DICTIONARY OF
INORGANIC CHEMISTRY

DICTIONARY
OF
INORGANIC CHEMISTRY

By

Satish Anand
Raj Kumar

ANMOL PUBLICATIONS PVT. LTD.
NEW DELHI - 110 002 (INDIA)

ANMOL PUBLICATIONS PVT. LTD.

Regd. Office: 4360/4, Ansari Road, Daryaganj,
New Delhi-110002 (India)
Tel.: 23278000, 23261597, 23286875, 23255577
Fax: 91-11-23280289
Email: anmolpub@gmail.com
Visit us at: www.anmolpublications.com

Branch Office: No. 1015, Ist Main Road, BSK IIIrd Stage
IIIrd Phase, IIIrd Block, Bangalore-560 085 (India)
Tel.: 080-41723429 • Fax: 080-26723604
Email: anmolpublicationsbangalore@gmail.com

Dictionary of Inorganic Chemistry

First Edition 1990

Reprint 1991, 1992, 1994, 2001, 2010

PRINTED IN INDIA

Printed at Atul Printers, Delhi.

Preface

This Dictionary, the first of its kind in India, is an effort to keep pace with continuing rapid developments and expanding vocabularies in Inorganic Chemistry. Throughout the process of compilation and editing this book our sincere effort was to make the dictionary serve as a ready reference for undergraduate and post-graduate chemistry students and for those appearing for competitive examinations, research scholars, teachers, and authors.

Every effort has been carefully made to write the entries in a clear and lucid style to provide both straight forward definitions and invaluable background information. At certain appropriate places, the line diagrams have been included whenever the meaning of a word can be best understood by means of a diagram.

The presentation throughout has been aimed at sustaining the interest of readers while enriching their vocabulary and comprehension of technical terms and expressions.

The dictionary will be of value to students of chemistry and to scientists and engineers working on science projects.

All comments from users on omissions of shortcomings will be most welcome.

Editors

A

A. The symbol for relative atomic mass. The A value for an element has been a measure of the mass of an average atom of the element compared to the mass of an atom of the isotope carbon-12.

Absorber. A body which is intended either to stop a radiation (it then absorbs the radiation particles) or to slow it down (in which case it absorbs the energy). In reactors, elements like cadmium or boron are used, which capture neutrons.

Absorption. Refers to a phenomenon which takes place when radiation passes through an absorber. Several effects may then appear, either separately or another :

(i) a slowing down, that is to say & lessening of the energy of the radiation (it can even come to a complete stop,

(ii) an attenuation, which means that only part of the radiation passes through the absorbing substance,

(iii) a transformation of the radiation, which acting upon the particles of matter of the absorber, induces the emissions of a secondary radiation.

Abundance. Refers to the measure of how much of a substance exists. It is often expressed as a percentage. Elements : The abundances of elements in the earth and in the sun have been shown in the tables below.

Earth elements	O	SI	Al	Fe	Ca	Na	K	Mg
% abundance	46.6	27.7	8.1	5.0	3.6	2.8	2.6	2.1

Sun elements	H	He	O	C	Si
% abundance	54.0	44.7	0.8	0.4	0.05

Isotopes : The abundances is isotopes in some commonly occurring elements have been shown in the following table (continued over page).

element	% abundance of each isotope	
Chlorine	$^{35}_{17}$Cl [75.5%]	$^{37}_{17}$Cl [24.5%]
Carbon	$^{12}_{6}$C [98.9%]	$^{13}_{6}$C [1.1%]
Magnesium	$^{24}_{12}$Mg [78.6%]	$^{25}_{12}$Mg [10.1%] $^{26}_{12}$Mg [11.3%]

If we know the percentage abundance of each isotope it becomes possible to calculate the A value for the element, *e.g.* :

$$^{35}_{17}\text{Cl} : {}^{37}_{17}\text{Cl} = 3 : 1$$

$$\text{A (Cl)} = \frac{(3 \times 35) + (1 \times 37)}{4} = 35.5$$

Accelerator. An apparatus which enables particles to be accelerated. From stationary particles the accelerator provides accelerated particles which possess a definite velocity. They will constitute the projectiles in the reaction. To accelerate them, the particles, which bear an electric change, are passed through a strong electric field.

There are two groups of accelerators :

(1) Linear accelerators, in which the speed of the particles is increased while always maintaining the same direction.

(2) Circular accelerators, in which the particle track is coiled up to save space, the acceleration being repeated at each revolution.

In the latter case, it is a magnetic field which guides the particles on a circular orbit or trajectory. In this way it is possible to make these particles acquire very high speeds, which almost reach the speed of light, *i.e.*, 300,000 km per second. This speed cannot be exceeded because of the principle of relativity.

Acceptor. As substance which, while normally not oxidized by oxygen or reduced by hydrogen, can be oxidized or reduced in presence of a substance which is itself undergoing oxidation or reduction.

Also an atom, molecule, or ion that is electron deficient and which can form a co-ordinate link with an electron donor. Thus in the complex ion $[Co(NH_3)_6]^{3+}$ the cobalt (III) ion is an acceptor and the ammonia the electron donor, π-acceptors have been molecules or atoms which accept electrons into π, p or d orbitals.

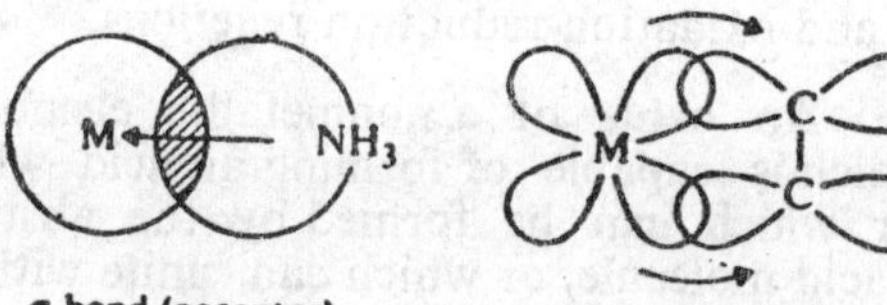

σ bond (acceptor) π bond (acceptor)

Acid. A substance which releases hydrogen ions (H^+) when it is added to water. The hydrogen ion is solvated, *i.e.*, a water molecule adds on to it, to give the oxonium ion (H_3O^+). Acidic solutions have a pH or less than 7.

Acids :

*turn blue litmus red.

*give carbon dioxide when added to carbonates.

*give hydrogen when added to certain metals.

*neutralize alkalis.

The Lowry-Bronsted theory defines an acid as a substance that exhibits a tendency to release a proton, and a base as a substance that tends to accept a proton. Thus, when an acid releases a proton, the ion formed is the conjugate base of the acid. Strong acids (*e.g.* HNO_3) react completely with water to give H_3O^+, *i.e* HNO_3 is stronger than H_3O^+ and the conjugate base NO_3 is weak. Weak acids (*e.g.* C I COOH and C_6H_5COOH) are only partly dissociated because H_3O^+ is a stronger acid than the free acids and the ions CH_3COO^- and $C_6H_5COO^-$ are moderately strong bases.

A further extension of the idea of acids and bases was made in the Lewis theory (G. N. Lewis, 1923). In this, a Lewis acid is a compound or atom that can accept a pair of electrons and a Lewis base is one that can donate a electron pair. This definition encompasses 'traditional' acid-base reactions. In

$$HCl + NaOH \rightarrow NaCl + H_2O$$

the reaction is essentially

$$H^+ + :OH^- \rightarrow H:OH$$

i.e. donation of an electron pair by OH^-. But it also includes reactions that do not involve ions, *e.g.*

$$H_3N: + BCl_3 \rightarrow H_3NBCl_3$$

in which NH_3 is the base (donor) and BCl_3 the acid (acceptor). The Lewis theory establishes a relationship between acid-base reactions and oxidation-reduction reactions.

Acid Anhydride. An oxide of a nonmetallic element of an organic radical which is capable of forming an acid when united with water, or which can be formed by the abstraction of water from the acid molecule, or which can unite with basic oxides to form salts.

Acid-base Reactions. Acids react with bases to form a salt and water only. Some examples are shown here.

Hydrochloric acid $HCl(aq)$	+	Sodium hydroxide $NaOH(aq)$	→	Sodium chloride $NaCl(aq)$	+	water $H_2O(l)$
Nitric acid $2HNO_3(aq)$	+	Copper(II) oxide $CuO(s)$	→	Copper(II) nitrate $Cu(NO_3)_2(aq)$	+	water $H_2O(l)$

Acid salts. Salts in which only some of the replaceable hydrogen atoms in an acid molecule have been replaced by a metal. Some examples are shown below :

Na_2CO_3	Sodium carbonate	$NaHCO_3$	Sodium hydrogencarbonate
Na_2SO_4	Sodium sulphate	$NaHSO_4$	Sodium Hydrogensulphate
Na_3PO_4	Sodium phosphate	Na_2HPO_4	Sodium hydrogenphosphate

Acid Salts (acidic salt). Refers to a salt in which there has been only partial replacement of the acidic hydrogen of an acid by metal or other cations. For polybasic acids the formulae have been of the type $NaHSO_4$ (sodium hydrogensulphate) and $Na_3H(CO_3)_2.2H_2O$ (sodium sesquicarbonate. For monobasic acids like HF the acid salts are of the form KHF_2 (potassium hydrogen fluoride).

Acidic Oxides. Most oxides of nonmetals react with water to from acidic solutions. Because of this they are termed as acidic oxides. Examples :

oxide		acid	
Carbon dioxide	CO_2	Carbonic acid	H_2CO_3
Sulphur dioxide	SO_2	Sulphurous acid	H_2SO_3
Sulphur trioxide	SO_3	Sulphuric acid	H_2SO_4

Actinides. Refers to a series of elements whose atomic number (number of protons in the nucleus) varies from 89 to 101.— These are successively, actinium, thorium, protoactinium, uranium, neptunium, plutonium, americium, curium, berkelium, californium, einsteinium, fermium, mendelevium. Elements beyond uranium are not to be found in nature ; they are artificially produced. They have been made up to element 104. This has been made by Russian workers at Dubha, by bombarding plutonium nuclei with neon nuclei accelerated in a cyclotron. The nuclei fuse and four neutrons are emitted to give a nucleus of atomic number 104. This element decays by fission with a half-life of 0.3 sec.

As with the lanthanoids, the first member actinium, has no electrons (AC [Rn] $6d^17\ s)^2$ but other members also show deviations from the smooth trend of electron filling expected from simple considerations, *e.g.*, thorium Th [Rn]$6d^27s^2$, berkelium Bk [Rn]$5f^86d^17s^2$.

The actinides are all radioactive and their chemistry is often extremely difficult to study. In general, artificial methods using high-energy bombardment are used to generate them. The outer electron configurations as are follows :

89 actinium (Ac) $6d^17s^2$,

90 thorium (Th) $6d^27s^2$

91 protacnium (Pa) $5f^26d^1s^2$

92 uranium (Ur) $5f^36d7s^2$

93 neptunium (Np) $5f^57s^2$ (or $5f^46d^17s^2$)

94 plutonium (Pu) $5f^6 7s^2$
95 americiun (Am) $5f^7 7s^2$
96 curium (Cm) $5f^7 6d^1 s^2$
97 berkelium (Bk) $5f^8 6d 7 s^2$ (or $5f^9 7s^2$)
98 Californium (Cf) $5f^{10} 7s^2$
99 einsteinium (Es) $5f^{11} 7s^2$
100 ferminium (Fm) $5f^{12} 7s^2$
101 mendelevium (Md) $5f^{13} 7s^2$
102 nobelium (Nb) $5f^{14} 7s^2$
103 lawrencium (Lw) $5f^{14} 6d^1 s^2$

The first four members (Ac to Ur) occur naturally. All are radioactive and this makes investigation difficult because of self-heating, short lifetimes, safety precautions, etc. Like the lanthanoids, the actinoids show a smooth decrease in atomic and ionic radius with increasing proton number. The lighter members of the series (up to americium) have f-electrons that can participate in bonding unlike the lanthanoids. Consequently, these elements resemble the transition metals in forming co-ordination complexes and displaying variable valency. As a result of increased nuclear charge the heaiver members (curium to lawrencium) tend not to use their inner f-electrons in forming bonds and resemble the lanthanoids in forming compounds containing the M^{3+} ion. The reason for this is pulling of these inner electrons towards the centre of the atom by the increased nuclear charge. Note that actinium itself does not have a of electron, but it is usually classified with the actionoids because of its chemical similarities.

Actinium. The first element of the Actinide series. It was discovered in 1899 by Debierne. It is known to have two isotopes, whose mass are 227 and 228; they are both radioactive, and can be produced artificially, by bombarding radium 226 with neutrons; radium 227 is obtained which by means of beta disintegration, gives actinium 227.

It occurs in minute quantities in uranium ores and has been used as a source of alpha particles.

Symbol : Ac; m.p. 1050°C; b.p. 3200°C (calc); r.d. 10.07 (calc.); p.n. 89; stablest isotope ^{227}Ac (half-life 21.6 years).

Actinium X. This is really a radium (atomic mass 223). Its atomic number is in fact 88. It is so-called because it belongs to the radioactive series of actinium.

Actinoids. See actinides.

Actinon. A radioactive gas, deriving from the radioactivity of actinium X, which splits up to produce an alpha particle and actinon. It is the isotope of radon of mass 219 and atomic number 86.

Actino-uranium. A name occassionally used to describe uranium 235 (qv). This derives from the fact that uranium 235 is the parent of the radioactive family of actinium. All the elements of this family are produced from uranium 235.

Activation. A process by means of which a body is rendered radioactive. Take for instance a stable body (not radioactive) : if it is bombarded (usually with a neutron) other bodies will be formed having a different mass and atomic number. These latter bodies may be unstable. Suppose neutrons are directed on to titanium. A titanium isotope is obtained which is having one neutron more; its mass therefore passes from 28 to 29. But this isotope is unstable and emits an alpha ray upon disintegrating. It is then said that titanium has been activated. Bombardment can also be effected with the help of other particles such as protons, deuterous, alpha particles, etc. The radioactivity of a body thus formed is called induced radioactivity. Depending on the particles used, alpha, beta and gamma rays, etc, are emitted.

Activity. A synonym of radioactivity (*e.g.* 'uranium has an alpha activity'). Also, the number of atoms which disintegrate in one second. The unit of activity is the curie.—In a radioactive source whose activity is 1 curie, 37 billion atoms disintegrate per second.

Activity Series (displacement series; electromotive series). An arrangement of the metals in the order of their tendency to react with water and acids, so that each metal displaces from solution those below it in the series and is displaced by those above it. The arrangement of the more common metals is : potassium, sodium, magnesium, aluminium, zinc, iron, tin, lead, hydrogen, copper, mercury, silver, platinum gold.

Affinity. The tendency of an atom or compound to react or combine with atoms or compounds of different chemical constitution. For example, paraffin hydrocarbons were so named because they are quite unreactive, the word "paraffin" meaning very little affinity." The hemoglobin molecule has a much greater affinity for carbon monoxide than for oxygen. The free energy decrease is a quantitative measure of chemical affinity.

Agate. A variety of chalcedony that forms in rock cavities and has a pattern of concentrically arranged bands of layers that lie parallel to the cavity walls. These layers are frequently alternating tones of brownish-red. Moss agate does not show the same banding and is a milky chalcedony containing moss-like or dendritic patterns formed by inclusions of manganese and iron oxide. Agates are used in jewellary and for ornamental purposes.

Aggregate. A collect term denoting any mixture of such particulates as sand, gravel, crushed stone, or cinder used in Portland cement formulations, road-building, paving compositions, animal husbandry, trickle filters, horticulture, etc.

Air. The mixture of gases which surrounds the earth. The average composition of pure air have been shown below. This composition varies from place to place and also varies with altitude. Other components

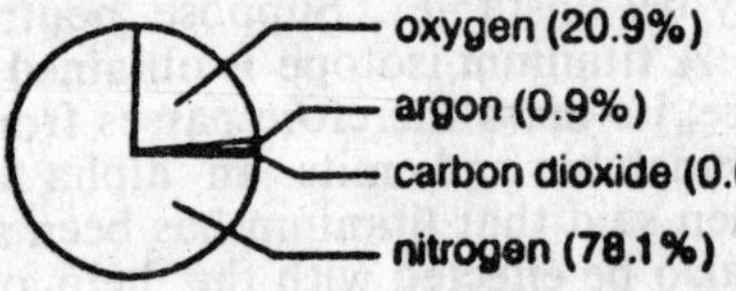

neon, helium, krypton, xenon—in small constant amounts.
water—in very variable amounts.

Air also contains pollutants, some of which have been shown below.

Pollutant	sources
Sulphur dioxide	Burning coal, oil.
Carbon monoxide	Engines, cigarettes.
Oxides of nitrogen	Car engines.
Soot	Fires, engines.
Pollen	Trees, flowers
Dust	Volcanoes.

Air is vital for life. The oxygen is necessary forrespiration and the carbon dioxide needed for photo-synthesis.

The composition of dry air, by volume, is

nitrogen 78.08%
oxygen 20.95%
argon 0.93%
carbon dioxide 0.03%

neon 0.0018%
helium 0.0005%
krypton 0.0001%
xenon 0.00001%.

Air also contains a variable amount of water vapour, as well as dust, pollen, and small amounts of other gases.

Air Gas. See producer gas.

Air Hardening. A method of hardening steel by allowing it to cool naturally in air, or with an air blast (air quenching). It is applicable only to certain alloy steels, in which the critical cooling rate is less than the normal rate of cooling in air.

Alabaster. A fine-grained form of gypsum, $CaSO_4 \cdot 2H_2O$, often showing veining and mottling and resembling marble. Soft and easily carved and polished. Also applied to a form of calcite, $CaCO_3$, which is harder than true alabaster.

Alchemy. The predecessor of chemistry, practiced from as early as 500 B.C. through the 16th century. Its two principal goals were transmutation of the baster metals into gold and discovery of a universal remedy. Modern chemistry grew out of alchemy by gradual stages.

Alcoa Process. A more efficient method of producing aluminium from bauxite which requires one-third less electric power than the Hall process. Alumina is reacted with chlorine, the resulting aluminium chloride yielding the metal and chlorine on electrolysis. No fluorine is required in the process. Prototype plants are under development.

Alkali. A base which is soluble in water. They are usually metal hydroxides, *e.g.* sodium hydroxide, but ammonia solution is also an alkali. Their reactions are affected by their strength and their concentration.

Oven cleaners, household ammonia and some paint strippers contain alkalis.

Alkalis :

(i) turn red litmus blue.

(ii) neutralize acids.

(iii) have a pH above 7.

(iv) react with acids to produce a salt and water only.

Alkali Metals. (group I elements). A group of soft reactive metals having an electronic configuration consisting of a noble-gas structure plus one outer electron. The alkali metals are lithium (Li), sodium (Na), potassium (K), rubidium (Rb), caesium (Cs), and francium (Fr).

The elements all readily form positive ions M^+ and consequently have been highly reactive (particularly with any substrate that is oxidizing). As the group has been descended there has been a gradual decrease in ionization potential and an increase in the size of the atoms; the group shows several smooth trends which follow from this. There has been a general decrease in the following: melting points, heats of sublimation, lattice energy of salts, hydration energy of M^+, ease of decomposition of nitrates and carbonates, and heat of formation of the "-ide" compounds (fluoride, hydridge, oxide, carbide, chloride).

Hydrolysis of the oxides or direct reaction of the metal with water gives rise to the formation of the hydroxide ion.

Salts of the bases MOH are known for all acids and these are generally white crystalline solids. The ions M^+ are hydrated in water and remain unchanged in most reactions of alkali metal salts.

Due to the ease of formation of the ions M^+ there have been very few coordination compounds of the type ML_n^+ apart from solvated species of very low correlation times. The group I elements are known to form a variety of organometallic compounds. Organo—alkali metal compounds—particularly the lithium compounds—are widely used in synthetic organic chemistry.

Francium is formed only by radioactive decay and in nuclear reactions; all the isotopes of francium have short half-lives.

Alkaline-earth Metals (group II elements). A group of moderately reactive metals, which are harder and less volatile than the alkali metals. The term alkaline earth strictly refers to the oxides, but is often used loosely for the elements themselves. The electronic configurations have been all those of a noble-gas structure with an additional two electrons in the outer s orbital. The elements have been beryllium (Be), magnesium (Mg), calcium (Ca), strontium (Sr), barium (Ba), and radium (Ra). The group shows an increasing tendency to ionize to the divalent state M^{2+}. The chemistry of the heavier members of the group has been largely that of divalent ions.

The group is showing a topical trend towards metallic character as the group is descended. The group also displays a smooth trend in the solubilities of the sulphates ($MgSO_4$ is soluble, $CaSO_4$ sparingly soluble, and $BaSO_4$ very insoluble). The trend to increasing metallic character has been also shown by the increase in thermal stabilities of the carbonates and nitrates with increasing relative atomic mass.

The elements all burn in air (beryllium must be finely powdered) to give the oxide MO (covalent in the case of beryllium) and for barium the peroxide, BaO_2 in addition to BaO. Oxides, CaO, SrO and BaO, react with. The metals Ca, Sr and Ba all react readily with water to give the hydroxide:

$$M + 2H_2O \rightarrow M^{2+} + 2OH^- + H_2$$

Because of its higher polarizing power, beryllium forms a rangeof complexes in which the beryllium atom should be treated as an electron acceptor (*i.e.* the vacant p orbitals are being used). Complexes such as etherates, acetylethanoates, and the tetrafluoride (BeF_4^{2-}) are formed, all of which are tetrahedral.

All isotopes of radium has been radioactive and randium was once widely used for radiotherapy.

Alkoxide. An organic compound having an ion of the type RO^-, where R is an alkyl group. Alkoxides are prepared by the reaction of metallic sodium on an alcohol. Sodium ethoxide ($C_2H_5O^-Na^+$) is an example.

Allobar. A mixture of the different isotopes of an element in proportions which differ from the natural isotopic composition.

Allotropes. Some elements exist in different forms in the same physical state. The chemical properties are the same but the physical properties are different. The best example is carbon. There are two allotropes : diamond and graphite. Diamond is hard and colourless, graphite is flakey and black. Other elements which exist in different allotropic forms are sulphur, phosphorus and tin.

Alloy. An alloy is a mixture which is made up of two or more metals or which contains metals and non-metals.

Alloys are much more widely used than pure metals because, by bringing two or more elements together in the right proportions, specific properties can be obtained. For example, aluminium is quite a soft metal but when a small amount of copper is mixed in the alloy duralumin is produced which is strong enough to be used in aircraft frames. Some common examples are: brass, bronze, duralumin, pewter, solder and steel.

Alloys may be completely homogeneous mixtures or may contain small particles of one phase in the other phase.

Alloy Elements. Elements which are added to give special properties to alloys, such as resistance to corrosion, heat, creep, etc. (*e.g.* the addition of chromium, tungsten, molybdenum, vanadium, etc., to Fe alloys).

Alloy, Fusible (low-melting alloy; fusible metal). An alloy melting in the range of about 51-260°C, usually containing bismuth, lead, tin, cadmium, or indium. Eutectic alloys are the particular compositions that have definite and minimum melting points compared with other mixtures of the same metals. The compositions of a few fusible alloys are given below :

System	Composition	Eutectic Temperature (°C)
Cd—Bi	60 Bi 40 Cd	144
In—Bi	33.7 Bi 66.3 In	72
	67.0 Bi 33 In	109
Pb—Bi	56.5 Bi 43.5 Pb	125
Sn—Bi	58 Bi 42 Sn	139
Pb—Sn—Bi	52 Bi-16 Sn-32 Pb	96
Pb—Cd—Bi	52 Bi-8 Cd-40 Pb	92
Sn—Cd—Bi	54 Bi-20 Cd-26 Sn	102
In—Sn—Bi	58 Bi-17 Sn-25 In	79
Pb—Sn—Cd—Bi	*50 Bi-10 Cd-13.3 Sn-26.7 Pb	70
In—Pb—Sn—Bi	49.4 Bi-11.6 Sn-18 Pb-21 In	57
In—Cd—Pb—Sn—Bi	44.7 Bi-5.3 Cd-8.3 Sn-22.6 Pb-19.1 In	47

*Wood's metal.

Alloy Steel. A steel having up to 10% of elements such as chromium, molybdenum, nickel, etc., usually with a low percentage of carbon. These added elements improve hardenability, wear resistance, toughness, and other properties. This term includes low-alloy steels, in which the alloy content does not exceed 5%, but does not include stainless.

Alnico (Trademark). The term used for any of a group of very hard brittle alloys which are used to make powerful permanent magnets. They have nickel, aluminium, cobalt, and copper in various proportions. †Iron, titanium, and niobium can also be present. They are having a high remanence and coercive force.

Alpha Particles. These are helium atoms without their electrons. They are produced in many nuclear reactions. They have a fairly short range in air (usually less than 10 cm) and are easily stopped by thin sheets of paper or foil.

Alstonite. A mixed carbonate of Ba and C.

Alum. Refers to a type of double salt. Alums have been the double sulphates which are obtained by crystallizing mixtures in the correct proportions. They have the general formulae :

$$M_2SO_4M'_2(SO_4)_3\ 24H_2O$$

Where M is a univalent metal or ion, and M' is a trivalent metal. Thus, aluminium potassiums sulphate (called potash alum, or simply alum is

$$K_2SO_4.Al_2(SO_4)_3.24H_2O$$

Aluminium ammonium sulphate (called ammonium alum) has been as follows .

$$(NH_4)_2SO_4Al_2(SO_4)_3\ 24H_2O$$

The name 'alum' originally came from the presence of Al^{3+} as the trivalent ion, but is also applied to other salts containing trivalent ions, thus :

Chromium (III) potassium sulphate (chrome alum)

$$K_2SO_4.Cr_2(SC_4)_3.24H_2O$$

Alum, Burnt (alum, dried) $AlNHO_4(SO_2)_2$ or AIK $(SO_4)_2$. Aluminium ammonium sulfate or aluminium potassium sulfate heated just sufficiently to drive off the water of crystallization.

Properties : White, odourless powder; sweetish taste. Absorbs moisture on exposure to air. Soluble in hot and slowly soluble in cold water; insoluble in alcohol.

Use : Medicine (astringent).

Alumina. See aluminium oxide.

Aluminate. See aluminium hydroxide.

Aluminium. Alluminium is the most abundant metal in earth's crust (approximately 8 per cent by mass). Its alloys are widely used at home and in industry.

Clay, shale, and granite all contain aluminium compounds but the metal is difficult to extract from them. It is obtained by the electrolysis of bauxite dissolved in cryolite, using graphite electrodes.

The metal has valency 3. Although it is a very reactive metal, it, and its alloys, are resistant to corrosion because of a layer of oxide on the surface of the metal. It is subject to anodizing.

It has the electronic structure of neon plus three additional outer electrons. There are numerous minerals of aluminium; it is the most. Commercially important minerals are bauxite (hydrated Al_2O_3), corundum (anhydrous Al_2O_3), cryolite (Na_3AlF_6), and clays and mica (aluminosilicates). The metal is produced on a massive scale by the Hall-Heroult method in which alumina, a non-electrolyte, is dissolved in molten cryolite and electrolysed. The bauxite contains iron, which would contaminate the product. The aluminium atom is much bigger than boron and its ionization potential is not particularly high. Consequently aluminium forms positive ions Al^{3+}. However, it also has non-metallic chemical properties. Thus, it is amphoteric and also has a number of covalently bonded compounds.

Unlike boron, aluminium does not form a vast range of hydrides. The ion AlH_4^- is widely used in the form as $LiAlH_4$ as a vigorous reducing agent.

The reaction of aluminium metal with oxygen is very exothermic but at ordinary temperatures an impervious film of the oxide protects the metal from further attack. There is only one oxide, Al_2O_3 (alumina), but a variety of polymorphs and hydrates are known. Aluminium metal will react with alkalis releasing hydrogen and producing initialy $Al(OH)_3$ then $Al(OH)_4^-$.

Aluminium reacts readily with the halogens; in the case of chlorine thin sheets will enflame. The fluoride has a high melting point (1290°C) and is ionic. The other halides are dimers in the vapour phase (two halogen bridges). Aluminium also forms a sulphide (Al_2S_3)), nitride (AlN), and carbide (Al_4C), the latter two at extremely high temperatures.

A number of very reactive aluminium alkyls are also known, some of which are important as polymerization catalysts.

Symbol : Al; m.p. 652.8°C; b.p. 1800°C; r.d. 2.702; p.n. 13; r.a.m. 26.27.

Aluminium Acetate (aluminium ethanoate, $Al(OOCCH_3)_3$). A white solid which is soluble in water. It is usually obtained as the dibasic salt, basic aluminium ethanoate $(Al(OH)(C_2H_2O_2)_2$. It is obtained by dissolving aluminium hydroxide in ethanoic acid. It finds use extensively as a mordant in dyeing and as a size for paper and cardboard products.

Aluminium Brass. An alloy having 76% copper, 21.5 to 22.25% zinc, 1.75 to 2.50% aluminium.

Uses: Condenser, evaporator and heat exchanger tubes and ferrules.

Aluminium Bromide ($AlBr_3$). A white solid soluble in water and many organic solvents.

Aluminous Cement. A hydraulic cement closely resembling Portland cement, but composed chiefly of calcium aluminate.

Aluminium Chloride ($AlCl_3$). A white covalent solid that fumes in moist air :

$$AlCl_3 + 3H_2O \rightarrow Al(OH_3) + 3\ HCl$$

It is obtained by heating aluminium in dry chlorine or dry hydrogen chloride. Vapour-density measurements reveal that its structure is a diamer; it consists of Al_2Cl_6 molecules in the vapour. Aluminium chloride finds use in Friedel-Crafts reactions in organic preparations.

Aluminium Fluoride (AlF_3). A white crystalline solid. Its primary use has been as an additive to the cryolite (Na_3AlF_6) electrolyte in the production of aluminium.

Aluminium Hydride, AlH_3: A white polymeric substance (H_2SO_4 plus $LiAlH_4$ in THF). A useful reducing agent (*e.g.* RCN gives RCH_2NH_2); reacts with alcohols to give alkoxy hydrides, $(RO)_nAlH_{3-n}$; forms complexes, *e.g.* $AlH_3.2NMe_3$ and complex hydrides $(AlH_6)^{3-}$ (NaH plus $NaAlCl_4$) and the more important tetrahedral tetrahydroaluminate ion. $[AlH_4]^-$, $LiAlH_4$ LiH plus $AlCl_3$ in ether) is a very important reducing agent, soluble in ethers (*e.g.* RCOOH gives RCH_2OH) and starting point for the formation of Al derivatives, *e.g.* RNA_2 gives $Li[Al(NHR_4]$. Alkoxy derivatives, *e.g.* $Li[AlH(OR)_3]$ and $Na[AlH(OMe)_2(OEt)$ are good and often specific reducing agents.

Aluminium Hydroxide ($Al(OH)_3$). A white powder which is prepared as a colourless gelatinous precipitate by adding ammonia solution or a small amount of sodium hydroxide solution to a solution of an aluminium salt. It is an amphoteric hydroxide. It finds use as a foaming agent in fire extinguishers and as a mordant in dyeing.

Aluminium Nitrate ($Al(NO_3)_3\ 9H_2O$). A hydrated white crystalline solid which is prepared by dissolving freshly prepared aluminium hydroxide in nitric acid.

Aluminium Oxide (alumina, Al_2O_3). A white powder that is almost insoluble in water. Due to its amphoteric nature it will react with both acids and alkalis. Aluminium oxide occurs naturally as bauxite, corundum, and white sapphire; it is manufactured by heating aluminium hydroxide. It finds use in the extraction by electrolysis of aluminium, as an abrasive (corundum), in furnace linings (because of its refactory properties), and as a catalyst (*e.g.* in the dehydration of alcohols).

Aluminium Potassium Sulphate (potash alum, $Al_2(SO_4)_3.K_2SO_4.24H_2O$ $AlK\ (SO_4)_2.12H_2O$). A white solid, which is soluble in water but insoluble in alcohol. It is prepared by mixing equimolecular quantities of solutions of ammonium and aluminium sulphate followed by crystallization. It finds use as a mordant for dyes, as a waterproofing agent, and as a tanning additive.

Aluminium Sulphate ($Al_2(SO_4)_3.18H_2O$). A white crystalline solid. It is used as a size for paper, a precipitating agent in sewage treatment, a foaming agent in fire control, and as a fireproofing agent.

Aluminium Trimethyl. See trimethyl-aluminium.

Alumino Ferric. Impure $Al_2(SO_4)_3$ having some Fe^2 and Fe^3. Used in water and sewage purification.

Alumstone. See alunite.

Alundun. An artificial form of corundum (α-Al_2O_3), which is made by fusing calcined bauxite in an electric furnace, and allowing the molten product to cool rather rapidly.

Alundum is used for highly refractory bricks, crucibles, refractory cement and muffles; also for small laboratory apparatus used as high temperatures (combustion tubes, pyrometer tubes, etc.)

Alunite, Alumstone. Mineral $KAl_3(SO_4)_2(OH)_6$. Occurs, usually in volcanic rocks, as white, grey or pinkish masses resembling limestone or marble. Used commercially as a source of potash alum and of potassium sulphate.

Amalgam. An amalgam is an alloy which contains mercury. Zinc amalgam is used for teeth fillings. Amalgamation was once used to extract gold and silver from crushed rock to earth which contained the metals.

Americium. Artificial element with atomic number 95 and atomic mass 241. Each of its nuclei contains 95 protons and 146 neutrons. It was produced in 1945 by Seaborg, James and Morgan by bombarding uranium 238 with helium ions. In this way, plutonium 241 is obtained, which, through beta decay, gives americium. Americium 241 has an average radioactive life of 475 years. It can equally well be produced by the repeated bombardment of neutrons on plutonium 241.

It is an element of actinoid series of metals. A transuranic element, it is not found naturally on earth but is synthesized from plutonium. Americium-241 has been used in gamma-ray radiography.

Symbol: Am; m.p. 294°C; b.p. 9607°C; r.d. 13.67; p.n. 95; stablest isotope ^{243}Am (half-life 7.4×10^3 years).

Americium Compounds. Americium is a typical actinide showing similar chemistry to uranium.

$$AmO_2^{2+} \xrightarrow{+1.60} AmO_2^{+} \xrightarrow{+0.7} Am^{4+} \xrightarrow{+2.8}$$

$$Am^{3+} \xrightarrow{-2.36} Am \text{ (volts in 1 M acid).}$$

Aqueons Am(III) is oxidized only with difficulty, Am^{4+} disproportionates to Am^{3+} and AmO_2^{+}. Americium(II) halides (Cl, Br, I) could be prepared from the metal and HgX_2 at 300—400°C, all of the trihalides are known and tan AmF_4 is formed from fluorine and AmO_2 AmO_2 is the stable oxide in air, Am_2O_3 is also known and mixed oxides, *e.g.* $Li_6Am''O_6$ and Li_3AmO_4 are prepared by solid state reactions using oxides, peroxides and oxygen. Oxidation to aqueous AmO_2^{2+} has been with peroxydisulphate.

Ammine. A coordination compound formed by the union of ammonia with a metallic substance, in such a way that the nitrogen atoms are linked directly to the metal. See cobaltammine and coordination compound. Note the distinction from amines, in which the nitrogen is attached directly to the carbon atom.

Ammonia (NH_3). A colourless gas with a pungent odour. It is very soluble in water giving an alkaline solution. Ammonia was once made from coal but now over 90% comes from the Haber process.

Ammonia is also produced by bacteria found on the roots of leguminous plants (peas and beans). When protein decomposes, ammonia is released. Both of these are important sources of plant food.

Ammonia is a covalent compound. It has a characteristic reaction with hydrogen chloride to give dense white fumes of ammonium chloride :

$$NH_3(g)+HCl(g)\rightarrow NH_4Cl(s)$$

Ammonia can be oxidized to nitric acid and this is the major source of nitric acid today. It is easily liquefied and it is usually carried in this state from place to place in tankers.

Ammonia is very soluble in water (a saturated solution at 0°C contains 36.9% of ammonia): the aqueous solution is alkaline and contains a proportion of free ammonia. Ammonia is also soluble in ethanol.

The compound does not burn readily in air but ignites, giving a yellowish brown flame, in oxygen. It will react, with atmospheric oxygen in the presence of platinum or a heavy metal catalyst—a reaction used as the basis of the commercial manufacture of nitric acid, which involves the oxidation of ammonia to nitrogen monoxide and then to nitrogen dioxide.

Ammonia is used commercially in the manufacture of fertilizers, mainly ammonium nitrate, urea, and ammonium sulphate. It is used to a smaller extent in the refrigeration industry. Liquid ammonia has been an excellent solvent for certain substances, which ionize in the solutions to give ionic reactions similar to those occurring in aqueous solutions.

Ammoniacal. Describing a solution in aqueous ammonia.

Ammonia Solution (NH_4OH) (formerly called Ammonium hydroxide). A solution of ammonia in water.

Ammonium Carbonate ($(NH_4)_2)CO_3$). A white solid that crystallizes as plates or prisms. This salt is manufactured from ammonium chloride and calcium carbonate. Commercial ammonium carbonate is used in baking powders, smelling salts, and in the dyeing and wool-scouring industries.

Ammonium Chloride (salt ammoniac, NH_4Cl). A white crystalline solid having a characteristic saline taste. Ammonium chloride can be manufactured by the action of ammonia on hydrochloric acid. It sublimes on heating because of the equilibrium :

$$NH_4Cl(s) = NH_3(g) + HCl(g)$$

Ammonium chloride has been used in galvanizing, as a flux for soldering, in dyeing and calico printing, and in the manufacture of Leclanche and 'dry' cells.

Ammonium Ion. The ion NH_4^+. formed by coordination of NH_3 to, H^+.

Ammonium compounds contain the ammonium ion as does a solution of ammonia gas in water. The ion is positively

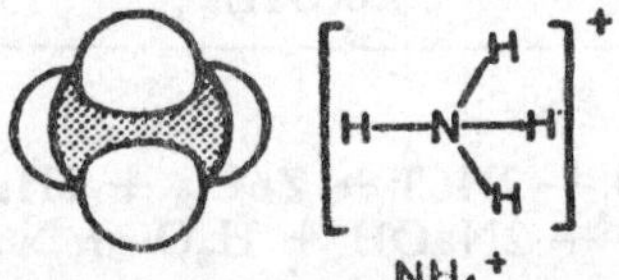

NH_4^+.

charged and the hydrogen atoms are arranged around the nitrogen atom in a tetrahedral way.

Ammonium Nitrate (NH_4NO_3). A colourless crystalline solid that is usually manufactured by the action of ammonia on nitric acid. It finds use in the manufacture of explosives and, because of its high nitrogen content, as a fertilizer.

Ammonium Sulphate $(NH_4)_2(SO_4)$. A colourless crystalline solid that is soluble in water. It is manufactured by the action of ammonia on sulphuric acid. It is the most important ammonium salt because of its widespread use as a fertilizer.

Amorphous. Describing a solid substance that has no 'long-range' regular arrangement of atoms; *i.e.* is not crystalline. Glasses are also amorphous.

Amphiboles. A large group of rock-forming metasilicate minerals. They have a structure of silicate tetrahedra linked to form double endless chains, in contrast to the single chains of the *pyroxenes to which they are closely related. They are present in many igneous and metamorphic rocks. The amphiboles show a wide range of compositional variation but conform to the general formula : $X_{23}Y_5Z_8O_{22}$ $(OH)_2$, where X=CA, Na, K. Mg or Fe^{2+}. Y=Mg.$Fe^{2+}Fe^{3+}$, Al, Ti or Mn; and Z=Si or Al. The hydroxyl ions may be replaced by F. Cl, or O.Most amphiboles are monoclinic, including cummingtonite, (Mg Fe^{2+}), Si_8O_{22}) $(OH)_2$ tremolite, Ca_2Mg_5 (Si_8O_{22}) $(OH,F)_2$; actinolite, $Ca_2(Mg.Fe^{2+})_5$ (Si_8O_{22}) $(OH,F)_2$; *hornblende, $NaCa_2$ (Mg,Fe^{2+}, $Fe^{3+}Al)_5$ $(Si, Al)_8$ $O_{22})$ $(O.H\ F)_2$; ednite, $NaCa_2$ $(Mg.Fe^{2+})_5$ $(Si_2$, $AlO_2)$ $(OH,F)_2$; and riebeckite, Na_2,Fe^{3+} (Si_8O_{22}) $(OH,F)_2$. Antiphyllite, $(Mg,Fe^{2+})^7$ (Si_8O_{22}) $(OH,F)_2$, and gedrite, (Mg, $Fe^{2+})_6$ Al $(Si,Al)_8O_{22})$ $(OH,F)_2$, are orthorhombic amphiboles.

Amphoteric (of oxides and hydroxides). Insoluble oxides and hydroxides which show both basic and acidic properties are said to be 'amphoteric'. The examples shown here will react with both acids and alkalis to form salts.

Examples

Al_2O_3	PbO	ZnO
$Al(OH)_3$	$Pb(OH)_2$	$Zn(OH)_2$

For example :

$$ZnO + 2HCl \rightarrow ZnCl_2 + 2H_2O$$
$$ZnO + 2NaOH + H_2O \rightarrow Na_2Zn(OH_4)$$
(Sodium Zincate)

The solid hydroxides of these metals all dissolve in sodium hydroxide solution.

A.M.U. See atomic mass unit.

Angstrom Symbol : Å A unit of length defined as 10^{-10} metre. The angstrom is sometimes still used for expressing wavelengths of light or ultraviolet radiation or for the sizes of molecules, although the nanometre is preferred.

Anhydrides. These are substances which react with water to produce acids. All acidic oxides are anhydrides, but some will give rise to one acid only whilst others will produce two.

Examples

Anhydride	Acid
SO_3	H_2SO_3
NO_2	HNO_3, HNO_2

Anhydrite $CaSO_4$. A natural calcium sulfate usually occurring as compact granular masses and resembling marble in appearance. Differs from gypsum in hardness and lack of hydration.

Anhydrous. Containing no water. The term usually denotes salts with no water of crystallization.

For example :

Anhydrous copper(II) sulphate $CuSO_4$

Anhydrous sodium carbonate Na_2CO_3

The hydrated salts are :

$$CuSO_4.5H_2O$$

$$Na_2Cl_3.10H_2O$$

The term is also used about liquids which are perfectly dry, *e.g.* anhydrous ether.

Annealing. Heating and cooling metals causes changes in their properties, especially their strength. When a metal is heated to a high temperature and then cooled very slowly the metal becomes soft and malleable. This is because the crystals in the metal have been allowed to grow very large. This slow-cooling is called 'annealing'.

Anodizing. Aluminium has a protective layer of oxide (Al_2O_3). This layer oxide can be thickened by making the piece of aluminium the anode (+) in an electrolysis cell. Dilute sulphuric acid is used as the electrolyte. Aluminium atoms give up electrons and they react with the water :

$$2Al(s) + 3H_2O(l) \rightarrow$$

$$Al_2O_3(s) + 6H^+(aq) + 6 \text{ electrons}$$

Aluminium oxide is formed and this can be dyed to produce attractive finishes to products, *e.g.* coloured milk-bottle tops.

Antibonding Orbital. See orbital.

Antimatter. Matter which contains only antiparticles (qv.). For example, a proton has a positive charge. The anti-proton has the same charge, but it is negative. The mass of the antiproton and that ofthe proton are equal. If a quantity of matter and the equivalent quantity of antimatter make contract over a sufficiently long period, and interaction of each particle with its corresponding antiparticle would be obtained. There would then be annihilation of matter with production of gamma rays. There is at present no evidence which would allow the conclusion that antimatter exists in the world in the form of atoms, like the matter, in an appreciable quantity. Antimatter is only observed in the form of isolated antiparticles, which are produced either by cosmic radiation or by large accelerators.

Antimonic. Designating an antimony (IV) compound.

Antimonous. Designating an antimony (III) compound.

Antimony. A toxic element which exists in two allotropic forms; the most stable is a brittle silvery metal. Arsenic belongs to group V of the periodic table. It occurs in many minerals, principally stibnite (Sb_2S_3). It is used in alloys and semiconductor devices.

Symbol : sb; m.p. 630.7°C; b.p. 1640°C; r.d. 6.69; p.n. 51; r.a.m. 121.7.

Antimony (III) Chloride (antimony trichloride, $SbCl_3$). A white deliquescent solid which is also known as butter of antimony. It is obtained by direct combination of antimony and chlorine. It readily gets hydrolysed by cold water to form a white precipitate of antimony (III) chloride oxide (antimonyl chloride, SbOCl) :

$$SbCl_3 + H_2O \rightleftharpoons SbOCl + 2HCl$$

Antimony (III) Chloride Oxide. See antimony (III) chloride.

Antimonyl Chloride. See antimony (III) chloride.

Antimoney (III) Oxide (antimony trioxide, Sb_2O_3). A white insoluble solid. It is an amphoteric oxide having a strong tendency to act as a base. It is prepared by direct oxidation by air, oxygen, or steam and is formed when antimony (III) chloride gets hydrolysed by excess boiling water.

Antimoney (V) Oxide (antimony pentoxide, Sb_2O_5). A yellow solid which is prepared by the action of concentrated nitric acid on antimony or by the hydrolysis of antimony (V) chloride. Although an acidic oxide, it is only slightly soluble in water.

Antimony Pentoxide. See antimony (V) oxide.

Antineutrino. An antiparticle which corresponds to a neutrino. The neutrino has no charge, nor has the antineutrino. Its rest mass is zero. Having neither a rest mass nor charge, it is difficult to detect.

Antineutron. An antiparticle corresponding to the neutron. Its mass is equal to that of the proton; its electric charge is zero. It is to be observed in cosmic radiation. It is produced in very high energy nuclear showers. K has also been studied with the large accelerators, where it is produced by sending an antiproton against a target which contains protons.

Antiparticle. An antiparticle corresponds to each existing particle. This fact has been verified experimentally for the great majority of particles. Particles and antiparticles have the same mass and opposite electric charges. They have the same spin and their magnetic moments are oppositely directed in regard to the direction of the spin, eg the proton and antiproton. Other quantum numbers are equally in opposition for particle and antiparticle. These are the baryon number, the lepton number and strangeness number. Some particles are at the same time their own antiparticles.

Antiparticle	proton	antiproton
Charge	+	—
Baryon number	+1	−1
Strangeness number	0	0
Lepton number	0	0

Antiproton. An antiparticle corresponding to the proton. It has the same mass as the proton, but a negative electric charge. Its existence, foreseen by theory, was experimentally shown in 1955 at the California Institute of Technology, in the United States. To produce an antiproton, it is necessary to produce a proton at the same time, for the number of nucleons less the number of antinucleons must remain constant throughout the reaction. Such a reaction is expressed by the following formula :

proton	+	proton	gives	proton
1	+	1	=	1
+antiproton	+	proton	+	proton
−1	+	1	1	1.

Apatite. $Ca_5(PO_4)_3F$. A phosphate mineral containing fluoride and phosphate anions. Used for the production of phosphorus and its compounds including superphosphates. Hydroxyapatite, $Ca_5(OH)(PO_4)_3$ is isostructural with apatite. These compounds are of importance as components of teeth. The mineral contains useful quantities of uranium.

Aq. This abbreviation is used to denote an aqueous solution, *e.g.* NaCl(aq).

Aqua Regia. A mixture of concentrated nitric acid and three parts of hydrochloric acid. It dissolves all metals including gold, hence the name. The mixture is having chlorine and NOCl.

Aquation. The process of replacement of other ligands by water, *e.g.*

$$[CO(NH_3)_4(H_2O)(NO_3)^{2+} \rightarrow (CO(NH_3)_4(H_2O_2]^{3+}.$$

The process of complexing by water molecules.

Aqueous Solution. This is a solution where water is the solvent.

Aquo Ions. Hydrated metal ions in aqueous solution: also applied to hydrated ions in complexes, *e.g.* $[Co(H_2O)_6]^{2+}$. Aquo ions can lose protons and act as acids, *e.g.*

$$[Fe(H_2O)_6]^{3+} \rightarrow [Fe(H_2O)_5(OH)]^{2+}$$

Aragonite. A rock-forming anhydrous mineral form of calcium carbonate, $CaCO_3$. It is much less stable than calcite, the commoner form of calcium carbonate, form which it may be distinguished by its greater handless and specific gravity. Over time aragonite undergoes recrystallization to calcit. Aragonite occurs in cavities in limestone, as a deposit in limestone caverns, as a precipitate around hot springs and geysers, and in high-pressure low-temperature metamorphic rocks; it is also found in the shells of a number of molluscs and corals and is the main constituent of pearis. It is white or colourless when pure but the presence of impurities may tint it grey, blue, green, or pink.

Argentic Compounds. Compounds of silver in its higher (+2) oxidation state, *e.g.* argentic oxide is silver(II) oxide (AgO).

Argentite. A sulphide ore of silver, Ag_2S. It crystalizes in the cubic system but most commonly occurs in massive form. It is dull grey-black in colour but bright when first cut and occurs in veins associated with other silver minerals. Important deposits occur in Mexico, Peru, Chile, Bolivia, Norway, and Czechoslovakia.

Argsntous Compounds. Compounds of silver in its lower (+1) oxidation state, *e.g.* argentous chloride is silver(I) chloride.

Argon. A rare gas which is contained in air in a 1 per cent proportion. Its mass is 40, but there is an isotope of mass 41, formed by the absorption of a neutron by argon 40. Argon 41 is radioactive, and it cannot be subjected to chemical reaction, because it is inert. This gas also constitutes a danger in piles in which the air is irradiated by neutrons. This circumstance necessitates the taking of special precautions to avoid contamination by argon 41.

Arsenic. A brittle grey metalloid. Its compounds are very poisonous— especially the oxide (As_2O_2). It is a group v element which occurs as allotropes Arsenic compounds are used in pesticides. The grey allotrope sublines at 613°C.

Arsenic. A toxic metalloid element which is existing in several allotropic forms, the most stable has been a brittle grey metal. It belongs to group V of the periodic table. Arsenic occurs native and in several ores including mispicket (FeSAs), real-gar (AsS), and orpiment (As_2S_3). It finds use in semiconductor devices, alloys, and gun shot. Various compounds have been used in medicines and agricultural insecticides and poisons.

Symbol : As; m.p. 817°C (grey) at 3 MPa pressure; **sublimes at 613°C (grey); r.d. 5.73 (grey; p.n. 33; r.a.m. 74.92.**

Arsenis (III) Chloride (arsenious chloride, $AsCl_3$). It is a poisonous oily liquid which fumes in moist air due to hydrolysis with water vapour :

$$AsCl_3+3H_2O=As_2O_3+6HCl$$

Arsenic (III) chloride is covalent and exhibits nonmetallic properties.

Arsenic Hydried. See arsine.

Arsenic (III) Oxide (arsenious oxide, As_2O_3). A colourless crystalline solid which is very poisonous (0.1 g would be a lethal dose). Analysis of the solid and vapour states reveals a dimerized structure of As_4O_6.

Arsenic (V) Oxide (arsenic oxide, As_2O_5). It is a white amorphous deliquescent solid. Which is an acidic oxide. It is prepared by dissolving arsenic (III) oxide in hot concentrated nitric acid, followed by crystallization then heating to 210°C.

Arsenious Chloride. See arsenic tri-chloride.

Arsenious Oxide. See arsenic (III) oxide.

Arsine (arsenic hydride, AsH_3) A poisonous colourless gas having an unpleasant smell. It decomposes to arsenic and hydrogen at 230°C. It is produced in the analysis for arsenic (Marsh's test).

Artificial Radioactivity. Radioactivity which is induced by bombarding stable nuclei with high-energy particles. For example :

$$27/13Al+n+24/11Na+He$$

represents the bombardment of aluminium with neutrons to produce an isotope of sodium. All the transuranic elements (atomic numbers 93 and above) have been artificially radioactive because they do not occur in nature.

Asbestos. A naturally occurring silicate which was widely used in the past for fireproof clothing and insulating materials. Its use today is restricted because of the dangers of contracting asbestosis, a fatal lung disease. It was much used in schools as bench mats and centre pieces for gauzes. These are now made of material which does not contain asbestos.

Association. A reversible chemical combination due to any of the weaker classes of chemical bonding forces. Thus the combination of two or more molecules due to hydrogen bonding, as in the union of water molecules with one another, or of acetic acid molecules with water molecules, it called association; also combination of water or solvent molecules with molecules of solute or with ions, *i.e.*, hydrate formation or solvation. Formation of complex ions or chelates, as copper ion with ammonia or copper ion with 8-hydroxyquinoline, are other examples. Aqueous solutions of soaps or synthetic detergents are often called association colloids.

Astatine. A radioactive element which belongs to the halogen group It occurs in minute quantities in uranium ores. 20 short-lived radioisotopes are known, all alpha-particle emitters.

Symbol : At; m.p. 302°C; b.p. 337°C; p.n. 85; stablest isotope^{210} At (half-life 8.3 hours).

Atmosphere

1. A unit of pressure. Although the pressure of the air which surrounds us varies from place to place and from time to time its value usually only varies slightly and is always approximately one atmosphere.

2. The mixture of gases which surrounds a planet, sun or moon. The atmosphere on earth has been air.

Atom. The smallest indivisible particle of an element that can exist, the building bricks with which everything is made. Atoms are small particles; 100 million placed end-to-end would measure 1 cm. They are made up of even smaller subatomic particles :

Protons (p) Neutrons (n) Electrons (e) These particles are found in the centre of the atom—the nucleus. These particles move round the nucleus.

The atom as a whole is electrically neutral although the protons and electrons carry electrical charges. These charges are equal in size but opposite in sign, hence the number of protons always equals the number of electrons. Atoms which lose or gain electrons are called ions. All atoms of the same element contain the same number of protons and have the same atomic number, but atoms of the same element can contain different numbers of neutrons.

An atom can be thought of as the smellest part of an element that can take part in a chemical reaction.

The electrons surrounding the nucleus are grouped into shells—i.e. main orbits around the nucleus. Within these main orbits there may be sub-shells.

The electronic structure of an atom refers to the way in which the electrons are arranged about the nucleus, and in particular the energy levels that they occupy. Each electron can be characterized by a set of four quantum numbers, as follows :

(1) The principal quantum number n gives the main energy level and has values 1, 2, 3, etc. (the higher the number, the further the electron from the nucleus). Traditionally these levels, or the orbits corresponding to them, are referred to as shells and given letters K, L, M, etc. The K-shell is the one nearest the nucleus.

(2) The orbital quantum number l, which governs the angular momentum of the electron. The possible values of l are (n−1), (n−2),...1, 0. Thus, in the first shell (n=1) the electrons can only have angular momentum zero (l=0). In the second shell (n=2). the values of l can be 1 or 0, giving rise to two subshells of slightly different energy. In the third shell (n=3) there are three subshells, with l=2, 1, or 0. The subshells are denoted by letters s(l=0), p(l=1), d(l=2)) f(l=3). The orbital quantum number is sometimes called the azimuthal quantum number.

(3) The magnetic quantum number m, which governs the energies of electrons in an external magnetic field. This can take values of +l, +(l−1),... l, 0,−1,...−(l−1), −l. In an s-subshell (*i.e.* l=0) the value of m=0. In a p-subshell (l=1), m can have values +1, 0, and −1; *i.e.* there are three p-orbitals in the p-subshell, usually designated p_x, p_y, and p_z. Under normal circumstances, these all have the same energy level.

(4) The spin quantum number m, which gives the spin of the individual electrons and can have the values $+\frac{1}{2}$ or $-\frac{1}{2}$

According to the Pauli exclusion principle, no two electrons in the atom can have the same set of quantum numbers. The numbers define the quantum state of the electron, and explain how the electronic structures of atoms occur.

Atomic Mass. Atoms have different masses if they contain different numbers of protons, neutrons and electrons. Because these particles are so small the mass of an atom is tiny. It is not usual to refer to the mass of an atom in grams or kilograms so we normally compare the mass of one atom with a standard mass. This is called the relative atomic mass (A).

Atomic mass of an atom is the mass of an atom of the isotope of carbon whose mass number is 12 is used as a reference. This consists of 1 nucleus containing 6 protons and 6 neutrons, and of 6 electrons: as it is due, above all, to the 12 nucleons (protons and neutrons) of the nucleus, the mass of the atom is fixed by agreement at 12. The atomic mass of other elements is not a whole number, but each nucleus is characterised by an integral mass number.

Atomicity. The number of atoms in a molecule of an element. With the inert gases there has been only one atom in the molecule and the atomicity is 1. All other gases have an atomicity of 2, *e g.* oxygen (O_2), nitrogen (N_2), except ozone (O_3). Molecules in solid sulphur have an atomicity of 8 (S_6).

Atomic Number (Z). The number of protons in the nucleus of an atom. All atoms of the same element have the same atomic number, *e.g.* sodium atoms all contain 11 protons. In a neutral atom the number of electrons equals the atomic number.

Atomic Orbital The energy levels in an atom may be described by the four quantum numbers. In wave mechanics, the energy of a particular system may be described by the Schrodinger equation and the wave function $\psi_{n/m}$ may be used to represent a solution of the wave equation in terms of these three numbers. Wave functions may be used to describe electron distribution and are thus sometimes referred to as atomic orbitals.

The wave function ψ_{100} (n=1, l=0, m=0) can be shown to correspond to a spherical electronic distribution around the nucleus and is an example of an s orbital. Solutions of other wave functions may be described in terms of p and q orbitals.

Atomic Radii. Half the closest distance of approach of atoms in the structure of the elements. This is easily defined for regular structures, *e.g.* A close-packed metals, but is less easy to define in elements with irregular structure *e.g.* The values may differ betwee allotropes *e.g.* C-C 1.54Å in diamond and 1.42Å in planes of graphite). Atomic radii are very different from ionic and covalent radii.

Atomic Volume. The relative atomic mass of an element divided by its density.

Atomism. A physical theory founded on the idea of the atom, and in general, of the concept of matter which embraces the hypothesis of distinct particles, in opposition to the concept of continuous matter.

Attenuation. Refers to the lessening of the intensity of a beam of particles. It has been produced by absorption (capture by an absorbent medium) or by scattering (change of direction in a scattering medium), both of which can occur simultaneously in the same medium.

Attenuation Factor. The ratio of the intensity of incident radiation to the intensity of transmitted radiation. Only some of the radiation which enters a certain thickness of matter comes out of it again The strength of the radiations has been attenuated by passing through matter; eg if 20 per cent of the particles pass through matter, the attenuation factor is 1/0.2=5.

Attenuator. A medium which induces attenuation.

Aufbau Principle A principle which governs the order in which the atomic orbitals have been filled in elements of successive proton number; *i.e.* a statement of the order of increasing energy. The order is as follows :

$1s^2, 2s^2, 2p^6, 3s^2, 3p^6, 4s^2, 3p^{10}, 4p^6, 5s^2,$
$4d^{10}, 5p^6, 6s^2, 5d^{10}, 6p^6, 7s^2, 5f^{14}, 6d^{10}$

(the superscript shows the maximum number of electrons for each level).

Note

(1) Hund's rule of maximum multiplicity applies; *i.e.* degenerate orbitals are occupied singly before spin pairing occurs.

(2) the unexpected position of the d-levels, which give rise to the first, second, and third transition series.

(3) the unusual position of the f-levels, giving rise to the lanthanoids and actinoids.

Auric Chloride. See gold (III) chloride.

Aurous. Designating a compound of gold (I).

Auger Effect. Refers to a phenomenon by means of which an excited atom may be de-excited by emitting an electron. For instance, the atom may be excited by absorption of an X-ray and may also be de-excited by re-emitting an X-ray. The Auger effect and de-excitation by means of X radiation are therefore two phenomena competing with one another.

Avalanche. A phenomenon in which a charged particle (usually an electron), colliding with the atoms contained in a gas, knocks out electrons of the gas (ionisation). These electrons, in turn, reproduce the same phenomenon. In this way there is a multiplication of free electrons and, at the same time, formation of ionised atoms through the loss of electrons. This process develops particularly in the presence of a high electric field. The electric field, in fact, exercises forces of opposing directions upon the negatively charged electrons and upon the positively charged ions. It tends to separate them and in this way to prevent their recombining. This is also called Townsend's avalanche (after the physicist Townsend who studied this phenomenon).

Avogadro Constant/number (L). The number of atoms that are contained in exactly 12 g of the carbon-12 isotope. More generally it is the number of particles present in a mole of substance. $L = 6 \times 10^{23}$ mol^{-1}.

Avogadro's Hypothesis/law. This stated that under the same conditions of temperature and pressure, equal volumes of gases contain the same number of molecules, *e.g.*, 10 dm^3 of oxygen and 10 dm^3 of hydrogen both contain the same number of molecules. They contain twice as many molecules as 5 dm^3 of chlorine—provided that all the volumes were measured at the same temperature and pressure.

Azomide. See hydrazoic acid.

Azimuthal Quantum Number. See atom.

Azurite, $Cu_3(OH)_2(CO_3)_2$. Blue basic carbonate of copper. An artificial form is known as Blue Verditer or Blue Ashes and is used as an artists' colour. It is also used as a gem stone.

B

Babbit Metal. Any of a group of related alloys used for making bearings. They consist of tin containing antimony (about 10%) and copper (1—2%), and often lead. The original alloy was invented in 1839 by the US inventor Isaac Babbit (1799-1862).

Back Bonding. Overlap of a filled orbital, generally a d orbital, of an acceptor with a vacant orbital d, p, or π (bonding or antibonding) of a ligand. Formally relieves the build-up of negative charge on the central atom. Particularly important in complexes in low oxidation states, *e.g.*, metal carbonyls.

Back Donation. Refers to a form of chemical bonding in which a ligand forms a sigma bond to an atom or ion by donating a pair of electrons, and the central atom donates electrons back by overlap of its d-orbitals with empty p- or d-orbitals on the ligand.

Baddeleyite. One form of ZrO_2 and a principal mineral source of Zr Normally contains some Hf which must be removed if the Zr is to be used in reactors.

Baking Powder. Refers to a synthetic leavening agent which is widely used in the baking industry. There are several types, all of which are composed of a carbonate, a weak acid or acidic compound, and a filler. A typical composition has been sodium bicarbonate, tartaric acid or monobasic calcium phosphate, and cornstarch. Ingredients sometimes used are ammonium carbonate and potassium bitartrate. Upon contact with moisture and heat, the active ingredients react to evolve carbon dioxide, which "raises" the dough in the early minutes of heat exposure, thus producing a stable solid foam. Wheat flour gluten has been sufficiently elastic to retain the bubbles of carbon dioxide.

Baking Soda. See sodium bicarbonate.

Ball Mill. Refers to a device commonly used in the chemical industry for reducing the size of social material. Ball mills are having slowly rotating steel or stone-lined drums, which contain steel balls, flint, or pebbles. The material is crushed by the tumbling action of the contents of the drum.

Balmer Series. In 1885 Balmer observed that the wavelengths (or frequencies) of certain lines in the spectrum of hydrogen were simply related to each other, and could be expressed by a general formula. The group of lines is termed the Balmer series. Other series were later discovered in the spectrum of hydrogen by Lyman, Paschen, Brackett and P fund.

All these series could be expressed by the equation

$$v=109\ 678.8/n^2-109\ 678/m^2,$$

wherein v is the wave-number, and n and m are integers which can only have certain definite values for each series. Thus, in the Lyman series, $n=1$, whilst each line is expressed by letting $m=2, 3, 4$, etc. In the Balmer series, $n=2$; in the Paschen series, $n=3$; in the Brackett series, $n=4$; and in the Pfund series, $n=5$.

Banana Bond. Informal name for the type of electron-deficient bond holding the B—H—B bridges in boranes and similar compounds.

Band Theory of Solids. In molecules, molecular orbitals arise by combination of atomic orbitals, the ideas of delocalized electrons can also be applied to large assemblies of atoms in solids. The quantum mechanical treatment of these electrons gives rise to the band theory of solids, in which there are bands of electron levels, sometimes overlapping, *e.g.*, metals, sometimes with a forbidden energy gap, *e.g.*, insulator oxides, between them.

Barterite. A nonferrous alloy having 88.5% copper, 5% nickel, 5% tin, 1.5% silicon. Sp. gr. 8.80; m.p. 1070°C. It offers good resistance to sulfuric acid in all dilutions up to 60%, sea water, moist sulfurous atmospheres, and mine waters.

Barft Process. A process which is formerly used for protecting iron by heating it in steam, to form a layer of tri-ircn tetroxide (Fe_3O_4).

Barite. See barytes.

Barium. Symbol Ba. A silvery-white reactive element which belongs to group II of the periodic table; a.n. 54; r.a.m. 137.33; r.d. 3.51; m.p. 725°C : b.p. 1640°C. It is found as the minerals barytes ($BaSO_4$) and witherite. Extraction is by high-temperature reduction of barium oxide with aluminium or silicon in a vacuum, or by electrolysis of fused barium chloride. The metal finds use as a getter in vacuum systems. It oxidizes readily in air and reacts with ethanol and water. Soluble barium compounds have been extremely poisonous. It was first identified in 1774 by Karl Scheele, and was extracted by Humphry Davy in 1808.

Barium Bicarbonate. Barium hydrogencarbonate.

Barium Carbonate ($BaCO_3$). A white insoluble salt that is found naturally as the mineral witherite. Barium carbonate can be readily precipitated by adding an alkali carbonate to a barium salt solution. On heating it decomposes reversibly with the formation of the oxide and carbon dioxide :

$$BaCO_3 \longrightarrow BaO + CO_2$$

It finds use as a rat poison.

Barium Chloride. A white compound, $BaCl_2$. The anhydrous compound is having two crystalline forms; an α form (monoclinic; r.d. 3.856), which transforms at 962°C to a β form (cubic; r.d. 3.917; m.p. 963°C; b.p. 1560°C). There is also a dihydrate, $BaCl_2 \cdot 2H_2O$ (cubic; r.d. 3.1), which loses water at 113°C. It may be prepared by dissolving barium carbonate (witherite) in hydrochloric acid and crystallizing out the dihydrate. The compound is used in the extraction of barium by electrolysis.

Barium Hydrogencarbonate (barium bicarbonate, $Ba(HCO_3)_2$). A compound that is found only in aqueous solution. It is prepared by the action of cold water containing carbon dioxide on barium carbonate, to which it reverts on heating :

$$BaCO_3+CO_2+H_2O \longrightarrow Ba(HCO_3)_2$$

Barium Hydroxide (baryta). A white solid, $Ba(OH)_2$ which is sparingly soluble in water. The common form is the octahydrate, $Ba(OH)_2, 8H_2O$; monoclinic; r.d. 2.18; m.p. 78°C. It can be prepared by adding water to barium monoxide or by the action of sodium hydroxide on soluble barium compounds and finds use as a weak alkali in volumetric analysis.

Barium Oxide (BaO). A white powder which is prepared by heating barium in oxygen or by thermal decomposition of barium carbonate. It finds use in the manufacture of lubricating-oil additives.

Barium Peroxide. A dense off-white solid. BaO_2 which is prepared by carefully heating barium oxide in oxygen; r d. 4.96; m.p. 450°C. It is used as a bleaching agent. With acids, hydrogen peroxide is formed and the reation is used in the laboratory preparation of hydrogen peroxide.

Barium Sulphate ($BaSO_4$). A white solid that is found naturally as the mineral barytes. Barium sulphate is very insoluble in water and can be prepared easily as a precipitate by adding sulphuric acid to barium chloride. Barium sulphate is an important industrial chemical. Under the name or 'blanc fixe' it finds use as a pigment extender in surface coating compositions. It also finds use in the glass and rubber industries.

Barytes (heavy spar). A mineral form of barium sulphate ($BaSO_4$).

Barytes, Heavy Spar. A common mineral barium sulphate, $BaSO_4$; it often forms well-developed orthorhombic crystals. Pure specimens of barytes are colourless or white, but brownish or bluish tints may be introduced by impurities. The mineral is used as a source of barium compounds; when finely ground it is used as a pigment and in oil-well drilling muds.

Basalt. Refers to a fine-grained basic igneous rock. It is composed chiefly of calcium-rich plagioclase feldspar and pyroxene; other minerals present may be olivine, magnetite, and apatite. Basalt is the commonest type of lava.

Base. Refers to a compound that releases hydroxyl ions, OH^-, in aqueous solution. Basic solutions are having a pH greater than 7. In the Lowry—Bronsted treatment a base is a substance that tends to accept a proton. Thus OH^- is basic as it accepts H^+ to form water, but H_2O is also a base (although somewhat weaker) because it can accept a further proton to form H_3O^+. In this treatment the ions of classical mineral acids such as SO^{-2}_4 and NO_3^- are weak conjugate bases of their respective acids. Electron-pair donors, like trimethylamine and pyridine, have been examples of organic bases.

Base Metal. Refers to a common relatively inexpensive metal, such as iron or lead, that corrodes, oxidizes, or tarnishes on exposure to air, moisture, or heat, as distinguished from precious metals, fuch as gold and silver.

Basic.

1. Describing a compound that is a base.
2. Describing a solution containing an excess of hydroxide ions : alkaline.

Basic Lining. A furnace lining containing basic compounds that decompose under furnace conditions to give basic oxides. The usual basic linings contain calcium and magnesium oxides or carbonates.

Basic-oxygen Process (BOP) Process). A high-speed method of making high-grade steel. It originated in the Linz-Donnewitz (L—D) process. Molten pig iron and scrap are charged into a tilting furnace, similar to the Bessemer furnace except that it has no tuyeres. The charge is converted to steel by blowing high-pressure oxygen onto the surface of the metal through a water-cooled lance. The excess heat produced enables up to 30% of scrap to be incorporated into the charge. The process has largely replaced the Bessemer and open-hearth processes.

Basic Salt. Refers to a compound that can be regarded as being formed by replacing some of the oxide or hydroxide ions in a base by other negative ions. Basic salts are thus mixed salt-oxides (*e.g.* bismuth(III) chloride oxide, BiOCl) or salt-hydroxides (*e.g.* lead (II) chloride hydroxide, Pb(OH)Cl).

Basic Slag. A slag produced in the manufacture of steel. It contains a variable amount of tricalcium phosphate, calcium silicate, lime and oxides of iron, magnesium and manganese. Used as a fertilizer for its phosphorus and lime.

Basic Stains. See staining.

Basis Metal. In electroplating, the metal that is being coated, it constitutes the cathode and may be any of a large number of metals.

Batch Process. Refers to a manufacturing process in which the reactants are fed into the process in fixed quantities (batches), rather than in a continuous flow. At any particular instant all the material, from its preparation to the final product, has reached a definite stage in the process. Backing a cake is an example of a batch process.

Battery Limits. Refers to that portion of a chemical plant in which the actual processes are carried out, as distinguished from storage buildings, offices and other subordinate structures, called offsites.

Bauxite. A mineral hydrated form of aluminium hydroxide , the principal ore of aluminium.

Bayer Process. Process for making alumina from bauxite. Bauxite is mixed with hot concentrated sodium hydroxide, which dissolves the alumina and silica. The silica is precipitated and the dissolved alumina is separated from the solids, diluted, cooled and then is crystallized as aluminium hydroxide. The aluminum hydroxide is calcined to give anhydrous alumina, which is then shipped to reduction plants.

Bearing Metals, Bearing Alloys. Alloys used in thin linings as bearings. Examples are tin-base Babbitt metals, Cu (70 % Pb (30%) alloy, Sn(6.25%)-Al alloys, and gun-metal (90:10 bronze).

Becquerel Symbol By. The SI unit of activity. The unit has been named after the discoverer of radioactivity A. H. Becquerel (1852-1908).

Bell Metal. A type of bronze which finds use in casting bells. It consists of 60-85% copper alloyed with tin, often with some zinc and lead included.

Benefication. A process used in extractive metallurgy whereby an ore, either metallic or nonmetallic, has been concentrated in preparation for further processing (smelting). Calcination is often an important step in beneficiation; others are physical separation of high-grade ore from impurities (gangue), br screening, washing, milling or magnetic means. A process foy removing sulfur from coal by chemical comminution has been developed.

Benfield Process. Removal of carbon dioxide from fuel gases, such as those obtained by gasifying coal in the Lurgi process, can be effected by counter-current scrubbing of the gases by hot potassium carbonate solution.

Benitoite, $BaTiSi_3O_9$. A silicate containing a cyclic Si_3O_9 species.

Bend Bond. See banana bond.

Berkelium. A radioactive transuranic element of the actinoid series of metals, not found naturally on Earth. Several radioisotopes have been synthesized.

Symbol : Bk ; p.n. 97 ; stablest isotope ^{247}Bk (half-life 1400 years).

Berthollide Compound. Solid phases showing a range of composition.

Beryl, $Be_3Al_2Si_6O_{18}$. A beryllium mineral containing hexagonal rings of six linked SiO_4 tetrahedra. Transparent forms of beryl are rare; when coloured green by Cr^{3+} they are known as emerald and when blue-green as aquamarine. Used for the extraction of Be and the manufacture of Be refractories.

Beryllate. A compound which is formed in solution when beryllium metal, or the oxide or hydroxide, dissolves in strong alkali. The reaction (for the metal) is often written

$$Be+2OH^-(aq)\rightarrow BeO_2^{2-}(aq)+H_2(g)$$

Beryllia. See beryllium oxide.

Beryllides. Intermetallic compounds made by chemically combining beryllium with such metals as zirconium and tantalum.

Beryllium. Symbol Be. A grey metallic element of group II of the periodic table; a.n. 4; r.a.m. 9.012; r.d. 1.85; m.p. 185°C; b.p. 2970 °C. Beryllium occurs as beryl (3BeO Al_2O_3) $6SiO_2$ and chrysoberyl ($BeO.Al_2O_3$). The metal is extracted from a fused mixture of BeF_2/NaF by electrolysis or by magnesium reduction of BeF_2. It has been used to manufacture Be—Cu alloys which have been used in nuclear reactors as reflectors and moderators because of their low absorption cross section. Beryllium oxide is used in ceramic and in nuclear reactors. Berylliums and its compounds have been toxic and can bring about serious lung diseases and dermatitis. The metal is resistant to oxidation by air because of the formation of an oxide layer, but will react with dilute hydrochloric and sulphuric acids. Beryllium compounds show high covalent character. The element was isolated independently by F. Wohler and A. A. Bussy in 1828.

Beryllium Bicarbonate. Beryllium hydrogencarbonate.

Beryllium Carbonate ($BeCO_3$). An unstable solid which may be prepared by prolonged treatment of a suspension of beryllium hydroxide with carbon dioxide. The resulting solution gets evaporated and filtered in an atmosphere of carbon dioxide.

Beryllium Chloride ($BeCl_2$). A white solid which is prepared by passing chlorine over a heated mixture of beryllium oxide and carbon. It is a poor conductor in the fused state and in the solid from consists of a covalent polymeric structure. It is readily soluble in organic solvents but is hydrolysed in the presence of water to give the hydroxide. The anhydrous salt finds use as a catalyst.

Beryllium Hydrogen Carbonate. Beryllium bicarbonate, $Be(HCO_3)_2$). A compound which is obtained in solution by the action of carbon dioxide on a suspension of the carbonate, to which it reverts on heating :

$$BeCO_3+CO_2+H_2O=Be(HCO_3)_2$$

Beryllium Hydroxide. A white crystalline compound, $Be(OH)_2$ which is precipitated from solutions of beryllium salts by adding alkali. Like the oxide, it is amphoteric and dissolves in excess alkali to give beryllates.

Beryllium Oxide. (BeO). A white solid which is prepared by heating beryllium in oxygen or by the decomposition of beryllium hydroxide or carbonate. Beryllium oxide is insoluble in water but it shows basic properties by dissolving in acids to form beryllium salts :

$$BeO+2H^+\rightarrow Be^{2+}+H_2O$$

However, beryllium oxide also resembles acidic oxides by reacting with alkalis to form beryllates :

$$BeO+2OH+H_2O\rightarrow Be(OH)_4^{2-}$$

It is thus an amphoteric oxide. Beryllium oxide has been used for the production of beryllium and beryllium-copper refractories, for high-output transistors, and for printed circuits. The chemical properties of beryllium oxide have been similar to those of aluminium oxide.

Beryllium Sulphate ($BeSO_4$). An insoluble salt which is prepared by the reaction of beryllium oxide with concentrated sulphuric acid. On heating, it breaks down to give the oxide. When heated in the presence of water, it readily dissolves to yield an acidic solution.

Bessemer Process. A process for converting pig iron to steel by oxidation of the impurities (C, Si, P, Mn) by blowing air through the molten metal. In the acid process for iron containing about 2.5% Si the Si is oxidized to a slag first followed by oxidation of C, carbon is then added as ferromanganese. In the basic process for iron rich in P the P is oxidized after the C. Plain carbon steels of low carbon content are produced.

Beta Decay. Refers to a type of radioactive decay in which an unstable atomic nucleus changes into a nucleus of the same masson ne but different proton number. The change involves the conversion of a neutron into a proton with the emission of an electron and an antineutrino ($n \rightarrow p + e^- + \nu$) or of a proton into a neutron with the emission of a position and a neutrino ($p \rightarrow n + e^+ + \nu$). An example has been the decay of carbon 14.

$${}_6^{14}C \rightarrow {}_7^{14}N + e^- + \nu^-$$

Beta-iron. Refers to a nonmagnetic allotrope of iron that exists between 768°C and 900°C.

Beta Particle. Negatively charged electrons.

Betatron. An electromagnetic device which is used for accelerating electrons (beta particles). Its action has been similar in principle to that of an electric transformer in which the secondary windings are replaced with focusing magnets. The electrons travel around the core in a vacuum tube placed between the magnets. At each revolution around the core the electrons pick up the same energy as the voltage that would have been induced in one turn of wire at that point. The betatron can generate electron beams up to 320 MeV. Invented by D.W. Kerst in 1940, it is used chiefly for basic physical research.

Betterton-kroll Process. A process which is used for obtaining bismuth and purifying desilverized lead that contains bismuth. Metallic calcium or magnesium is added to the molten lead to cause formation of high melting intermetallic compounds with bismuth. These separate as a surface scum and are skimmed off. The excess calcium and magnesium are removed from the lead by use of chlorine gas as mixed molten chlorides of lead or zinc. Bismuth of 99.995% purity is produced in this way.

Betts Process. An electrolytic process which is used for removing impurities from lead, in which pure lead gets deposited on a thin cathode of pure lead, from an anode containings as much as 10% of silver, gold, bismuth, copper, antimony, arsenic, selenium, and other impurities. The electrolyte has been lead fluosilicate and fluosilicic acid. The scrap anodes and the residues of impurities associated with them have been either recast into anodes or treated to recover antimonial lead, silver, gold, bismuth, etc.

Bi. Prefix used formerly in naming acid salts. For instance, sodium bisulphate ($NaHSO_4$) is sodium hydrogensulphate, etc.

Bicarbonate of Soda. See sodium hydrogencarbonate.

Bicarbonate. See hydrogencarbonate.

Bi-gas Process. A high-pressure operation which is used for the conversion of solid fuel into substitute natural gas (SNG) using to stages of gasification. Coal slurry is dried and the dried coal powder has been devolatilized and partially gasified at about 1200K. Char formed in the process is then gasified at about 1950 K. As all the ash is melted in the second stage the process is suitable for a wide range of coals.

Binary. Describing a compound or alloy formed from two elements.

Binary Acid. An acid in which the acidic hydrogen atom(s) have been bound directly to an atom other than oxygen. Examples are hydrogen chloride (HCl) and hydrogen sulphide (H_2S). Such compounds are sometimes called hydracids.

Bioinorganic Chemistry. Study of the mechanisms involved in the behavior of metal-containing molecules in living organisms, *e.g.*, biological transport of iron, the effect of copper on nucleic acid and nucleoproteins, molybdenum, and manganese complexes, etc.

Biotite. A component of igneous rocks and of soil, similar to mica. It is a silicate of magnesium, iron, potassium and aluminum.

Biotite, Black Mica. $K(Mg,Fe)_3(AlSi_3)O_{10}(OH,F)_2$. A constituent of many rocks including granite. Of no economic importance.

Birkeland-eyde Process. An industrial process which is used for fixing nitrogen (as nitrogen monoxide) by passing air through an electric arc:

$$N_2+O_2 \rightarrow 2NO$$

Bismuth. A brittle pinkish metallic element which belongs to group V of the periodic table. It is found native and in the ores Bi_2S_3 and Bi_2O_3. Bismuth is widely used in alloys, especially low-melting alloys. Compounds find use in cosmetics and medicines.

Symbol : Bi; m.p. 271.3°C; b.p. 1560°C; r.d. 9.75; p.n. 83; r.a.m. 208.98.

Bismuth Alloys. Alloys of bismuth with Pb, Sn and Cd are of special interest; they melt at low temperatures, in many cases below

100%°C. These alloys give well-defined castings, and with sufficient Bi present expand on cooling.

Name	Bi	Pb	Sn	Cd	m.p. °C
Newton's	50	31	19	—	98
Rose's	50	28	22	—	100
D'Arcet's	50	25	25	—	97
Lichtenberg's	50	30	20	—	98
Wood's	50	25	12.5	12.5	71
Lipowitz's	50	27	13	10	71

These find use for fire protective devices, fuses, solders, etc. The alloys of bismuth with lead and antimony find use for stereotype plates.

Bismuthates(V). Formed from, *e.g.*, Na_2O_2 and Bi_2O_3 (impure $NaBiO_3$). Never found pure, they are very strong oxidizing agents (e.g. Mn^{2+} to MnO_4^- in acid solution).

Bismuth (III) Carbonate Dioxide. (Bismuthyl carbonate, $Bi_2O_2CO_3$). A white solid which is prepared by mixing solutions of bismuth nitrate and ammonium carbonate. It contains the $(BiO)^+$ ion.

Bismuth(III) Chloride. (Bismuth trichloride, $BiCl_3$). A white deliquescent solid which may be prepared by direct combination of bismuth and chlorine. Bismuth(III) chloride dissolves in excess dilute hydrochloric acid to form a clear liquld, but if diluted it gives rise to a white precipitate of bismuth(III) chloride oxide (bismuthyl chloride, BiOCl) :

$$BiCl_3 + H_2O \rightarrow BiOCl + 2HCl$$

Bismuth(III) Chloride Oxide. †See bismuth(III) chloride.

Bismuth(III) Nitrate Oxide. (Bismuthyl nitrate, $BiONO_3$). A white insoluble solid, often referred to as bismuth subnitrate. It gets precipitated when bismuth(III) nitrate is diluted and is having the $(BiO)^+$ ion.

Bismuth Trichloride. See bismuth(III) chloride.

Bismuthyl Compound. A compound having the $(BiO)^+$ ion or BiO grouping.

Bismuthyl Chloride. See bismuth(III) chloride.

Bismuthyl Nitrate. See bismuth(III) nitrate oxide.

Bismuth Subcarbonate. (Bismuth oxycarbonate; bismuth carbonate, basic) $(BiO)_2CO_3$ or Bi_2O_3 CO_2, with one-half H_2O. It is white, odorless powder; tasteless; insoluble in water and alcohol; soluble in nitric or hydrochloric acid with effervescene.
It is prepared by adding ammonium carbonate to a solution of a bismuth salt.

Bisulphate. See hydrogensulphate.

Bisulphite. See hydrogensulphite; aldehydes.

Bittern. The solution of salts remaining when sodium chloride gets crystallized from sea water.

Bitumen. A mixture of solid or semisolid hydrocarbons which is obtained from coal, oil, etc.

Bituminous Coal. See coal.

Biotite. An important rock-forming silicate mineral, a number of the mica group of minerals, in common with which it is having a sheetlike crystal structure. It is usually black, dark brown, or green in colour.

Bivalent (or divalent). Having a valency of two.

Black Cyanide. A mixture containing 45% calcium cyanide, made from calcium cyanamid by heating it with salt and carbon.

Blackdamp (Choke Damp). Air left depleted in oxygen following the explosion of firedamp in a mine.

Black Lead. Graphite.

Black Powder (blasting powder). A low explosive composed of potassium nitrate, charcoal and sulfur. Typical proportions are 75, 15 and 10%.

Black Rouge. Iron oxide, black.

Black Sand. A deposit of dark minerals having a high specific gravity found in stream beds and on beaches. Magnetite and ilmenite has been usually present, as also monazite and other minerals.

Blance Fixe. Precipitated $BaSO_4$ used as a fine white pigment in coating paper. It does not absorb ink.

Blast Furnace. A furnace for producing iron from iron (III) oxide. A mixture of the ore with coke and flux is heated by pre-heated air blown into the bottom of the furnace. The flux is often calcium oxide (from limestone). Molten pig iron is run off from the bottom of the furnace.

Blast-furnace Gas. By-product gas from smelting iron ore, obtained by the passage of the hot air over the coke in the blast furnaces. A typical gas will analyze 12.9% carbon dioxide, 26.3% carbon monoxide, 3.7% hydrogen, 57.1% nitrogen.

Bleach Bath. A bath used in photography to convert developed silver back to silver halide prior to removal (colour) or sulphide tonning. $K_2Fe(CN)_6$ and KBr is a commonly used bleach bath.

Bleaching Earths. There are two types : (1) fuller's earth possessing absorptive properties in the natural state, (2) activated clay in which the absorptive powers which are low in the raw state are enhanced by heat or chemical treatment (usually a mild acid leach).

Bleaching Powder. A white solid that can be regarded as a mixture of calcium chlorate(I) (calcium hypochlorite), calcium chloride, and calcium hydroxide. It may be prepared on a large scale by passing a current of chlorine through a titled cylinder down which is passed calcium hydroxide. Its bleaching powder arises from the formation, in the presence of air containing carbon dioxide, of the oxidizing agent chloric(I) acid (hypochlorous acid, HClO) :

$$Ca(ClO)_2.Ca(OH)_2.CaCl_2 + 2CO_2 \rightarrow$$
$$2CaCO_3 + CaCl_2 + 2HClO$$

Blende. A naturally occurring metal sulphide, *e.g.* zinc blende ZnS.

Blister Copper. Copper (96—99% purity) produced by the reduction and smelting of copper ores.

Block. See periodic table.

Blue Print Paper. Light sensitive paper (turns blue) impregnated with $[Fe(CN)_6]^{3-}$ and iron(III) cit-rate or oxalate. The paper is fixed by washing with water. Brown print paper is having a metal (Pt, Pd, Au, Hg, Cu) salt and Fe(III).

Blue Vitriol. Copper(II) sulphate pentahydrade ($CuSO_4$ $5H_2O$).

Boehmite. Refers to a mineral form of a mixed aluminium oxide and hydroxide. AlO.OH.

Boghead Coal. Refers to a variety of bituminous or subbituminous coal which resembles cannel coal in appearance and behaviour during combustion.

Bohr Theory. The theory published in 1913 by the Danish physicist Niels Bohr (1885—1962) to explain the line spectrum of hydrogen. He assumed that a single electron of mass *m* travelled in a circular orbit of radius *r*, at a velocity *v*, around a positively charged nucleus. The *angular momentum of the electron would then be *mvr*. Bohr proposed that electrons could only occupy orbits in which this angular momentum had certain fixed values, $h/2\pi$, $2h/2\pi$, $3h/\pi ... nh/2\pi$, where h is the Planck constant. This means that the angular momentum is quantized, *i e.* can only have certain values, each of which is a multiple of *n*. Each permitted values of *n* is associated with an orbit of different, radius and Bohr assumed that when the atom emitted or absorved radiation of frequency *v*, the electron jumped from one orbit to another, the energy emitted or absorbed by each jump is equal to *hv*. This theory gave good results in predicting the series of lines observed in the *hydrogen spectrum but not with any other spectrum. The idea of quantized values of angular momentum water later explained by the wave nature of the electron. Each orbit has to have a whole number of wavelengths around it; *i.e.* $n\lambda = 2\pi r$, where λ is the wavelength and *n* a whole number. The wavelength of a particle is given h/mv, so $nh/mv = 2\pi r$, which leads to $mvr = nh/2\pi$. Modern atomic theory does not allow subatomic particles to be treated in the same way as large objects, and Bohr's reasoning is somewhat discredited. However, the idea or quantized angular momentum has been retained.

Boiler Scale. A rocklike deposit occurring on boiler walls and tubes in which hard water has been heated or evaporated. Consists largely of calcium carbonate, calcium sulfate of similar materials, depending on the mineral content of the water.

Bond. The linkage between atoms in molecules and between molebules and ions in crystals.

Bond Energy. The energy required to break a particular bond and, precisely, to produce the products in particular electronic states. Bond energies of multiple bonds are generally greater than those of single bonds.

Average bond energies at 298 K

Bond	ΔH(kJ mol^{-1})
C—C	348
C=C	612
C=C (benzene)	518
C≡C	837
C—H	412
C—Cl	338
Cl—Cl	242

H—H	436
I—Cl	211
N≡N	944
N—N	388
O=O	496
O—H	463

Bonderizing. A process of phosphate coating* of metals.

Bonding Orbital. See orbital.

Bone Ash. An ash which is composed principally of tribasic calcium phosphate, but containing minor amounts of magnesium phosphate, calcium carbonate and calcium fluoride. Noncombustible.

Bone Black (bone char; bone charcoal). Black pigment made by carbonizing bones.

Bone China. Ceramic tubleware of high quality in which a small percentage of bone ash gets incorporated.

Bone Meal. A product made by grinding animal bones. The fertilizer grade contains from 43.55% tricalcium phosphate. 20-25% phosphoric acid, and 4-5% ammonia.

Boracite, $Mg_7B_{16}O_{30}Cl_2$. An important boron mineral.

Boral. A composite material consisting of boron carbide crystals in aluminium, with a cladding of commercially pure aluminium.

Borane. Formally BH_3, *i.e.* borane pyridine is $H_3BNC_5H_5.BH_3$ is only stable when complexed.

Borane Anions. The anions $B_nH^x{}_n{}^-$ formally formed from boron

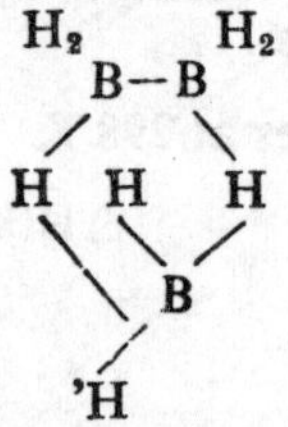

$[B_3H_8]^-$

hydrides and having structures containing bonding similar to that of the boron hydrides.

The simplest borane anion is tetrahedral BH^- Borane anions with polyhedral (closo) structures include $[B_9H_9]^{2-}$, $[B_{10}H_{10}]^{2-}$, $[B_{12}H_{12}]^{2-}$ (NaB_3H_8 at 200—230°C under vacuum).

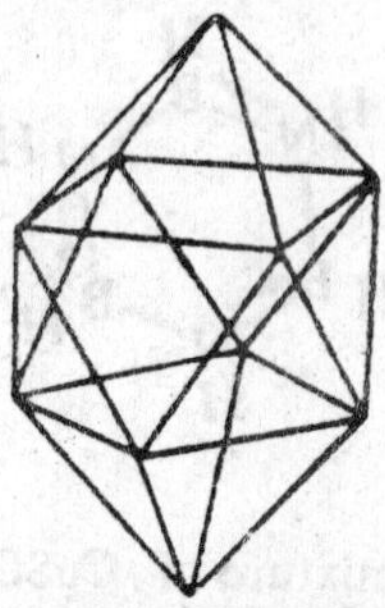

$[B_{10}H_{10}]^{2-}$

Boranes (Boron Hydrides). Any of a group of compounds of boron and hydrogen, many of which can be prepared by the action of acid on magnesium boride (MgB_2). The boranes are a remarkable group of compounds in that their structures cannot be described using the conventional two electron-covalent bond model (see : electron-deficient compound).The simplest example is diborane (B_2H_6). Other boranes include B_4H_{10}, B_5H_9, B_5H_{11}, B_6H_{10}, and $B_{10}H_4$. The larger borane molecules have open or closed polyhedra of boron atoms.

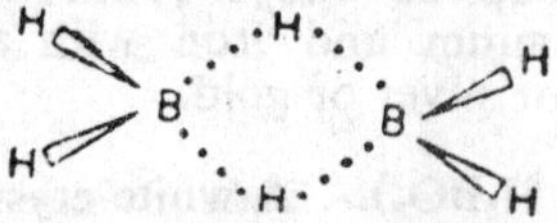

Diborane the simplest of the bonares.

Borax. $Na_2[B_4O_5(OH)_4],8H_2O$. Occurs naturally as kernite

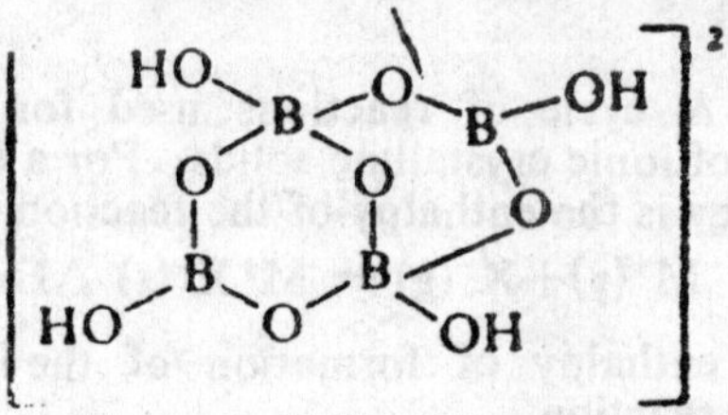

($Na_2B_4O_5(OH)_4,2H_2O$) and tincal ($Na_2B_4C_5(OH)_4.8H_2O$) and also present in the brines of Searles Lake, U.S.A.

Borazine, $B_3N_3H_6$. A cyclic compound prepared by heating B_2H_6 and NH_3 to 200°C or from $LiBH_4$ and NH_4Cl at 200°C. It is also known as borazole. Forms π-bonded complexes with metals. Substituted derivatives are formed from alkylboranes or alkylamines.

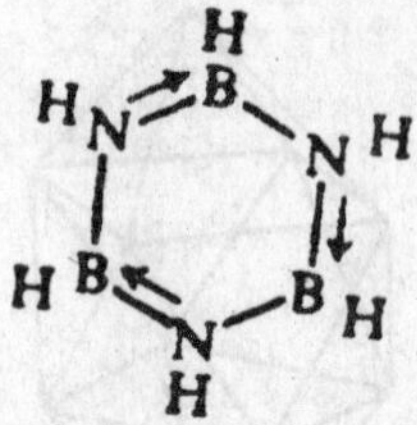

Borazole. See borazine.

Bordeaux Mixture. A mixture of $CuSO_4$ and $Ca(OH)_2$ made up in water and used as a fungicide.

Borate. Any of a wide range of ionic compounds that have negative ions containing boron any oxygen. Lithium borate, for example, contains the simple anion $B(OH)_4^-$.

"Borazon"[245] Trademark for a boron nitride formed at very high pressures (85,000 atm) and temperatures (1800°C) from mixtures of boron and nitrogen or from ordinary hexagonal boron nitride in the presence of catalysts such as lithium, calcium, magnesium, or their nitrides.

Borcher's Metal. A group of alloys of chromium with nickel and cobalt, or of chromium and iron with a small proportion of molybdenum and/or silver or gold.

Boric Acid (boracic acid, H_3BO_3). A white crystalline solid which in solution it is a very weak acid. Boric acid is used as a mild antiseptic eye lotion Pyrex 'glass'.

Boride. A compound of boron, especially one with a more electropositive element.

Born-haber Cycle. A cycle of reactions used for calculating the lattier energies of ionic crystalline solids. For a compound MX, the lattice energy is the enthalpy of the reaction

$$M^+(g)+X^-(g) \rightarrow M^+X^-(s) \ \Delta H_L$$

The standard enthalpy of formation of the ionic solid is the enthalpy of the reaction

$$M(s)+\tfrac{1}{2}X_2(g) \rightarrow M^+X^-(s) \ \Delta H_f$$

The cycle involves equating this enthalpy (which can be measured) to the sum of the enthalpies of a number of steps proceeding from the elements to the ionic solid. The steps are :

(1) Atomization of the metal :

$$M(s) \rightarrow M(g) \Delta H_L$$

(2) Atomization of the nonmetal :

$$\frac{1}{2}X_2(g) \rightarrow X(g) \ \Delta H_2$$

(3) Ionization of the metal :

$$M(g) \rightarrow M^+(g)+c \ \Delta H_3$$

This is obtained from the ionization potential

(4) Ionization of the nonmetal :

$$X(g)+e \rightarrow X^-(g) \ \Delta H_4$$

This is the electron affinity.

(5) Formation of the ionic solids :

$$M^+(g)+X^-(g) \rightarrow M^+X^-(s) \Delta H_L$$

Equating the enthalpies gives :

From which ΔH_L can be found. It is named after the German phyicist Max Born (1882-1970) and the chemist Fritz Haber (1868-1934).

Bornite. An important ore of copper which is composed of a mixed copper-iron sulphide, Cu_5FeS_4.

Boron. A hard rather brittle metalloid element of group II of the periodic table. It is having the electronic structure is $^2 2s^2 p^1$. Boron is of low abundance (0.0003%) but the natural minerals occur in very concentrated form as borax ($Na_2B_4O_7.10H_2O$) and colemanite ($Ca_2B_6O_{11}$).

The element is obtained by conversion to boric acid followed by dehydration to B_2O_3 then reduction with magnesium. High-purity boron for semiconductor applications is obtained by conversion to boron trichloride, which can be purified by distillation, then reduction using hydrogen.

Boron Carbide. A black solid, B_4C, soluble only in fused alkali; it is extremely hard. It is manufactured by the reduction or boric oxide with petroleum coke in an electric furnace. It is used largely as an abrasive.

Boron Fibre. A vapour-deposited filament made by deposition of boron on a heated tungsten wire. Tensile strength 350,000 to 450,000 psi; elastic modules 55 million psi;

Boron Hydrides. Six hydrides of boron ranging from B_2H_6 to $B_{10}H_{24}$, which have complex molecular structures.

Boron Nitride (BN). A compound formed by heating boron in nitrogen (1000°C). It has two crystalline forms.

Boron Oxide (B_2O_3.) A glassy hygroscopic solid that eventually forms boric acid. It forms various salts but also exhibits some amphoteric properties.

Boron Tribromide (BBr_3). A colourless liquid.

Boron Trichloride (BCl_3). A fuming liquid which may be by passing dry chlorine over heated boron. It is rapidly hydrolysed by water :

$$BCl_3+3H_2O \rightarrow 3HCl+H_3BO_3$$

Boron Trifluoride (BE_3). A colourless fuming gas which is prepared by heating a mixture of boron oxide, calcium fluoride, and concentrated sulphuric acid.

Borosilicate. Refers to any of a large number of substances in which BO_3 and SiO_4 units are linked to form networks with a wide range of structures.

Bosch Process. The reaction

$$CO+H_2O \rightarrow CO_2+H_2$$

Using water gas over a hot catalyst, used to make hydrogen for the Haber process.

Brackett Series. See hydrogen spectrum.

Brass. A group of alloys consisting of copper and zinc. A typical yellow brass might contain about 67% copper and 33% zinc.

Brazing. A method of joining metals that involves the use of a filler metal or alloy, at above 430°C. The filler alloys may be silver/copper/zinc 660-870°C or nickel/boron/silicon /chromium, suitable for superalloys in the range above 1100°C; silver-cadmium alloy is also used.

Braunite. Brown Mn_2O_3 containing some SiO_2. Used as a source of manganese and as a colourizing agent for bricks and pottery.

Brazilite, ZrO_2. A variety of baddeleyite.

Brazing Metal. See solder.

Bredt's Rule. A double bond cannot be introduced at the bridgehead carbon of a bridged bicyclic or polyclic system with small or medium-sized rings.

Breeze. Coke particles less than one-half inch in diameter. This occurs to the extent of about 100 pounds per ton of coal processed.

Breunnerite. Impure magnesite ($MgCO_3$) containing 5-30% iron carbonate. Also known as giobertite and mesitite.

Bridge. An atom joining two other atoms in a molecule.

Brillouin Zones. Electronic theory of metals divides the electronic states of a metal into a series of broad energy levels known as Brillouin zones.

Brine. Any solution of sodium chloride and water, usually containing other salts also. Concentration range from 3% (ocean) to 20% or more.

Brin Process. A process formerly used for making oxygen by heating barium oxide in air to form the peroxide and then heating the peroxide at higher temperature (>800°C) to produce oxygen

$$2BaO_2 \rightarrow 2BaO + O_2$$

Britannia Metal. A silvery alloy consisting of 80−90% tin, 5−15% antimony, and sometimes small percentages of copper, lead, and zinc. It finds use in bearings and some domestic articles.

Bromate. A salt or ester of a bromic acid.

Bromic (I) Acid (hypobromous acid). A yellow liquid, HBrO. It is a weak acid but a strong oxidizing agent.

Bromic(V) Acid. A colourless liquid, $HBrO_3$, made by adding sulphuric acid to barium bromate. It is a strong acid.

Bromide. See halide.

Bromine Symbol Br. A halogen element : a.n. 35; r.a.m. 70.909; r.d. 3.13; m.p. −7.2°C; b.p. 58.78°C. It is a red volatile liquid at room temperature, having a red brown vapour. Bromine is obtained from brines by displacement with chlorine. Large quantities are used to make 1, 2-dibromoethane as a petrol additive. The element was discovered in 1826 by Antoine Balard.

Bromine Trifluoride (BrF_3). A colourless fuming liquid which is prepared by direct combination of fluorine and bromine.

Bronze. A alloy of copper and tin, usually containing from 1 to 10% tin; special types contain from 5 to 10% aluminium (aluminium bronze); fractional percentages of phosphorus (phospher bronze) as deoxidizer; or low percentages of silicon (silicon bronze).

Bronze Blue. Any of a number of varieties of iron-blue pigments.

Brucite $Mg(OH)_2$. Natural magnesium hydroxide. Properties: Colour white, gray, greenish; luster pearly or waxy, Sp. gr. 2.39; Mohs hardness 2.5.

Burnable Poison. A neutron absorber (or poison), such as boron, which, when purposely incorporated in the fuel or fuel cladding of a nuclear reactor, gradually "burns up" (is changed into nonabsorbing material) under neutron irradiation.

By-product. A substance obtained during the manufacture of a main chemical product. For example, propanone is not now manufactured from propan-l-ol, but is obtained as a by-product in the manufacture of phenol.

C

Cadmium. A transition metal which is obtained as a by-product during the exraction of zinc. It is used to protect other metals from corrosion, as a neutron absorber in nuclear reactors, and in alkali-type batteries.

Symbol: Cd; mp. 320°C; b.p. 765°C; r.d. 8.7; p.n. 48; r.a.m. 112.41.

Cadmium Red, Orange, Scarlet. Pigments consisting of CdS(Se) prepared from a solution of $CdSO_4$ with added BaS and Se.

Cadmium Sulphate, $3CdSO_4.8H_2O$. The most important Cd salt.

Cadmium Sulphide, CdS. Occurs naturally as a green ockite and obtained as yellow precipitate from Cd^{2+} (aq) and H_2S. Used as a yellow pigment.

Cadmium Yellow. Natural cadmium sulphide. Used as a pigment.

Caesium. A soft silvery highly reactive element of the alkali metal group. It occurs in several silicate minerals, including pollucite ($CsAlSi_2O_6$) Caesium is used in photocells, as a catalyst, in the caesium atomic clock, and recently in ion-propulsion systems for spacecraft.

Symbol: Cs; m.p. 28.5°C; b.p. 690°C; r.d, 1.87; p.n. 55; r.a.m. 132 91.

Caesium Chemistry. Caesium is an element of group I, electronic configuration $6s^1$. It forms a single series of components in the +1 oxidation state; most compounds are predominently ionic. It also forms a series of lower oxides, *e.g.*, Cs_7O, $Cs_{11}O_3$ containing cluster cations.

Caesium Chloride, CsCl. A typical alkali halide formed from Cs_2CO_3 and HCl. The structure is that adopted by many

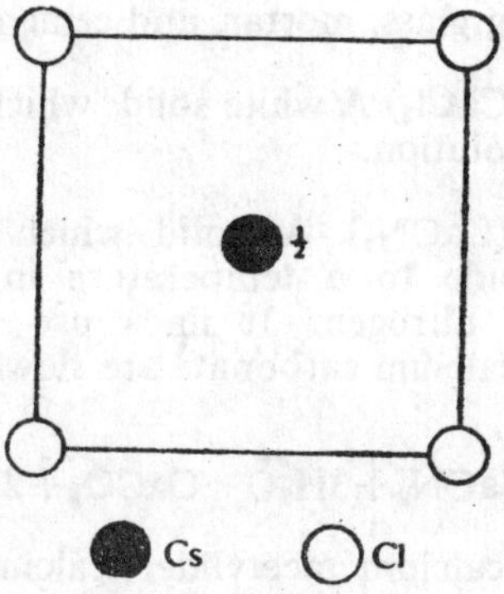

compounds AX with a large radius ratio of A : X. The Cl^- are at the corners of a cubic unit cell and the Cs^+ at the unit ce centre, each ion has a co-ordination number eight. Adopted by CsCl, CsI, AgLi, HgTl, etc.

Calamine, $ZnCO_3$ (British), $ZnSiO_4,H_2O$ (U.S.). Medicinal calamine is a basic zinc carbonate coloured with Fe_2O_3. Used either in lotions or as a powder for sunburn, sore skin and dermatitis.

Calcination. The formation of a calcium carbonate deposit from hard water.

Calcinite. A mineral form of potassium hydrogencarbonate, $KHCO_3$.

Calcite. A mineral form of calcium carbonate occurring in limestone and marble.

Calcium. Symbol Ca. A soft grey metallic element belonging to group II of the periodic table. Calcium compounds are common in the earth's crust *e.g.* limestone and marble ($CaCO_3$), gypsum ($CaSO_4.2H_2O$), and fluorite (CaF_2). The elem : is extracted by electrolysis of fused calcium chloride and is used as a getter in vacuum systems and a deoxidizer in producting nonferrous alloys.

Calcium Acetylide. See calcium dicarbide.

Calcium Bicarbonate. See calcium hydrogencarbonate.

Calcium Carbide. See calcium dicarbide.

Calcium Carbonate. ($CaCO_3$) A white solid which occurs naturally in two crystalline forms: calcite and aragonite. These minerals make up the bulk of such rocks as marble, limestone, and chalk. Calcium carbonate also occurs in the mineral dolomite ($CaCO_3 . MgCO_3$). It is sparingly soluble in water. Calcium carbonate is a basic raw material in the Solvay process and is used for making glass, mortar, and cement.

Calcium Chloride. ($CaCl_2$) A white solid which is readily soluble in water and the solution.

Calcium Cyanamide ($CaCN_2$). A solid which is prepared by heating calcium dicarbide to a temperature in excess of 800°C in an atmosphere of nitrogen. It finds use as a fertilizer because ammonia and calcium carbonate are slowly formed when water is added :

$$CaCN_2 + 3H_2O \rightarrow CaCO_2 + 2NH_3$$

Calcium Dicarbide (calcium acetylide, calcium carbide, CaC_2) A colourless solid when pure. It is produced on a large scale by heating calcium oxide with coke at a temperature in excess of 2000°C in an electric-arc furnace. Water is then added to give ethyne (C_2H_2), an important industrial organic chemical:

$$CaC_2 + 2H_2O \rightarrow C_2H_2 + Ca(OH)_2$$

Calcium Fluoride. A white crystalline solid, CaF_2; r.d. 3.2; m.p. 1360°C; b p. 2500°C. It occurs naturally as the mineral fluorite for fluorspar) and is the main source of fluorine.

Calcium Hydrogencarbonate (calcium bicarbonate, $Ca(HCO_3)_2$.] A solid formed when water containing carbon dioxide dissolves calcium carbonate :

$$CaCO_3 + H_2O + CO_2 \rightarrow Ca(HCO_3)_2$$

Calcium hydrogencarbonate is a cause of temporary hard water.

Calcium Hydroxide (**slaked lime**). A white solid, $Ca(OH)_2$, which dissolves sparingly in water. It is manufactured by adding water to calcium oxide, a process that evolves much heat and is known as slaking.

Calcium Nitrate A white deliquescent compound, $Ca(NO_3)_2$, that is very soluble in water. It can be prepared by neutralizing nitric acid with calcium carbonate and crystallizing it from solution as the tetrahydrate $Ca(NO_3)_2.4H_2O$, which exists in two monoclinic crystalline forms (a r.d.) 1.9; β r.d. 1.82). There is also a tribydrate, $(CaNO_3)_2.3H_2O$.

Calcium Octadecanoate (calcium stearate). An insoluble white salt, $Ca(CH_3)(CH_2)_{16}COO)_2$, which is formed when soap is mixed with water containing calcium ions and is the scum produced in hard-water regions.

Calcium Oxide (quicklime). A white solid compound, CaO, which is formed by heating calcium in oxygen or by the thermal decomposition of calcium carbonate to a temperature above 500°C.

$$CaCO_3(s) \rightleftharpoons CaO(s) + CO_2(g)$$

Calcium Phospahte (V). A white insoluble powder, $Ca_3(PO_4)_2$ which is found naturally in the mineral apatite $(Ca_5(PO_4)_3(OH,F,Cl)$ and as rock phosphate. It is also the main constituent of animal bones. It can be prepared by mixing solution, having calcium ions and hydrogenphosphate ions in the presence of an alkali :

$$HPO_4^{2-} + OH^- \rightarrow PO_4^{3-} + H_2O$$
$$3Ca^{2+} + 2PO_4^{3-} \rightarrow Ca_3(PO_4)_2$$

It is used extensively as a fertilizer.

Calcium Stearate. See calcium octadecanoate.

Calcium Sulphate. A white solid compound, $CaSO_4$. It occurs naturally as the mineral anhydrite, and as the dihydrate, gypsum, $CaSO_4.2H_2O$. Calcium sulphate is sparingly soluble in water.

Calgon. Tradename for a substance often added to detergents to remove unwanted chemicals that have dissolved in water and would otherwise react with soap to form a scum. Calgon consists of complicated polyphosphate molecules, which absorb dissolved calcium and magnesium ions. The metal ions get trapped within the Calgon molecules.

Caliche. An impure commercial form of sodium nitrate.

Californium. Symbol Cf. A radioactive metallic transuranic element which is belonging to the actinoids; a.n. 98; mass number of the most stable isotope 251 (half-life about 700 years). Nine isotopes are known; californium-252 is an intense neutron source, which makes it useful in neutron activation analysis and potentially useful as a radiation source in medicine. The element was first produced by G.T. Seaborg and associates in 1950.

Calomel. See mercury(I) chloride.

Calorizing. The process for coating or impregnating a surface of steel, Fe, Cu, etc., with Al. Articles so treated show a high resistance to many corrosive materials, and are not easily oxidized at high temperatures.

Calx. A metal oxide obtained by heating an ore to high temperatures in air.

Carat

1. A measure of fineness (purity) of gold. Pure gold is described as 24-carat gold. 14-carat gold contains 14 parts in 24 of gold, the remainder usually being copper.
2. A unit of mass equal to 0.200 gram which is used to measure the masses of diamonds and other gemstones.

Carbide Formers. Alloying elements which are added to steel (*e.g.* Mn, Cr, Mo, W, Ti, V, Nb) which stabilizer carbides and confer wear resistance when present in a coarse form or creep resistance when present in a fine coherent form.

Carbides. Derivatives of carbon with elements of lower electronegativity. Carbides are formed by direct interaction of the elements or metal plus suitable hydrocarbon. The most electropositive elements from salt-like carbides which are essentially ionic. Be_2C and Al_4C_3 contain C^{4-} and give CH_4 on hydrolysis. Group I, II, Cu group, Zn, Cd, and Al, La, Pr, Th, form acetylides containing $[C_2]^{2-}$ ions (HC_2H with water). Other lanthanide and actinide carbides contain C_2 units but give hydrocarbons in addition to ethyne on hydrolysis.

Carbon, C. At. no. 6, at. wt. 12.011, m.p. 3550°C, b.p. 4830°C. Occurs in both free and combined (CO_2, $(CO_3)^{2-}$, etc.) states. The crystalline forms are diamond (metastable) and graphite. Active carbon is widely used in industry. Steel is an iron-carbon system. Carbon fibres, produced by pyrolysis, confer great strength on plastic materials. ^{12}C is the standard of atomic mass, ^{13}C has nuclear spin of $\frac{1}{2}$ and can be used for n.m.r. measurements, ^{14}C is radioactive and is used for radiocarbon dating since it is formed in the atmosphere and absorbed only by living materials. Carbon burns in oxygen and reacts with halogents.

Carbonate. A salt of carbonic acid (containing the ion CO_3^{2-}).

Carbonate Minerals. A group of common rock-forming minerals having the anion CO_3^{2-} as the fundamental unit in their structure. The most important carbonate minerals are calcite, dolomite, and magnesite.

Carbonation. The solution of carbon dioxide in a liquid under pressure.

Carbon Black. A fine carbon powder which is made by burning hydrocarbons in insufficient air. It is used as a pignment and a filler (*e.g.* for rubber).

Carbon Cycle. One of the major cycles of chemical elements in the environment. Carbon (as carbon dioxide) is taken up from the atmosphere and incorporated into the tissues of plants in photosynthesis. It may then pass into the bodies of animals as the plants are eaten. During the respiration of plants, animals, and organisms that bring about decomposition, carbon dioxide is returned to the atmosphere. The combustion of fossil fuels (*e.g.* coal and peat) also releases carbon dioxide into the atmosphere.

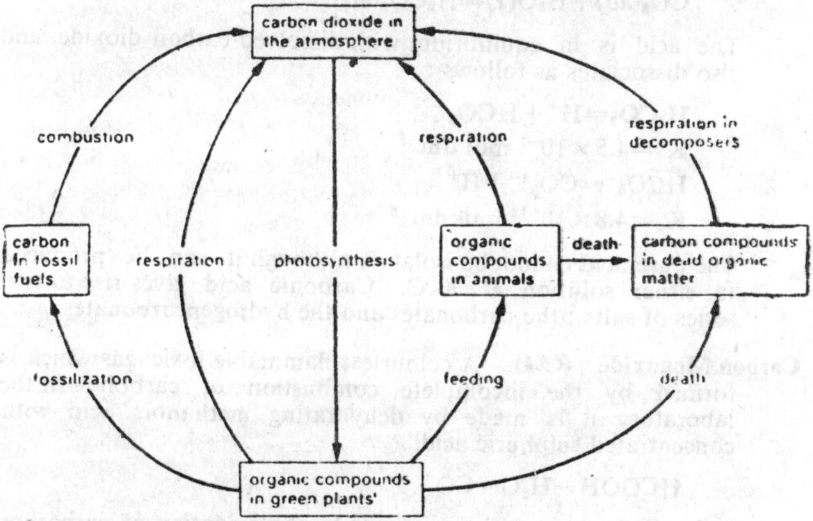

The carbon cycle in nature

Carbon Dating. ^{14}C is continuously formed in the atmosphere by the reaction

$$^{14}N + {}^{1}n \rightarrow {}^{14}C + {}^{1}H$$

The ^{14}C exchanges with ^{12}C in living organisms, but exchanges ceases on death. The radioactive ^{14}C content decays with a half-life of 3760 years. Hence the age of a once living material may be established by determining the amount of ^{14}C.

Carbon Dioxide (CO_2). A colourless odourless nonflammable gas which is formed when carbon burns in excess oxygen. It is a so produced by respiration. Carbon dioxide occurs in the atmosphere (0.03% by volume) and is converted in plants to carbohydrates by photosynthesis. In the laboratory it is prepared by the action of dilute acid on metal carbonates. Industrially, it is obtained as a by-product in certain processes, such as fermentation or the manufacture of lime. The main uses are as a refrigerant (solid carbon dioxide, called dry ice) and in fire extinguishers and carbonated drinks.

Carbon Fibres. Fibres of carbon in which the carbon is having an oriented crystal structure. Carbon fibres are made by heating textile fibres and are used in strong composite materials for use at high temperatures.

Carbonic Acid. A dibasic acid, H_2CO_3 formed in solution when carbon dioxide is dissolved in water.

$$CO_2(aq)+H_2O(l)\rightleftharpoons H_2CO_3(aq)$$

The acid is in equilibrium with dissolved carbon dioxide, and also dissociates as follows :

$$H_2CO_3\rightleftharpoons H^++HCO_3^-$$
$$K_a=4.5\times 10^{-7}\ \text{mol dm}^{-3}$$
$$HCO_3^-\rightleftharpoons CO_3^{2-}+H^+$$
$$K_a=4.8\times 10^{-11}\ \text{mol dm}^{-3}$$

The pure acid cannot be isolated, although it can be produced in either solution at 30°C. Carbonic acid gives rise to two series of salts : the carbonates and the hydrogencarbonates.

Carbon Monoxide (CO). A colourless flammable toxic gas which is formed by the incomplete combustion of carbon. In the laboratory it is made by dehydrating methanoic acid with concentrated sulphuric acid.

$$HCOOH-H_2O\rightarrow CO$$

Industrially, it could be produced by the oxidation of carbon or of natural gas, or by the water-gas reaction. It is a powerful reducing agent and is used in metallurgy.

Carbon Suboxide. See tricarbon dioxide.

Carbonyl. A complex in which C=O ligands are coordinated to a metal atom. A common example has been tetracarbonyl nickel(0), $Ni(CO)_4$.

Carbonylate ions. Ions having metal carbonyls. Anions may be prepared by the action of a base or an alkali metal on the carbonyl (e.g. $Mn_2(CO)_{10}$ plus pyridine or Na gives $(Mn(CO_5)^-)$. Cations may be prepared from a Lewis acid and a co-ordinating group on the carbonyl halide (e.g. $Mn(CO)_5Cl$ plus $AlCl_3$ plus CO gives $[Mn(CO_6]^+)$. These ions generally have a noble gas electronic structure (both $[Mn(CO)_5]^-$ and $[Mn(CO)_6]^+$ have the Kr electronic structure for Mn).

Carbonyl Chloride (Phosgene). A colourless gas, $COCl_2$, with an odour of freshly cut hay. It is used in organic chemistry as a chlorinating agent, and was formerly used as a war gas.

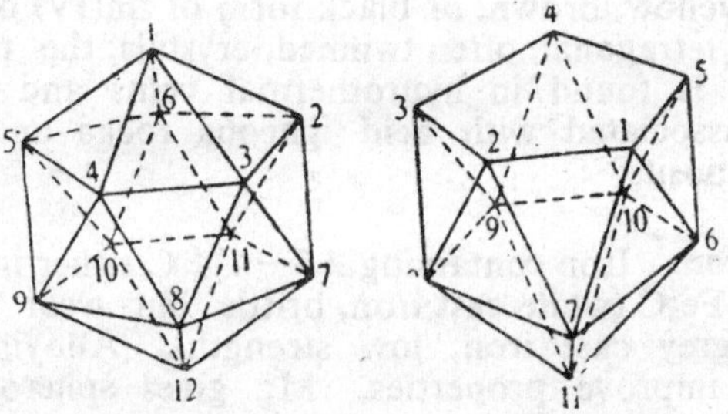

closo-$B_{10}C_2H_{12}$ (C at 1.2-or 1,7-)

nido-$[B_9C_2H_{11}]^{2-}$ (C at 1,2-or 3,5-)

Carboranes. Derivatives of the borane anions in which a CH group has formally replaced a BH^-, Prepared by the action of acetylenes and borane derivatives. The series contain closo (polyhedral) carboranes, *e.g.* 1,2- and 1,7-$B_{10}C_2H_{12}$ (based on an icosahedron), $B_7C_2H_9$ and nido (open) carboranes, e.g. $B_7C_2H_{13}$. Nido carborane anions, *e.g.* $[B_9C_2H_{11}]^{2-}$ form metal complexes, *e g.* $[(B_9B_2H_{11})_2\ Fe]^{2-}$ through the open pentagonal faces of the ions (dicarbollide derivatives). Similar derivatives are formed by, *e.g.*, the $[B_7C_2H_9]^{2-}$ ion (dicarbazapides).

Carborundum. Formally the trade name for SiC produced by the Carborundum Co. but in practice also used for SiC from other sources.

Carburizing. See case hardening.

Carnallite, KCl, $MgCl_2,6H_2O$. A principal component of many salt deposits. Used as a source of potassium salts.

Carnotite. A radioactive mineral consisting of hydrated uranium potassium vanadate $K_2(UO_2)_2(VO_4)_2\cdot nH_2O$. It varies in colour from bright yellow to lemon- or greenish-yellow. It is a source of uranium, redium, and vanadium.

Caro's Acid. See peroxosulphuric(VI) acid.

Case Hardening. A process for producing a hard, resistant surface on a tough core for steel components such as gear wheels, splines. The component is manufactured from a low carbon (tough) steel. It is then heated to about 910°C in an atmosphere rich in carbon, *e.g* in a hydrocarbon gas or packed in charcoal. The carbon diffuses into the surface of the steel to a depth of about 1 mm, forming a high carbon surface which is subsequently quenched to martensite for maximum hardness and wear resistance.

Cassiterite. A yellow, brown, or black form of tin(IV) oxide, SnO_2, that forms tetragonal, often twinned, crystals; the principal ore of tin. It is found in hydrothermal veins and metasomatic deposits associated with acid igneous rocks and in alluvial (placer) deposits.

Cast Iron, Pig Iron. Iron containing 2.5—4% C either in the form of cementite, Fe_3C (white cast iron, brittle but wear resistant) or graphite (grey cast iron, low strength). Alloying metals are added to improve properties. Mg gives spheroidal graphite (S.G.) cast irons which are almost as strong as steel but have good casting properties. SG cast iron may be heat treated to give strength or toughness. P gives ease of casting. Malleable cast irons are cast as white cast iron and then annealed.

Cathode Rays. In a low pressure gas discharge, electrons of varying velocities travel towards the cathode. These electrons are termed cathode rays.

Cathodic Protection. A method of metal protection based upon the fact that there is no dissolution of metal ions from the cathode of an electrochemical cell, since the cathode is more negative than the electrode potential of the particular metal. Cathodic protection is used to prevent corrosion of submerged or underground metal work such as ships, piers, boiler installations or underground pipe-lines.

Caustic. Describing a substance that is strongly alkaline (*e.g.* caustic soda).

Caustic Potash. See potassium hydroxide.

Caustic Soda. See sodium hydroxide.

Celestine, $SrSO_4$. Native strontium sulphate, a major source of Sr compounds. Mainly colourless but some specimens are sky-blue in colour.

Cement. A powdered mixture of calcium silicates and aluminates, which is obtained by heating limestone ($CaCO_3$) with clay, and grinding the result. When mixed with water reactions take place with the water (hence the name hydraulic cement), and a hard solid aluminosilicate is formed.

Cement Fondu. A hydraulic cement composed chiefly of calcium aluminate.

Cement, High Alumina. A dark-coloured cement produced by heating a mixture of bauxite and limestone to about 1850 K. Varying amounts of iron oxides and silica are also present.

This cement is very resistant to attack by sulphates, sea water and acid waters. It is also used with crushed firebrick to produce refractory concrete.

Cement, Rapid-hardening. Portland or other which has a shorter setting time than ordinary Portland cement. This is achieved by slight changes in composition or kiln conditions or by the addition of calcium chloride to the water used.

Cement, Rapid-hardening. Portland or other cements which harden rapidly so that concrete may be used in three to four days.

Cement, Supersulphated. A cement made from granulated blast furnace slag of fairly high alumina content, anhydrite or burnt gypsum and about 5% Portland cement. It has a high sulphate resistance and is used largely for work underground and for work in sea water.

Cementation Process of Steel Manufacture. An obsolete process of steel manufacture involving direct interaction of wrought iron and charcoal in an air-free furnace for days followed by welding and hammering.

Cementite (Fe_3C). A constituent of certain cast irons and steels. The presence of cementite increases the hardness of the metal.

Cerium. Symbol Ce. A silvery metallic element belonging to the lanthanoids; a.n. 58; r.a.m. 140.12; r.d. 6.77 (20°C); m.p. 798°C; b.p. 3433°C. It occurs in allanite, bastnasite, cerite, and monazite. Four isotopes occur naturally : cerium—136,—138,—140, and—142; fifteen radioisotopes have been identified. It was discovered by M.H. Klaproth in 1803.

Carmet. A composite material consisting of a ceramic in combination with a sintered metal, used when a high resistance to temperature, corrosion, and abrasion is needed.

Cerussite. An ore of lead consisting of lead carbonate, $PbCO_3$. It forms well-shaped orthorhombic crystals.

Chalcedony. A fibrous form of SiO_2 with a structure similar to that of quartz. Similar to onyx and agate. Pure samples are white but chalcedony is generally coloured by impurities.

Chalcocite, Copper Glance, Cu_2S. A black copper ore.

Chalconens. The elements oxygen, sulphur, selenium, tellurium, polonium of group VI. Chalcogenid contain the X^{2-} species.

Chalcogens. See group VI elements.

Chalconides. Binary compounds formed between metals and group VI elements; *i.e.* oxides, sulphides, selenides, and tellurides.

Chalcopyrite (Copper Pyrites). A brassy yellow mineral consisting of a mixed copper-iron sulphide, $CuFeS_2$, crystallizing in the tetragonal system; the principal ore of copper.

Chalk. A natural form of calcium carbonate ($CaCO_3$) formed originally by marine organisms. (Blackboard chalk is calcium sulphate, $CaSO_4$).

Charcoal. A form of carbon produced by slow combustion of wood in a deficiency of oxygen. Formerly used as a fuel and reducing agent. Used to absorb gases. Has a low-order graphite structure.

Charge Transfer Band or Spectrum. An absorption band which corresponds to transfer of an electron from one atom or group to another.

Charge Transfer Complexes. Complexes in which there is weak interaction, charge transfer, between the donor and the acceptor properties which are generally induced in the acceptor. Examples of charge transfer complexes include the complexes formed between aromatic derivatives and halogenes. The class is practically indistinguishable from true complexes. Charge (electron) transfer takes place in all complexes.

Cheddite. Any of a group of high explosives made from nitro compounds mixed with sodium or potassium chlorate.

Chelate. An inorganic complex in which a ligand is coordinated to a metal ion at two (or more) points, so that there is a ring of atoms including the metal. The process is known as chelation.

Chelate Compound. A compound in which co-ordinate links complete a closed ring, *e.g.* in beryllium acetylacetonate.

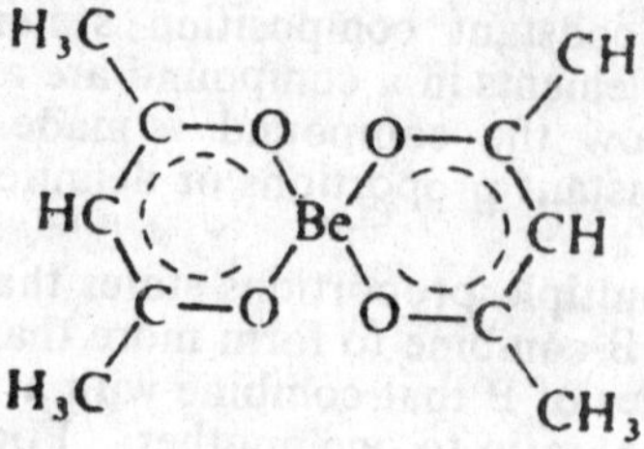

Chelate Effect. The enhanced stability of a complex containing chelate rings as compared with the complex containing monodentate ligands. E.g. ethylenediamine complexes.

are more stable than ethylamine complexes.

In complexes of chelates there are a number of types of isomerism which may occur. In a tris-(ethylenediamine) octahedral complex two optically active isomers occur (often denoted Δ and Λ).

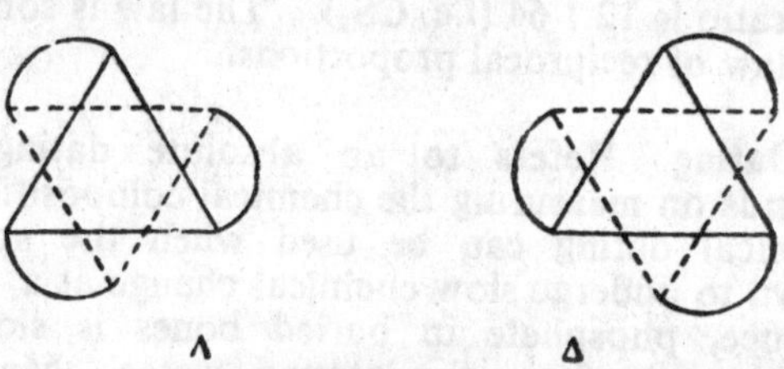

Λ Δ

A ligand such as ethylenediamine is not planar and has a spiral form (gauche) giving rise to further forms. When the direction of one C–C bond in one ethylenediamine is parallel to the 3-fold axis the isomer is termed the lel form, when it is inclined to the axis it is termed the ob form.

Chemical Bond. A link which is formed between atoms that leads to an aggregate of sufficient stability to be regarded as an independent molecular species. Chemical bonds include covalent bonds, electrovalent bonds, coordinate bonds, and metallic bonds. Some authors also include hydrogen bonds and van der Waals forces.

Chemical Combination. The combination of elements to give compounds. There are three laws of chemical combination :

(1) The law of constant composition states that the proportions of the elements in a compound are always the same, no matter how the compound is made. It is also called the law of constant proportions or definite proportions.

(2) The law of multiple proportions states that when two elements A and B combine to form more than one compound, then the masses of B that combine with a fixed mass of A are in simple ratio to one another. For example, carbon forms two oxides. In one, 12 grams of carbon is combined with 16 grams of oxygen (CO); in the other 12 g of carbon is combined with 32 grams of oxygen (CO_2). The oxygen masses combining with a fixed mass of carbon are in the ratio 16.32, *i.e.* 1 : 2.

(3) The law of equivalent proportion states that if two elements A and B each form a compound with a third element C, then a compound of A and B will contain A and B in the relative proportions in which they reac with C. For example, sulphur and carbon both form compounds with hydrogen. In methane 12 g of carbon react with 4 g of hydrogen. In hydrogen sulphide, 32 g of sulphur react with 2 g of hydrogen (*i.e.* 64 g of S for 4 g of hydrogen). Sulphur and carbon form a compound in which the C : S ratio is 12 : 64 (*i.e.* CS_2). The law is sometimes called the law of reciprocal proportions.

Chemical Dating. Refers to an absolute dating technique that depends on measuring the chemical composition of a specimen. Chemical dating can be used when the specimen has been known to undergo slow chemical change at a known rate. For instance, phosphate in buried bones is slowly replaced by fluoride ions from the ground water. Measurement of the proportion of fluorine present gives a rough estimate of the time that the bones have been in the ground. Another, more accurate, method depends on the fact that amino acids in living organisms have been L-optical isomers. After death, these racemize and the age of bones could be estimated by measuring the relative amounts of D- and L- amino acids present.

Chemical Reaction. Refers to a change in which one or more chemical elements or compounds (the reactants) form new compounds (the products).

Chert. See flent.

China Clay, Kaolin. A while powdery material arising from the decomposition of feldspars in granites. The clay is washed out with water and purified by settling. Used in paper and pottery manufacture and as a filler in textiles and paint.

China-stone, Cornish Stone. A granite rock having partially kaolinized feldspar. Used as a flux for ceramics

Chinese White. Zinc oxide.

Chlorate. A salt of a chloric acid.

Chloric (I) Acid (hypochlorous acid, HClO). A colourless liquid produced when chlorine is dissolved in water. It is a bleach and gives chlorine water its disinfectant properties.

Chloric (III) Acid (chlorous acid $HClO_2$). A pale yellow liquid which is produced by reacting chlorine dioxide with water. It is a weak acid and oxidizing agent.

Chloric (V) Acid (chloric acid, $HClO_3$). A colourless liquid having a pungent odour, formed by the action of dilute sulphuric acid on barium chlorate. It is a strong acid and has bleaching properties.

Chloric (VII) Acid (perchloric acid $HClO_4$). A colourless liquid that fumes strongly in moist air. It is made by vacuum distillation of a mixture of potassium perchlorate and concentrated sulphuric acid. In contact with organic material it is dangerously explosive.

Chloride. See halide.

Chloirnation

1. A chemical reaction in which a chlorine atom is introduced into a compound.

2. The treatment of water with chlorine to distinfect it.

Chlorine Symbol Cl. A *halogen element; a.n. 17; r.a.m. 35.453; d. 3.214 g dm^{-3}; m.p. —100.98°C; b.p—346° C. It is a poisonous greenish-yellow gas and occurs widely in nature as sodium chloride in seawater and as halite (NaCl), carnallite ($KCl.MgCl_2.6H_2O$), and sylvite (KCl). It is manufactured by the electrolysis of brine and also obtained in the *Downs process for making sodium. It has many applications, including the chlorination of drinking water, bleaching, and the manufacture of a large number of organic chemicals.

It reacts directly with many elements and compounds and is a strong oxidizing agent. Chlorine compounds contain the element in the 1, 3, 5 and 7 oxidation states. It was discovered by Karl Scheele in 1774 and Humphry Davy confirmed it as an element in 1810.

Chlorine Dioxide (ClO_2). An orange gas which is a powerful oxidizing agent and its explosive properties in the presence of a reducing agent were used to make one of the first matches. It is widely used in the purification of water and as a bleach in the flour and wood pulp industry. On an industrial scale an aqueous solution of chlorine dioxide is prepared by passing nitrogen dioxide up a tower packed with a fused mixture of aluminium oxide and clay, down which a solution of sodium chlorate flows.

Chlorine Monoxide See dichlorine oxide.

Chlorite. A chlorate (III) salt; *i.e.* a salt of chloric (III) acid (chlorous acid).

Chloroplatinic Acid. A reddish crystalline compound, H_2PtCl_6 made by dissolving platinum in aqua regia.

Chlorous Acid. See chloric(III) acid.

Chondrodite, $2Mg_2SiO_4$, $Mg(OH)_2$. Chondrodite gives its name to a series of silicates containing discrete SiO_4^{4-} ions. Other chondrodite minerals are norbergite Mg_2SiO_4, $Mg(OH)_2$, humite $3Mg_2SiO_4$, $Mg(OH)_2$ and clinohumite $4Mg_2SiO_4$, $Mg(OH)_2$. Fluoride can replace hydroxyl.

Chromate. A salt containing the ion CrO_4^{2-}.

Chrome. Having a chromium plated surface.

Chrome Alum, $KCr(SO_4)_2$, $12H_2O$. A typical alum used as a gelatine hardener*.

Chrome Green, Cr_2O_3. See chromium oxides.

Chrome Iron Ore. A mixed iron-chromium oxide, $FeO.Cr_2O_3$, used to make ferrochromium for chromium steels.

Chrome Red. A basic lead chromate, $PbO.PbCrO_4$, used as a red pigment.

Chrome Yellow. Lead chromate, $PbCrO_4$, used as a pigment.

Chromic. Outdated nomenclature for chromium (III) compounds.

Chromic Acid. A hypothetical acid, H_2CrO_4, konwn only in chromate salts.

Chromic Anhydride. See chromium (VI) oxide.

Chromic Compounds. Compounds having chromium in a higher (+3 or +6) oxidation state; *e.g.* chromic oxide is chromium (VI) oxide (CrO_3).

Chromic Oxide. See chromium (III) oxide.

Chromite, $FeCr_2O_4$. Brownish-black mineral with the spinel structure used as a source of Cr and its compounds.

Chromium Symbol Cr. A hard silvery transition element; a.n. 24; r.a.m. 52,000; r.d. 7.19; m.p. 1900°C; b.p. 2640°C. The main ore has been chromite ($FeCr_2O_4$). The metal is extracted by heating chromite with sodium chromate, from which chromium can be obtained by electrolysis. Alternatively, chromite can be heated with carbon in an electric furnace to give ferrochrome, which is used in making alloy steels. The metal is also used as a shiny decorative electroplated coating and in the manufacture of certain chromium compounds. The element was discovered in 1797 by Vauquelin.

Chromium (II) Oxide. A black insoluble powder, CrO which is prepared by oxidizing chromium amalgam with air. At high temperatures hydrogen reduces it to the metal.

Chromium (III) Oxide (chromic oxide, chromium sesquioxide, Cr_2O_3). A green powder that is isomorphous with iron (III) oxide and aluminium (III) oxide. Chromium (III) oxide is prepared by gently heating chromium (III) hydroxide or by heating ammonium dichromate. Chromium (III) oxide is used as a pigment in the paint and glass industries.

Chromium (IV) Oxide (chromium dioxide, CrO_2) A black solid which is prepared by heating chromium (III) hydroxide in oxygen at a temperature of 300°—350°C. Chromium (IV) oxide has been very unstable.

Chromium (VI) Oxide (chromium trioxide, chromic anhydride, CrO_3). A red crystalline solid formed when concentrated sulphuric acid is added to a cold saturated solution of potassium dichromate. Chromium (VI) oxide is readily soluble in water, forming a solution that contains several of the polychromic acids. On heating it decomposes to yield chromium (III) oxide. Chromium (VI) oxide finds use as an oxidizing reagent.

Chromium Potassium Sulphate. A red crystalline solid, $K_2SO_4\,Cr_2(SO_4)_3.24H_2O$; r.d. 1.91. It is used as a mordant.

Chromium Steel. Any of a group of stainless steels having 8—25% of chromium. A typical chromium, steel might contain 18% of chronium, 8% of nickel, and 0.15% of carbon. Chromium steels have been highly resistant to corrosion and are used for cutlery, chemical plant, ball beaiings, etc.

Chromous Compounds. Compounds having chromium in its lower (+2) oxidation state ; *e.g.*, chromous chloride is chromium (II) Chloride ($CrCl_2$).

Chromous Oxide. See Chromium (II) oxide.

Chromyl Chloride. (CrO_2Cl_2). A dark red covalent liquid which is prepared either by distilling a dry mixture of potassium dichromate and sodium chloride with concentrated sulphuric acid or by the action of concentrated sulphuric acid on chromium (VI) oxide dissolved in concentrated hydrochloric acid. Chromyl chloride gets hydrolysed by water and with solutions of alkalis it undergoes immediate hydrolysis to produce chromate ions. Chromyl chloride oxidizes methyl groups at the ends of aromatic side chains to aldehyde grouping (Etard's reaction).

Chrysotile. See serpentine.

Ciment Fondu. An aluminous cement which sets slowly but hardens rapidly. Resistant to sea water and to sulphates in solution. Also used as a refractory cement.

Cinnabar. A bright red mineral form of mercury (II) sulphide, HgS, crystallizing in the hexagonal system; the principal ore of mercury.

Clark Process. See hardness of water.

Clathrate Compounds. Molecular compounds formed by the inclusion of molecules of one type in holes in the lattice of another. Organic compounds, particularly nitro-compounds and quinol, form lattices which are able to take up small molecules such as the noble gases. One of the best known clathrate compounds is the nickel ammonia cyanide benzene clathrate. Use in the separation of *e.g.* gases.

Claude Process. A process which is used for liquefying air on a commercial basis. Air under pressure is used as the working substance in a piston engine, where it does external work and cools adiabatically. This cool air is fed to a counter-current heat exchanger, where it reduces the temperature of the next intake of high-pressure air. The same air is re compressed and used again, and after several cycles eventually liquefies. The process was perfected in 1902 by the French scientist Georges Claude (1870-1960).

Claudetite. A mineral form of arsenic(III) oxide. As_4O_6.

Clay. A fine-grained desposit consisting chiefly of clay minerals. In soil science clay is regarded as a soil with particles less than 0.002 mm in size.

Clay Minerals. Very small particles, chiefly hydrous silicates of aluminium, sometimes with magnesium and/or iron substituting for all or part of the aluminium, that are the major constituents of clay materials. The chief groups of clay minerals are: kaolinite, $Al_4Si_4O_{10}(OH)_6$, the chief constituent of kaolin; halloysite, $Al_4Si_4(OH)_8O_{10}$ $4H_2O$; illite, $KAl_4(SiAl)_8O_{18}$ $2H_2O$; montmoril lonite, $(Na\ Ca)_{0.33}(Al,Mg)_2Si_4O_{10}(OH)_2$ nH_2O, formed chiefly through alteration of volcanic ash: vermiculite, $(Mg.Fe, Al)_3(Al,Si)_4O_{10}(OH)_2.4H_2O$, used as an insulating material and potting soil.

Cleveite. A variety of uraninite ($UO_{2-2.5}$) containing up to 10% lanthanides and also appreciable quantities of He.

Coal. A brown or black carbonaceous deposit derived from the accumulation and alteration of ancient vegetation, which originated largely in swamps or other moist environments.

As the processes of coalification (*i.e.* the transformation resulting from the high temperatures and pressures) continue, there is a progressive transformation of the deposit: the proportion of carbon relative to oxygen rise and volatile substances and water are driven out. The various stages in this process are referred to as the ranks of the coal. In ascending order, the main ranks of coal are :lignite (or brown coal), which is soft, brown, and has a high moisture content; subbituminous coal, which is used chiefly by generating stations; bituminous coal, which is the most abundant rank of coal; semibituminous coal; semianthracite coal, which has a fixed carbon content of between 86% and 92%; and anthracite coal, which is hard and black with a fixed carbon content of between 92% and 98%.

Most deposits of coal were formed during the Carboniferous and Permian periods.

Coal Gas. A fuel gas produced by the destructive distillation of coal which is made by heating coal in the absence of air in local gas works. Typically, it contained hydrogen (50%), methane (35%), and carbon monoxide (8%). By-products of the process were coal tar and coke.

Cobalt. A transition metal occurring in Canada in association with nickel. It is used in alloys for magnets, cutting tools, and electrical heating elements.

Symbol : Co; m.p. 1495°C: b.p. 2880°C; r.d. 8 9; p.n. 27; r.a.m. 58.93.

Cobaltic Oxide. See cobalt(III) oxide.

Cobaltous Oxide. See cobalt(II) oxide.

Cobalt(II) Oxide. A pink solid, CoO; cubic; r.d. 6.45; m.p. 1795°C. The addition of potassium hydroxide to a solution of cobalt (II) nitrate gives a bluish-violet precipitate, which on boiling is converted to pink impure cobalt (II) hydroxide. On heating this in the absence of air, cobalt(II) oxide is formed. The compound is readily oxidized in air to form tricobalt tetroxide, CO_3O_4, and readily gets reduced by hydrogen to the metal.

Cobalt (III) Oxide (cobaltic oxide, CO_2O_3). A dark grey powder which is formed by the thermal decomposition of either cobalt(II) nitrate or carbonate in air. If heated in air, cobalt (III) oxide undergoes further oxidation to yield tricobalt tetroxide, CO_3O_4.

Cobalt Steel. Any of a group alloy steels having 5-12% of cobalt, 14—20% of tungsten, usually with 4% of chromium and 1—2% of vanadium. They have been very hard but somewhat brittle. Their main use is in high-speed tools.

Coinage Metals. A group of three malleable ductile metals forming subgroup IB of the periodic table : copper (Cu), silver (Ag), and gold (Au). Their outer electronic configurations have the form $nd^{10}(n+l)s$. They are generallyclassified with the transition elements.

Coke. A form of carbon which is prepared by the destructive distillation of coal. Coke is used for blast-furnaces and other metallurgical and chemical processes requiring a source of carbon. Lower-grade cokes, made by heating the coal to a lower temperature, are used as smokeless fuels for domestic heating.

Columbium. A former name for the element niobium.

Common Salt. Sodium chloride.

Complex. Any compound in which the bonding is by interaction of the electrons of the donor with empty orbitals of the acceptor. In some complexes the electron flow may take place in both directions simultaneously. The interaction may take place between charged or uncharged species.

Where the structure is known a complex species comprising the acceptor and its ligands is formulated within square brackets, *e.g.* $[Co(NH_3)_6]Cl_3$. Bridging ligands are designated

μ-L, *e.g.* $Fe_2(CO)_9$ is $[(OC)_3Fe(\mu\text{-}CO)_3Fe(CO)_3]$. Hapto designates the number of ligand atoms actually bonded to the acceptor, *e.g.* $[(h^5-C_5H_5)Mn(CO)_3]$ has 5 carbon atoms (plus 3 carbonyls) bonded to manganese.

Complex Ion. A complex ion is formed by the co-ordination of other ions or molecules to an ion to form a stable entity. Thus the Co^{3+} ion and ammonia give $[CO(NH_3)_6]^{3+}$ whilst the Fe^{3+} ion and cyanide ions give the $[Fe(CN)_6]^{3-}$ complex. Depending upon the stability of the complex a solution of a complex salt may or may not give the reactions of the individual components of the complex. It should be noted that hydrated salts generally contain aquo-complexes, *e.g.* $[Cu(H_2O)_4]^{2+}$ and that the anions of oxy-salts may be considered as being complexes formed by (mostly hypothetical) positive ions and oxide ions, *e.g.* the NO_3^- ion as N^{5+} and $3O^{2-}$.

Compound. A substance formed by the combination of elements in fixed propertions. The formation of a compound involves a chemical reaction; *i.e.* there is a change in the configuration of the valence electrons of the atoms. Compounds, unlike mixtures, cannot be separated by physical means.

Condy's Fluid. A mixture of calcium and potassium permanganates (manganate (VII) used as an antiseptic.

Configuration

1. Refers to the arrangement of electrons about the nucleus of an atom. Configurations are represented by symbols, which contain :

(1) An integer, which is the value of the principal quantum number (shell number).

(2) A lower-case letter representing the value of the azimuthal quantum number (1), *i.e.*

s means 1=0, p means 1=1,
d means 1=2, f means 1=3.

(3) A numerical superscript giving the number of electrons in that particular set; for example, $1s^2$, $2p^3$, $3d^5$.

The ground state electronic configuration (*i.e.* the most stable or lowest energy state) may then be represented as follows, for example, He, $1s^2$; N, $1s^2\ 2s^22p^5$. These have been identical to the 'electron box' models, which are often met. However, elements are commonly abbreviated by using an inert gas to represent the 'core', *e.g.* Zr, $[Kr]4d^25s^2$.

2. The arrangement of atoms or groups in a molecule.

Congeners. Elements that belong to the same group in the periodic table.

Conservation of Mass, Law of. Formulated by Lavoisier in 1774, this law states that matter cannot be created or destroyed. Thus in a chemical reaction the total mass of the products equals the total mass of the reactants (the term 'mass' must include any solids, liquids, and gases—including air—that participate).

Constantan. An alloy having an electrical resistance that varies only very slightly with temperature (over a limited range around normal room temperatures). It consists of copper (50—60%) and nickel (40—50%) and is used in resistance wire, thermocouples, etc.

Constant Proportions. See chemical combination.

Contact Process. An industrial process which is for the manufacture of sulphuric acid. Sulphur (IV) oxide and air are passed over a heated catalyst (vanadium (V) oxide or platinum) and sulphur (VI) oxide is produced:

$$2SO_2+O_2 \rightarrow 2SO_3$$

The sulphur (VI) oxide gets dissolved in sulphuric acid

$$SO_3+H_2SO_4 \rightarrow H_2S_2O_7$$

The resulting oleum is diluted to yield sulphuric acid.

Converter. The reaction vessel in the Bessemer process or some similar steel-making process.

Co-ordinate Bond. The linkage of two atoms by a pair of electrons, being provided by one of the atoms (the donor). The co-ordinate bond is formally identical with the covalent bond. An atom capable of accepting the electrons is the acceptor, the molecule donating the electrons is the donor or ligand.

Co-ordination Arrangements. The arrangements of atoms or ligands around a central atom depends upon the sizes of the atoms and the electronic configuration of the central atom (lone-pairs of electrons occupy positions in space). For details of the most common co-ordination arrangements see under the individual headings. Common arrangements are:

Co-ordination
number

2 linear $BeCl_2$, CO_2
bent H_2O

3 trigonal planar BF_3
pyramidal NH_3

4 tetrahedral CH_4.$[NiCl_4]^{2-}$
square planar $[PtCl_4]^{2-}$
5 trigonal bipyramidal PF_5
square pyramidal IF_5
6 octahedral NaCl.$[Co(NH_3)_6]^{3+}$
trigonal prismatic NiAs
7 pentagonal bipyramid ReF_7
face-centred octahedron $[NbOF_6]^{3-}$
face-centred trigonal prism $[TaF]_7^{-2}$
8 cubic CsCl
square anti-prism $[Mo(CN)_8]^{4-}$
dodecahedral $[Mo(CN)_8]^{4-}$
12 icosahedral $[Ce(NO_3)_6]^{2-}$
cubic close-packed perovskite

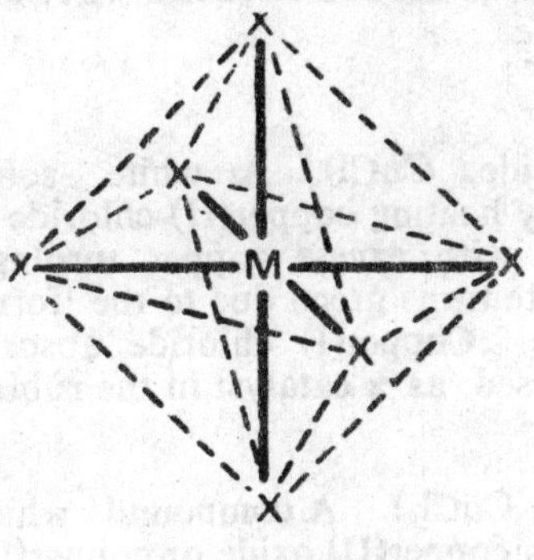

octahedral

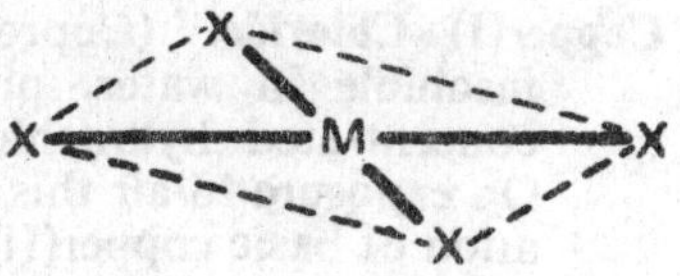

square planar

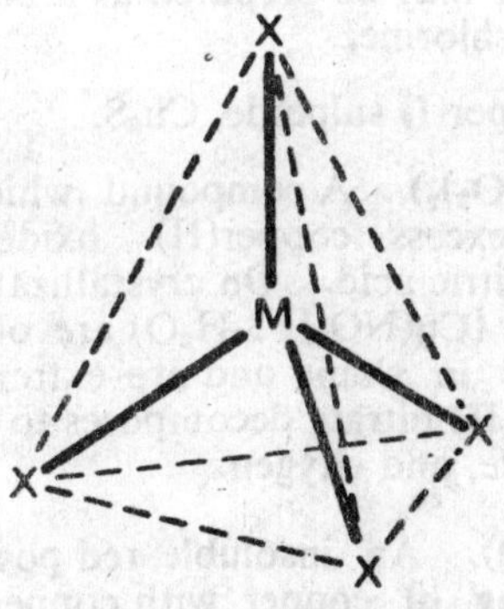

tetrahedral

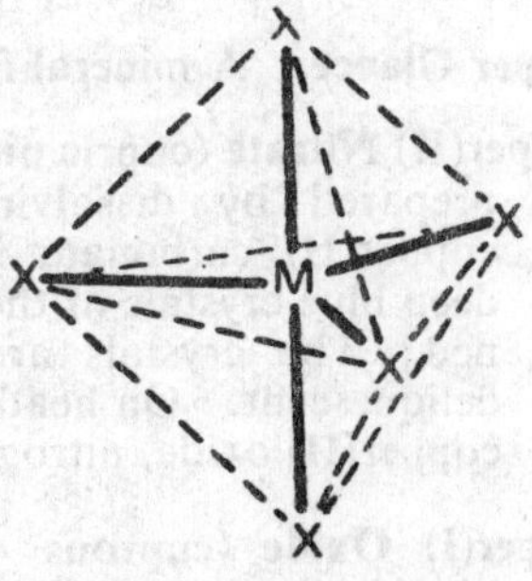

trigonal-bipyramid

Co-ordination Compound. A compound having a co-ordinate bond or bonds.

Co-ordination isomerism. Isomerism in co-ordination compounds in terms of the actual arrangement of the atoms.

Co-ordination Numbers. The number of groups or ions surrounding a particular molecule or ion in a crystal or in solution; in complexes the total number of co-ordinated and covalently bonded groups surrounding the central atom. For elements Li-F the maximum co-ordination number for covalent compounds is 4, but 6 co-ordination can occur in ionic compounds. For elements Na-Cl the maximum co-ordination number is 6.

Copper. A transition metal which is found in nature principally as the sulphide. It could be extracted by roasting the ore in a controlled air supply and purified by electrolysis of copper(II) sulphate solution using impure copper as the anode and pure copper as the cathode. Copper is used in electrical wires and in such alloys as brass and bronze.

Copperas. See iron(II) sulphate.

Copper(I) Chloride. (Cuprous chloride, CuCl). A white solid, insoluble in water, prepared by heating copper(II) chloride in concentrated hydrochloric acid with excess copper turnings. On exposure to air this precipitate turns green due to the formation of basic copper(II) chloride Copper(I) chloride absorbs carbon monoxide gas. It is used as a catalyst in the rubber industry.

Copper(II) Chloride. (Cupric chloride, $CuCl_2$). A compound which is prepared by dissolving excess copper(II) oxide or copper(II) carbonate in dilute hydro-chloric acid. On crystallization, emerald green crystals of the dihydrate ($CuCl_2 2\text{-}H_2O$) are obtained. The anhydrous chloride may be prepared as a brown solid by burning copper in excess chlorine.

Copper Glance. A mineral form of copper(I) sulphide, Cu_2S.

Copper(II) Nitrate (cupric nitrate, $Cu(NO_3)_2$). A compound which is prepared by dissolving either excess copper(II) oxide or copper(II) carbonate in dilute nitric acid. On crystallization, deep blue crystals of the trihydrate ($Cu(NO_3)_2\ 3\text{-}H_2O$) are obtained. The crystals are prismatic in shape and are extremely deliquescent. On heating, copper(II) nitrate decomposes to give copper(II) oxide, nitrogen(IV) oxide, and oxygen.

Copper(I) Oxide (cuprous oxide, CuO). An insoluble red powder which is prepared by the heating of copper with copper(II) oxide or the reduction of an alkaline solution of copper(II) sulphate. It is oxidized to copper(II) oxide when heated in air. It is used in the glass industry.

Copper(II) Oxide. A black insoluble solid, CuO; monoclinic; r.d. 6.3; m.p. 1326°C. It is obtained by heating either copper(II) carbonate or copper(II) nitrate. It decomposes on heating above 800°C to copper(I) oxide and oxygen.

Copper Pyrites. See chalcopyrite.

Copper(II) Sulphate (cupric sulphate, $CuSO_4$). A compound prepared as the hydrate by the action of dilute sulphuric acid on copper(II) oxide or copper(II) carbonate. On crystallization, blue triclinic crystals of the pentahydrate ($CuSO_4.5H_2O$) are formed. Industrially copper(II) sulphate is obtained by passing air through a hot mixture of dilute sulphuric acid and scrap copper. The solution formed is recycled until the concentration of the copper(II) sulphate is sufficient. The monohydrate ($CuSO_4 H_2O$) is formed at 100°C and the anhydrous salt at 250°C. Anhydrous copper(II) sulphate is white; it is extremely hygroscopic and turns blue on absorption of water.

Copper(II) sulphate is used as a wood preservative, a fungicide (in Bordeaux mixture), and in the dyeing and electroplating industries.

Corrosion. Chemical or electrochemical attack on the surface of a metal.

Corundum (emery, Al_2O_3). A naturally occurring form of aluminium oxide that sometimes contains small amounts of iron and silicon (IV) oxide. Ruby and sapphire are impure crystalline forms. It is used in various polishes, abrasives, and grinding wheels.

Covalent Bond. A bond formed by the sharing of an electron pair between two atoms. The covalent bond is conventionally represented as a line, thus H-Cl indicates that between the hydrogen atom and the chlorine atom there is an electron pair formed by electrons of opposite spin and implying that the binding forces are strongly localized between the two atoms.

Covalent Radius. The equilibrium distance between two atoms joined by a covalent bond of a given type (*i.e.* single, double or triple) is found to be, within close limits, constant. Thus the C—C bond distance of 1.64 A found is diamond is the same as that found in cyclic saturated hydrocarbons. In general the A—B covalent bond distance is the mean of the A—A and B—B bond distances. Thus, a set of covalent radii may be drawn up such that the sum of any two distances gives the distance between the corresponding atoms in a covalent molecule. The covalent radii are, in general, half the internuclear distances in the elements, except when the latter are very electronegative (*e.g.* fluorine). For values see table under radii.

Covellite, CuS. A native copper ore which can be ground to a blue powder. Has the structure $Cu_4^{I}Cu_2^{II}(S_2)_2S_2$.

Creep. The slow yielding of metals and alloys (generally at high temperatures) under a load which does not cause fracture. The process is diffusion controlled.

Cristobalite. A mineral form of silicon(IV) oxide, SiO_2.

Cryolite. A rare mineral form of sodium aluminofluoride, Na_3AlF_6, which crystallizes in the monoclinic system. It is used chiefly as a flux in the production of aluminium from bauxite.

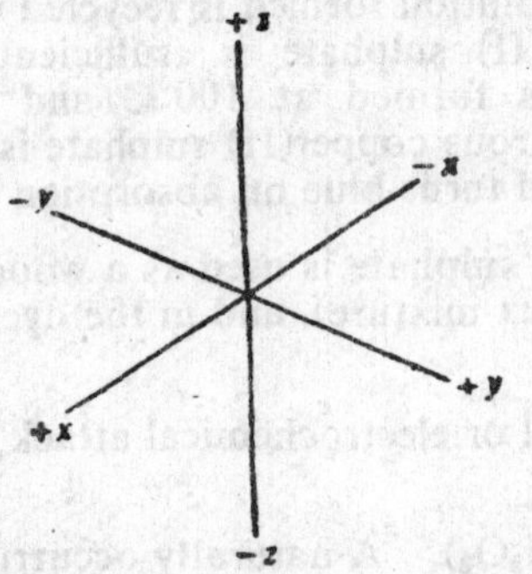

Crystal Field Theory. An approach to transition metal compounds which explains the u.v. and visible spectra (electronic) and magnetic properties in terms of a central ion surrounded by electron-rich ligands which split the d orbitals into a pattern determined by the geometry of the co-ordination, the distance of approach of the ligands and the tendency to covalent bonding.

Cupellation. Refers to a method of separating noble metals (*e.g.* gold or silver) from base metals (*e.g.* lead) by melting the mixture with a blast of hot air in a shallow porous dish (the cupel). The base metals are oxidized, the oxide being carried away by the blast of air or absorbed by the porous container.

Cuprammonium ion. The tetraamminecopper(II) ion $[Cu(NH_3)_4]^{2+}$.

Cupric Compounds. Compounds having copper in its higher (+2) oxidation state; *e.g.* cupric chloride is coper(II) chloride ($CuCl_2$).

Cupric Nitrate. See copper(II) nitrate.

Cupric Oxide. See copper(II) oxide.

Cupric Sulphate. See copper(II) sulphate.

Cuprite. A red mineral cubic form of copper(I) oxide, Cu_2O.

Cupronickel. A series of strong coppernickel alloys having up to 45% nickel. Those containing 20% and 30% nickel are very malleable, can be worked cold or hot, and are very corrosion-resistant. Cupronickel with 25% nickel is widely used in coinage.

Cuprous Chloride. See copper(I) chloride.

Cuprous Compounds. Compounds containing copper in its lower (+1) oxidation state; *e.g.* cuprous chloride is copper(I) chloride (CuCl).

Cuprous Oxide. See copper(I) oxide.

Curium, Cm. At.no. 96, m.p. 1340°C, b.p.~3540°C, ^{242}Cm (162 days) and ^{244}Cm (18 years) are formed by multiple neutron irradiation of ^{239}Pu; ^{248}Cm (4.7×10^5 years) is formed by decay of ^{252}Cf, the compounds are purified by ion exchange chromatography. The metal is formed by reduction of CmF_3 with Ba, it has the double-hexagonal cp structure. The metal is silvery in colour and is rapidly oxidized. ^{244}Cm on neutron irradiation is a potential source of higher actinides.

Curium Compounds. These are mainly in the +3 state and include all of the trihalides.

$$Cm^{4+} \xrightarrow{>2.8\text{volt}} Cm^{3+}$$

The +4 state is stabilized in aqueous solution by fluoride ions. CmO_2 and CmF_4 are formed by strong oxidation or the action of fluorine.

Cyanamide

1. An inorganic salt containing the ion CN_2^{2-}
2. A colourless crystalline solid, H_2NCN, made by the action of carbon dioxide on hot sodamide. It is hydrolysed to urea in acidic solutions.

Cyanamide Process. An industrial process for fixing nitrogen by heating calcium dicarbide in air.

$$CaC_2 + N_2 \rightarrow CaCN_2$$

The product, calcium cyanamide, hydrolyses to ammonia and can be used as a fertilizer. The process is expensive due to the high cost of calcium dicarbide. See also nitrogen fixation.

Cyanate. See cyanic acid.

Cyanide

1. A salt of hydrogen cyanide, containing the cyanide ion.
2. See nitrile.

Cyanide Process. A method of extracting gold by dissolving it in potassium cyanide (to form the complex ion $[Au(CN)_2]^-$). The ion can be reduced back to gold with zinc.

Cyanogen (C_2N_2). A toxic flammable gas prepared by heating mercury cyanide.

D

Dalton's Atomic Theory. A theory of chemical combination, first stated by the British chemist John Dalton (1766-1844) in 1803. It involves the following postulates :

1. Elements consist of indivisible small particles (atoms).
2. All atoms of the same element are identical; different elements have different types of atom.
3. Atoms can neither be created nor destroyed.
4. 'Compound elements' (*i.e.* compounds) are formed when atoms of different elements join in simple ratios to form 'compound atoms' (*i.e.* molecules).

Dalton also proposed symbols for atoms of different elements (later replaced by the present notation using letters).

Dating Techniques. Methods of estimating the age of rocks, palaeontological specimens, archaeological sites, etc. Relative dating techniques date specimens in relation to one another; for example, stratigraphy has been used to establish the succession of fossils. Absolute (or chronomteric) techniques give an absolute estimate of the age and fall into two main groups. The first depends on the existence of something that develops at a seasonally varying rate, as in dendrochronology and varve dating. The other uses some measurable change that occurs at a known rate, as in chemical dating radioactive (or radiometric) dating.

Dative Bond. Coordination bond.

Daughter

1. A nuclide which is produced by radioactive *decay of some other nuclide (the parent).

2. An ion or free radical which is produced by dissociation or reaction of some other (parent) ion or radical.

D.block Elements. The term used for the transition elements of the first, second, and third long periods, *i.e.* Sc to Zn, Y to Cd, and La to Hg. They are so called because in general they have inner d-levels with configurations of the type $(n-1)dxs^2$ where $x=1-10$.

D-d Transition. Electronic transition occuring between d orbitals which have been split by a crystal field. Generally occurs in the visible region but may occur in the u.v. or near i.r.

Deacon Process. A former process which was used for making chlorine by exidizing hydrogen chloride in air at 450°C using a copper chloride catalyst. It was patented in 1870 by Henry Deacon (1822—76).

De Braglic Wave. A wave associated with a particle, such as an electron or proton. In 1924, Louis de Broglie suggested that since electromagnetic waves can be described as particles (photons), particles of matter could also have wave properties. The wavelengh (λ) is having the same relationship to momentum (p) as in electromagnetic radiation :

$$\lambda=h/p$$

where h is the Planck constant.

Debye. A unit of electric dipole moment in the electrostatic system, used to express dipole moments of molecules. It is the dipole moment produced by two charges of opposite sign, each of 1 statcoulomb and placed 10^{-18} cm apart, and has the value 3.33564×10^{-30} coulomb metre.

Decay Symbol : da A prefix denoting 10. For example, 1 decametre (dam)=10 metres (m).

Decahydrate. A crystalline hydrate containing ten molecule of water per molecule of compound.

Decay. May be defined as the spontaneous transformation of one radioactive nuclide into a daughter nuclide or may not, with the emission of one or more particles or photons. The decay of N_0 nuclides to give N nuclides after time t is given by $N=N_a \exp(-\gamma t)$ were γ is called the decay constant or the disintegration constant. The reciprocal of the decay constant is the mean life. The time required for half the original nuclides to decay (*i.e.*, $N=\frac{1}{2}N_0$) is called the half-life of the nuclide. The same terms are applied to elementary particles that spontaneously transform into other particles. For example, a free neutron decays into a proton and an electron.

Deci. Symbol : d A prefix denoting 10^{-1}. For example, 1 decimetre (dm)$=10^{-1}$ metre (m).

Decomposition. Refers to a chemical reaction in which a compound is broken down into compounds with simpler molecules.

Defect. A discontinuity in a crystal lattice. A point defect is consisting either of a missing atom or ion creating a vacancy in the lattice (a vacancy is sometimes called a Schottky defect) or an extra atom or ion between two normal lattice points creating an interstitial. A Frenkel defect consists of a vacancy in which the missing atom or ion has moved to an interstitial position. If more than one adjacent point defect occurs in a crystal there may be a slip along a surface causing a line defect (or dislocation). Defects have been caused by strain or, in some cases, by irradiation. All crystalline solids are having an equilibrium number of point defects above absolute zero; this number increases with temperature. The existence of defects in crystals has been important in the conducting properties of semiconductors.

Definite Proportions. See chemical combination.

Degenerate. Describing different quantum states that have the same energy. For instance, the five d orbitals in transition-metal atom all have the same energy but different values of the magnetic quantum number m. Differences in energy occur if a magnetic field is applied or if the arrangement of ligands around the atom is not symmetrical. The degeneracy is then said to be 'lifted'.

Degradation. A chemical reaction involving the decomposition of a molecule into simpler molecules, usually in stages.

Delocalization. Refers to a spreading out of bonding electrons in a molecule over the molecule.

Delocalized Bond (non-localized bond). Refers to a type of bonding in molecules that occurs in addition to sigma bonding. The electrons forming the delocalized bond have been no longer regarded as remaining between two atoms; *i.e.*, the electron density of the delocalized electrons is spread over several atoms and may spread over the whole molecule.

Delta Bonding. Lateral overlap between two orbitals *e.g.* d orbitals) such that there are four regions of overlap.

Delta Rays, (**δ rays**). Rays emitted when α-particles are absorbed by matter δ-Rays consist of electrons with relatively low velocity compared with those of β-particles.

Delta-iron. See iron.

Dendrochronology. An absolute dating technique using the growth rings of trees. It depends on the fact that trees in the same locality show a characteristic pattern of growth rings resulting from climatic conditions. Thus it becomes possible to assign a definite date for each growth ring in living trees, and to use the ring patterns to date fossil trees or speciments of wood (*e.g.*, used for buildings or objects on archaeological sites) with life-spans that overlap those of living trees. The bristlecone pine (Pinus aristata), which lives for up to 5000 years, has been used to date specimens over 8000 years old. Fossil specimens accurately dated by dendrochronology have been used to make corrections to the *carbon-dating technique. Dendrochronology is also helpful in studying past climatic conditions. Analysis of trace elements in sections of rings can also provided formation on past atmospheric pollution.

Denitrification. The return of fixed nitrogen, generally nitrates and ammonium salts, to the atmosphere generally as N_2 and N_2O.

Desalination. The removal of salt from sea water for irrigation of the land or to provide drinking water. The process is normally only economic if a cheap source of energy, such as the waste heat from a nuclear powder station, can be used. Desalination using solar energy has the greatest economic potential since shortage of fresh water is most acute in hot regions. The methods employed include evaporation, often under reduced pressure (fiash evaporation); freezing (pure ice forms from freezing brine); reverse osmosis; electrodialysis; and ion exchange.

Detonating Gas. A mixture of H_2 and O_2 (molar ratio 2 : 1) produced by electrolysis of water and which explodes violently to reform H_2O on ignition.

Deuterated Compound. A compound in which one or more ^1H atoms have been replaced by deuterium (^2H) atoms.

Deuterium. D. At.no. 1,at.wt. 2– 01. An isotope of hydrcgen which, because of the relatively large difference in masses, gives markedly different physical properties including rates of reactions to deutero compounds—bracketed figures below are for hydrogen analogues. D_2, b.p. −249.7 (−252.8°C); D_2O, b.p. 101.4 (100.0°C), m.p. 3.8 (0.0°C), ionic product at 25°C 1.1×10^{-15} (1.0×10^{-14}). Ordinary hydrogen compounds contain 1 part in 4500 D. D_2O (heavy water) is obtained by preferential evaporation of H_2O or preferential electrolysis of H_2O. Deuterium compounds are used as tracers in chemical and biological investigations; as solvents, etc., when protons are under investigation (*e.g.* n.m.r.), D_2O is used as a moderator in atomic reactors.

Deuton, Deuteron $^2D^+$. The nucleus of the deuterium atom. Used in linear accelerators for bombarding other nuclei.

Devarda's Alloy, 45% Al, 50% Cu, 5% Zn. Used to detect and analyse for nitrates.

Developed Colours. See ingrain dyes.

Developers, Photographic. See aminophenol developers, colour development. Solutions which cause reduction of exposed silver halide grains to silver.

Deuterium Oxide. See heavy water.

Devitrification. Loss of the amorphous nature of glass as a result of crystallization.

Diagonal Relationship. There is a general trend in the periodic table for electronegativity to increase from left to right along a period and decrease down a group. Thus a move of one across and one down', *i.e.*, a diagonal move in the table, gives rise to effects that tend to cancel each other. There is a similar combined effect for size, which decreases along a period but increaseer down a group. This diagonal relationship gives rise to similarities in chemical properties, which are particularly noticeable for the following pairs : Li-Mg; Be-Al; B-Si.

Diamagnetism. The property of all substances that causes repulsion of the substance from a magnetic field. Originates in the interaction of the electronic charge and the field and is an additive property of the atoms and groups present. Order of magnitude 10^{-3} that of paramagnetism so that the effect is swamped by the presence of unpaired electrons.

Diamond. A crystalline form of carbon, and one of the hardest and most infusible substances known; d 3·520. In a crystal of diamond every carbon atom is linked tetrahedrally to four others. The structure has C—C 1·54A°, equal to that found in saturated hydrocarbons.

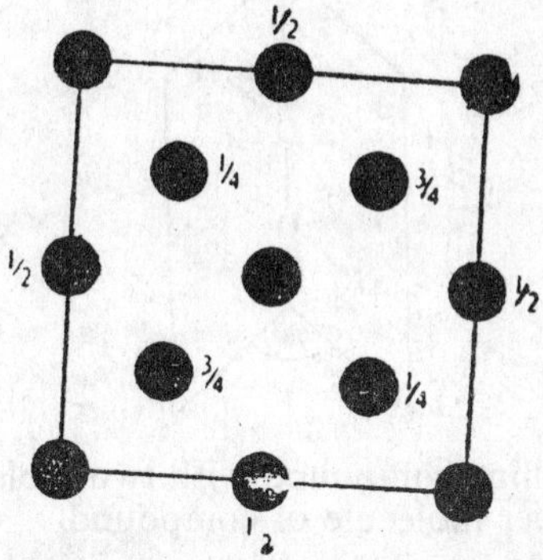

Crystal structure of diamond

Diaspore. A mineral form of a mixed aluminium oxide and hydroxide, AlO.OH.

Diatomic. Describing a molecule that consists of two atoms. Hydrogen (H_2), oxygen (O_2), nitrogen (N_2), and the halogens are examples of diatomic elements.

Diatomite, Kieselguhr. A hydrated silica, the remains of diatoms. The particles are hollow and the material is used as a filler and an absorbent (nitro-glycerine gives dynamite). Other uses are for polishing, filtration, decolourizing and for heat-insulation, Tripolite is a compact form.

Dibasic Acid. An acid that has two acidic hydrogen atoms in its molecules. Sulphuric (H_2SO_4) and carbonic (H_2CO_3) acids are common examples.

Dichroism. The property of some crystals, such as tourmaline, of selectively absorbing light vibrations in one plane while allowing light vibrations at right angles to this plane to pass through. Polaroid is a synthetic dichroic material.

Dichromate (VI). A salt containing the ion Cr_2O_7. Dichromates are strong oxidizing agents.

Dihedral Angle. The angle relating two parts of a molecule. In H_2O_2 the dihedral angle in the solid is 94°.

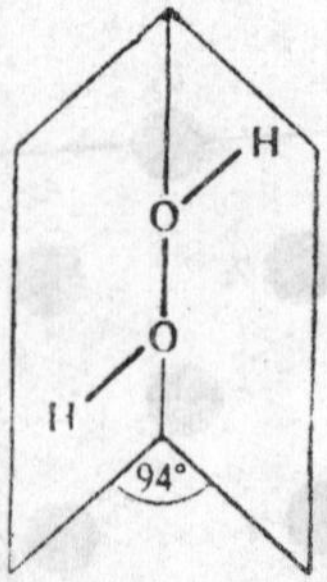

Dihydrate. A crystalline compound with two molecules of water of crystallization per molecule of compound.

Diiodine Hexachloride (iodine trichloride, I_2Cl_6). A yellow crystalline solid which is made by reacting excess chlorine with iodine. It is a strong oxidizing agent.

Dilead (II) Lead (IV) Oxide (red lead, Pb_3O_4). A powder which is obtained by heating lead (II) oxide or lead (II) carbonate hydroxide at 400°C. It is black when hot and red or orange when cold. On heating it decomposes to yield lead (II) oxide and oxygen. Dilead (II) lead (IV) oxide is used as a pigment and in glass making. In practice it tends to have less oxygen than denoted in the formula.

Dimer. A compound (or molecule) which is formed by combination association of two molecules. For example, aluminium chloride ($AlCl_3$) is a dimer (Al_2Cl_6) in the vapour.

Dinitrogen Oxide (nitrous oxide, N_2O). A colourless gas with a faintly sweet odour and taste. It is prepared commercially and in the laboratory by the careful heating of ammonium nitrate :

$$NH_4NO_3(s) \rightarrow N_2O(g) + 2H_2O(g)$$

The gas is used as a mild anaesthetic in medicine and dentistry, being marketed in small steel cylinders. It is sometimes called laughing gas because it induces a feeling of elation.

Dinitrogen Tetroxide (N_2O_4). A colourless gas that becomes a pale yellow liquid below 21°C and solidifies below-11°C. On heating, the gas dissociates to nitrogen dioxide molecules :

$$N_2O_4(g) \rightarrow 2NO_2(g)$$

Liquid dinitrogen tetroxide has good solvent properties and is used as a nitrating agent.

Dioxonitric (III) Acid. See nitrous acid.

Dioxygen O_2. Formally the oxygen molecule but used for dioxygen complexes and dioxygenyl, $(O_2)^+$, salts.

Dioxygenyl (O_2)$^+$. See dioxygen.

Dioxygenyl Compounds. Compounds containing the positive ion O_2^+, as in dioxygenyl hexafluoroplatinate O_2PtF_6—an organe solid that sublimes in vacuum at 100°C.

Diphosphane (disphosphine, P_2H_4). A yellow liquid that can be condensed out from phosphine in a freezing mixture. It ignites spontaneously in air.

Diphosphine. See diphosphane.

Dipole Moment. In a heteronuclear diatomic molecule, because of the difference in electronegativities of the two atoms, one atom acquires a small positive charge (q_+), the other a negative charge (q_-). The molecule is then said to have a dipole moment whose magnitude $\mu = qd$, where d is the distance of separation of the charges. With polyatomic molecules the net dipole moment is the vector sum of the dipole moments of the individual bonds within the molecule. Thus symmetrical molecules, *e.g.* CCl_4, may contain polar bonds but possess no net dipole moment. Measurements of dipole moments may be used to give information about the structure of complex molecules.

Disilane. See silane.

Dislocation. See defect.

Disodium Hydrogenphosphate (V) (disodium orthophosphate). A colourless crystalline solid, Na_2HPO_4) soluble in water and insoluble in ethanol. It is known as the dihydrate (r.d. 2 066), heptahydrate (r.d. 1.68), and dodecahydrate (r.d. 1.52). It may be prepared by titrating phosphoric acid with sodium hydroxide to an alkaline end point (phenolphthalein) and is used in treating boiler feed water and in the textile industry.

Disodium Orthophosphate. See disodium hydrogenphosphate (V).

Disodium Tetraborate-10 water (borex, $Na_2B_4O_7.10H_2O$). A white crystalline solid, sparingly soluble in cold water but readily soluble in hot water It occurs naturally as salt deposits in dry lake beds, especially in California. It is an important industrial material, being used in the manufacture of enamels and heat-resistant glass, as a paper glaze, etc.

Its full systematic name is sodium (I) heptaoxotetraborate (III)-10 water. In solution hydrolysis occurs :

$$B_4O_7^{2-} + 7H_2O = 2OH^- + 4H_3BO_3$$

The borax-bead test gives characteristic colours with certain cations.

Disproportionation. A process in which a compound of one oxidation state changes to compounds of two or more oxidation states, *e.g.* $2Cu^+ \rightarrow Cu + Cu_2^+$ in aqueous solution. Alternatively a redistribution of groups around a central atom, *e.g.*

$$2PF_4Cl \rightarrow PF_3Cl_2 + PF_5$$

Disulphur Dichloride (sulphur monochloride, S_2Cl_2). A red fuming liquid with a strong smell. It is prepared by passing chlorine over molten sulphur and is used to harden rubber.

Disulphuric (VI) Acid (pyrosulphuric acid). A colourless hygroscopic crystalline solid. $H_2S_2O_7$; r.d. 1.9; m.p. 35°C. It is commonly encountered mixed with sulphuric acid as it is formed by dissolving sulphur trioxide in concentrated sulphuric acid. The resulting fuming liquid, called oleum or nordhausen sulphuric acid, is produced during the contact process and is also widely used in the sulphonation of organic compounds.

Dithionate. A salt of dithionic acid, containing the ion $S_2O_6^{2-}$, usually formed by the oxidation of a sulphite using manganese (IV) oxide. The ion has neither pronounced oxidizing nor reducing properties.

Dithionic Acid. An acid, $H_2S_2O_6$, known in the form of its salts (dithionates).

Dithionite. See sulphinate.

Dithionous Acid. See sulphinic acid.

Divalent (bivalent). Having a valency of two.

D-lines. Two close lines in the yellow region of the visible spectrum of sodium, having wavelengths 589.0 and 589.6 nm. As they are prominent and easily recognized they are used as a standard in spectroscopy.

Dobereiner's. Triads of chemically similar elements in which the central number, when placed in order of increasing atomic weight, has an atomic weight approximately equal to the average of the outer two. Other chemical and physical properties of the central number also lie between those of the first and last numbers of the triad. Dobereiner noted this relationship in 1817; the triads are now recognized as consecutive numbers of a group of the periodic table; *e.g.* Ca, Sr, and Ba and Cl, Br, and I.

Dodecahedral Co-ordination. Co-ordination by eight ligands at the corner of a dodecahedron. Found in the $[Mo(CN)_8]^{4-}$ ion. A dodecahedron has twelve triangular faces and consists of two pentagonal pyramids sharing four corners with two faces at the open edges.

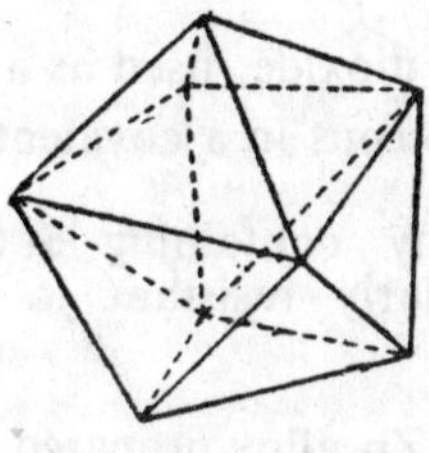

Dolomite. A mineral, $CaCO_3.MgCO_3$, used as an ore of magnesium.

Donor. An ion or molecule that provides a pair of electrons in forming a coordinate bond.

Doping. The name given to the practice whereby electrical, magnetic and other properties of an ionic solid may be changed by incorporation of an altervalent ion in the host lattice. E.g. the incorporation of small amounts of Li^+ or Ga^{3+} into the nickel oxide lattice results in a respective decrease or increase in conductivity of the latter. When the dopent is not incorporated directly into the lattice, but rather forms discrete aggregations within it the process is referred to as domain doping.

D-orbital. See orbital.

Double Bond. A covalent bond between two atoms that includes two pairs of electrons, one pair being the single bond equivalent (the sigma pair) and the other forming an additional bond, the pibond (π bond). It is conventionally represented by two lines, for example $H_2C{=}O$.

Double Decomposition (metathesis). A chemical reaction involving exchange of radicals, *e.g.*

$$AgNO_3(aq)+KCl(aq) \rightarrow KNO_3(aq)+AgCl(s)$$

Double Salt. A salt formed by crystallization from a solution of two or more components that has distinct properties in the solid but in solution behaves as a mixture of the components. E g. ion (II) ammonium sulphate, $FeSO_4(NH_4)_2,SO_46H_2O$. Double salts are either weak, easily dissociated complexes or the solid lattice has an extensive interaction, *e g.* hydrogen bonding, not found in the components.

Doublet. A pair of associated lines in certain spectra, *e.g.*, the two lines that make up the sodium D-lines.

Downs Process. An electrolytic process for making chlorine from fused sodium chloride. The anode is a central piece of graphite, surrounded by a cylindrical steel cathode. A grid and dome between the two keeps the products separate.

Dry Ice. Solid carbon dioxide, used as a refrigerant.

Duplet. A pair of electrons in a covalent chemical bond.

Duriron. An iron alloy containing Si (14%), Mn (2%), C (1%), S (0.1%). Particularly resistant to acids and used in chemical plant.

Dutch Metal. A Cu–Zn alloy prepared in leaf form in imitation of gold-leaf.

Dy. Dysprosium.

Duralumin (Trademark). A strong light-weight aluminium alloy containing 3–4% copper, with small amounts of magnesium, manganese, and sometimes silicon. It is widely used in aircraft bodies.

Dysprosium. A soft malleable silvery element of the lanthanoid series of metals. It occurs in association with other lanthanoids. One of its few uses is as a neutron absorber in nuclear reactors.

Symbol : Dy; m.p. 1412°C; b.p. 2562°C; r.d. 8.56; p.n. 56; r.a.m. 162.50.

E

Earth. The planet that orbits the sun between the planets Venus and Mars. The earth is having three layers : the gaseous atmosphere (see earth's atmosphere), the liquid hydrosphere, and the solid lithosphere. The solid part of the earth is also having three layers : the crust with a mean thickness of about 32 km under the land and 10 km under the seas; the mantle, which extends some 2900 km below the crust; and the core, part of which is believed to be liquid. The composition of the crust has been oxygen 47%, silicon 28%, aluminium 8%, iron 4.5%, calcium 3.5%, sodium and potassium 2.5% each, and magnesium 2.2%. Hydrogen, carbon phosphorus, and sulphur, have been all present to an extent of less than 1%.

Earth's Atmosphere. Refers to the gas that surrounds the earth. The composition of dry air at sea level has been : nitrogen 78.08%, oxygen 20.95%, argon 0.93%. carbon dioxide 0.03%, neon 0.0018%, helium 0.0005%, krypton 0.0001%, and xenon 0.0001%. In addition to water vapour, air in some localities is having sulphur compounds, hydrogen peroxide, hydrocarbons, and dust particles.

Ebonite. See vulcanite.

Edta (ethylenediamine tetracetic acid). A compound with the formula

$$(HOOC\ CH_2)_2N(CH_2)_2N(CH_2COOH)_2$$

It finds use in forming chelates of transition metals.

Effective Atomic Number Rule. Applied to transition metal carbonyls and to many organometallic derivatives. It is frequently found that the total number of electrons available to the central atom from its own electrons, by donation (lone pairs, etc. of electrons from π-bonding systems), by covalent bonding (each covalent bond to a substituent or another metal contributes 1 electron), and taking into account charge add up to the total number of electrons of the next noble gas. Does not apply to paramagnetic species and there are many other exceptions to the rule. E g. $ClMn(CO)_5$. Mn 7 valence electrons, 1 from covalent bond to Cl, 5×2 electrons from carbonyls. Total 18 electrons, krypton structure.

Efflorescence. Refers to the process in which a crystalline hydrate loses water, forming a powdery deposit on the crystals.

Eigen Function. In wave mechanics, the Schrodinger equation may be written using the Hamiltonian operator H as

$$H\psi = E\psi$$

Only certain energy values (E) will lead to solutions of this equation. The corresponding values of the wave functions are called Eigen functions or characteristic wave functions.

Einsteinium, Es. At no. 99, ^{253}Es (20 days) and ^{254}Es (270 days) may be formed by multiple neutron irradiation of Am, Pu and Cm; it is purified by ion exchange chromatography. The metal has not yet been prepared.

Einsteinium Compounds. The element shows oxidation states of +2 and +3, the +3 state being that of a typical tripositive actinide, $EsCl_3$, EsOCl and Es_2O_3 are known. $EsBr_3$ is reduced to $EsBr_2$ by hydrogen.

Electric-arc Furnace. A furnace which is used in melting metals to make alloys, especially in steel manufacture, in which the heat source is an electric arc. In the direct-arc furnace, such as the Heroult furnace, an arc gets formed between the metal and an electrode. In the indirect-arc furnace, like the Stassano furnace, the arc gets formed between two electrodes and the heat is radiated onto the metal.

Electrochemical Series. A series in which the elements are arranged in decreasing magnitude of their standard oxidation potentials. The use of the series allows some qualitative generalities to be drawn. Thus, for example, the higher the metal is in the series the greater the tendency for it to be oxidized. Similarly if a metal is added to a solution of another metal ion, which is below it in the electrochemical series, the metal ion will be displaced, *e.g.* zinc metal will displace copper from solutions of copper salts. The following series shows some of the more common elements in order (oxidation potentials in brackets) :

Li (3.04)	Na (2.71)	H_2 (0 00)
Cs (2.92)	Mg (2.38)	Cu (−0.34)
Rb (2.92)	H^- (2.23)	I^- (−0.54)
K (2.92)	Al (1.71)	Ag (−0.80)
Ba (2.90)	Mn (1.03)	Au (−1.42)
Sr (2.89)	Zn (0.76)	F^- (−2.87)
Ca (2.76)	Cr (0.74)	

Electrodeposition. Refers to the process of depositing one metal on another by electrolysis, as in electroforming and electroplating.

Electroforming. Refers to a method of forming intricate metal articles or parts by electrodeposition of the metal on a removable conductive mould.

Electrolytic Corrosion. Corrosion that takes place through an electrochemical reaction.

Electrolytic Cell. Refers to a cell in which electrolysis takes place *i.e.* one in which current is passed through the electrolyte from an external source.

Electrolytic Gas. Refers to the highly explosive gas which is formed by the electrolysis of water. It consists of two parts hydrogen and one part oxygen by volume.

Electrolytic Refining. The term used for the purification of metals by electrolysis. It is commonly applied to copper. A large piece of impure copper finds use as the anode with a thin strip of pure copper as the cathode. Copper (II) sulphate solution is the electrolyte. Copper dissolves at the anode : $Cu \rightarrow Cu^{2+} + 2e$, and is deposited at the cathode. The net result is transfer of pure copper from anode to cathode. Gold and silver in the impure copper form a so-called anode sludge at the bottom of the cell, which could be recovered.

Electromagnetic Spectrum. Refers to the range of wavelengths over which electromagnetic radiation extends. The longest waves (10^5–10^{-3} metres) have been radio waves, the next longest (10^{-3}—10^{-6} m) have been infrared waves, then comes the narrow band (4—7×10^{-7} m) of visible radiation, followed by ultraviolet waves (10^{-7}—10^{-9} m), X-rays (10^{-9}—10^{-11} m), and gamma rays (10^{-11}—10^{-14} m).

Electrometallurgy. The uses of electrical processes in the separation of metals from their ores, the refining of metals, or the forming or plating of metals.

Electromotive Series. See electrochemical series.

Electron. An elementary particle of negative charge (—1.602 192· C) and rest mass 9.109 558 kg. Electrons are present in all atoms in shells around the nucleus.

Electron Affinity. The energy released when an electron is added to a neutral atom in the gaseous state, *i.e.*

$$A(g)+e \rightarrow A^-(g)+\text{energy}$$

Electron affinities may be estimated using a Born-Haber cycle.

Electron affinities (kJ mol^{-1})

			H				He
			73				—54
Li	Be	B	C	N	O	F	Ne
57	—66	15	123	—31	141	333	—99
Na	Mg	Al	Si	P	S	Cl	
21	—67	26	135	60	169	348	
						Br	
						340	
						I	
						297	

Electron Alloys. See magnesium alloys.

Electron Compounds. Compounds which are formed by transition metals and one of the main group elements in which the structure of particular phases depends on the ratio of the total valency electrons available to the total number of atoms in the simplest empirical formula (making assumptions such that Fe, Co, Ni group elements have 0, Cu group 1 valence electrons). The ratios are 3 : 2 (β-phase bcc), 21 ; 13 (γ-phase complex), 7 : 4 (ε-phase, hcp)-

Electron-deficient Compounds. Compounds in which a filled-shell configuration cannot be written for each atom if it is regarded that each bond between two atoms is having two electrons. The bonding in these compounds can be rationalized in terms of multi-centre bonds in which two electrons may bind many atoms.

Electron Diffraction. When we allow beam of electrons to interact with atoms or molecules, because of the wave properties of the electrons, they get diffracted. The diffraction of a beam of electrons by the surface of a nickel single crystal was used by Davisson and Germer to demonstrate that electrons have wave properties. Electron diffraction finds use to study molecular structure, especially in the gas phase and the investigation of surfaces.

Electronegativity. A measure of the tendency of

H 2.1				
Li 1.0	C 2.5	N 3.3	O 3.5	F 4.0
Na 0.9	Si 1.8	P 2.1	S 2.5	Cl 3.0
				Br 2.8
				I 2.5

each atom in a stable molecule to attract electrons to itself. Thus in covalently bonded hydrogen chloride, HCl, the electrons are much closer to the chlorine than to the hydrogen, the chlorine being the more electronegative atom. Numerical value of electronegativity have been calculated by Pauling from bond dissociation energies, ionization energies and electron affinity values using the equations :

$$D_{A-B}=\tfrac{1}{2}(D_{A-A}+D_{B-B})+23(x_A-x_B)^2$$

and

$$125x_A=(\text{ionization energy}-\text{electron affinity})$$

where x is the electronegativity and D the bond dissociation energy. In general, electronegativity values increase from left to right across each row of the periodic table and decrease down each group. Some typical values are shown on previous page :

Electronic Energy Levels. See energy levels.

Electron Spin Resonance (ESR). A similar technique to nuclear magnetic resonance, but applied to unpaired electrons in a molecule (rather than to the nuclei). It is a powerful technique for studying free radicals and transition-metal complexes.

Electronvolt Symbol eV. Refers to a unit of energy equal to the work done on an electron in moving it through a potential difference of one volt. It finds use as a measure of particle energies although it is not an SI unit.

$$1 \text{ eV} = 1.602 \times 10^{-19} \text{ joule.}$$

Electroplating. Refers to the process of coasting a solid surface with a layer of metal by means of electrolysis (*i.e.* by electrodeposition).

Electropositive. Describing elements that tend to lose electrons and form positive ions. The alkali metals have been typical electropositive elements.

Electrovalent Compounds. Compounds in which the major binding force is electrostatic attraction between positive and negative ions. The lattice does not contain discrete molecules and the ions are packed together to occupy space most efficiently (generally close packing of anions). Electrovalent compounds are distinguished from covalent compounds by the conductivity of melts and solutions in polar solvents, low volatility, solubility in polar solvents.

Electrum

1. An alloy of gold and silver containing 55—88% of gold.
2. A German silver alloy containing 52% copper, 25% nickel, and 22% zinc.

Element. A substance which cannot be further divided by chemical methods. The basic substances which build up chemical compounds. An element is defined by its atomic number (nuclear charge and electronic configuration). The table on pp. 173—4 shows the elements and their normal, ground-state, electronic configurations.

Elementary Particles. The fundamental particles found in nature. Those known at present have been proton, neutron, electron, photon, positron, neutrino, meson, muon, hyperon, graviton and baryon.

Elements, Abundance of. The relative abundance of elements depends on whether measurements are made on the crust, meteorites, core, etc. For the earth's crust (E) abundance of the lighter elements (expressed relative to $Si = 10^6$ are :

	E	A
Na	105000	0.51
Mg	68000	11.25
Al	300000	1.07
Si	10^6	13.84
P	3430	0 08
S	1640	2.74
K	62000	0.06
Ca	80000	1.07
Ti	‵000	0.06
Mn	1950	0.13
Fe	126000	38.8
Ni	100	2.70
O	2.96×10^6	27.17

(A) represents the average % composition of the earth by weight.

Elinvar. Trade name for a nickel—chromium steel containing about 36% nickel, 12% chromium, and smaller proportions of tungsten and manganese. Its elasticity does not vary with temperature and it is therefore used to make hairspring for watches.

Emanation. The former name for the gas radon, of which there are three isotopes : Rn—222 (radium emanation), Rn—220 (thoron emanation), and Rn—219 (actinium emanation).

Emerald. The green gem variety of beryl : one of the most highly prized gemstones. The finest specimens occur in the Muzo mines, Colombia. Other occurrences include the Ural Mountains, the Transvaal in South Africa, and Kallgunan in India. Emeralds can also be successfully synthesized.

Emanation. An obsolete name for radon.

Embonic Acid. See pamoic acid.

Emerald. The grass-green variety of beryl ; contains some Cr^{3+}.

Emerald Green, Pairs Green, Schweinfurter green. Basic copper ethanoate arsenates(III) prepared from verdigris (or other basic copper salt). sodium arsenate(III) and ethanoic acid. Once used as pigments but now used in insecticldes for spraying fruit trees. Readily decomposed to soluble argenic compounds so use is very restricted.

Emery Impure α-Al_2O_3 (corundum) containing some iron oxide (magnetic). Used as an abrasive and polishing agent.

Enantiotropy. Refers to the existence of different stable allotropes of an element at different temperatures; sulphur, for example, exhibits enantiotropy.

†The phase diagram for an enantiotropic element is having a point at which all the allotropes can coexist in a stable equilibrium. At temperatures above or below this point, one of the allotropes will be more stable than the other(s).

Enclosure Compound. See clathrate.

Energy Band. Refers to a range of energies that electrons can have in a solid. In a single atom, electrons exist in discrete energy levels. In a crystal, in which large numbers of atoms have been held closely together in a lattice, electrons are influenced by a number of adjacent nuclei and the sharply defined levels of the atoms become bands of allowed energy; this approach to energy levels in solids have been often known as the band theory. Each band represents a large number of allowed quantum states. Between the bands have been forbidden bands. The outermost electrons of the atoms (*i.e.* the ones responsible for chemical bonding) form the valence band of the solid. This has been the band, of those occupied, that has the highest energy.

The band structure of solids accounts for their electrical properties. In order to move through the solid, the electrons have to change from one quantum state to another. This can only occur if there are empty quantum states close to the electrons. In general, if the valence band is full, electrons cannot change to new quantum states in the same band. For conduction to occur, the electrons have to be in an unfilled—

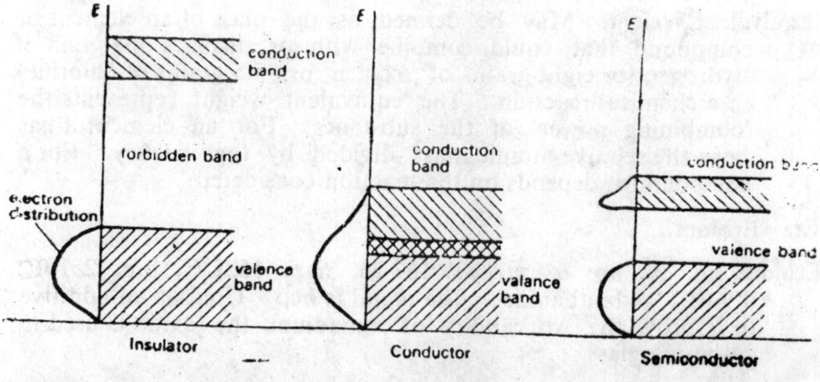

the conduction band. Metals are good conductors either because the valence band and the conduction band are only half-filled or because the conduction band overlaps with the valence band; in either case vacant states are available. In insulators the conduction band and valence band are separated by a wide forbidden band and electrons do not have enough energy to 'jump' from one to the other.

In intrinsic semiconductors the forbidden gap is narrow and, at normal temperatures, electrons at the top of the valence band can move by thermal agitation into the conduction band (at absolute zero, a semiconductor would act as an insulator). Doped semiconductors have extra bands in the forbidden gap.

Energy Level. One of the discrete energies that an atom or molecule, for instance, can have according to quantum theory. Hence in an atom there have been certain definite orbits that the electrons can be in, corresponding to definite electronic energy levels of the atom. Similarly, a vibrating or rotating molecule can be having discrete vibrational and rotational energy levels.

Engel's Salt. See potassium carbonate.

Enrichment. Refers to the process of increasing the abundance of a specified isotope in a mixture of isotopes. It is generally applied to an increase in the proportion of U—235, or the addition of Pu—239 to natural uranium for use in a nuclear reactor or weapon.

Epitaxy (Epitaxial Growth). Growth of a layer of one substance on a single crystal of another, such that the crystal structure in the layer has been the same as that in the substance. It finds used in making semiconductor devices.

Epsomite A mineral form of magnesium sulphate heptahydrate, $MgSO_4$ $7H_2O$.

Equivalent Weight. May be defined as the mass of an element or compound that could combine with or displace one gram of hydrogen (or eight grams of oxygen or 35.5 grams of chlorine) in a chemical reaction. The equivalent weight represents the 'combining power' of the substance. For an element it has been the relative atomic mass divided by the valency. For a compound it depends on the reaction considered.

Er. Erbium.

Erbium, Er. At. no. 68, at. wt. 167.26, m.p. 1522°C, b.p. 2510C, d 9·01. A lanthanide. The metal is hcp. Used as an additive to improve the workability of vanadium; the oxide is used in colouring glass.

Erbium Compounds. Erbium forms a single series of typical lanthanide compounds in the + 3 state

$$E^{3+} (f^{11} \text{ rose}) \dashrightarrow Er(-2{\cdot}30 \text{ volts in acid}).$$

Erdmann's Salt. $NH_4[Co(NA_3)_2(NO_2)_4]$.

Erg. A former unit of energy used in the e g.s. system. It is equal to 10^{-7} joule.

Essential Element. Any of a number of elements required by living organisms to ensure normal growth, development, and maintenance. Apart from the elements found in organic compounds (*i.e.* carbon, hydrogen, oxygen, and nitrogen), plants, animals, and microorganisms all require a range of elements in inorganic forms in varying amounts, depending on the type of organism. The major elements, present in tissues in relatively large amounts (greater than 0.005%), are calcium, phosphorus, potassium, chlorine, sulphur, and magnesium. The trace elements occur at much lower concentrations and thus requirements are much less. The most important are iron, manganese, zinc, copper, lodine, cobalt, selenium, molybdenum, chromium, and silicon. Each element may fulfil one or more of a variety of metabolic roles.

Ethylenediaminetetracetic Acid. See edta.

Europium Symbol Eu. It is a soft silvery metallic element belonging to the lanthanoids; a.n. 63; r.a.m. 151.96; r.d, 5.245 (20°C); m.p. 822°C, b.p. 1529°C. it is found in small quantities in bastanite and monazite. Two stable isotopes occur naturally; europium—and curopium—153, both of which have been neutron absorbers. It was discovered by Sir William Crookes in 1889.

Exa-Symbol : E A profix denoting 10^{18}.

Excitation. Refers to the process of producing an excited state of an atom, molecule, etc.

Excitation Energy. Refers to the energy which is required to change an atom, molecule, etc. from one quantum state to a state with a higher energy. The excitation energy (sometimes called excitation potential) refers to the difference between two energy levels of the system.

Excited State. Refers to a state of an atom molecule, or other system, with an energy greater than of the ground state.

Exclusion Principle. The principle, enunciated by Pauli in 1925, that no two electrons in an atom could have an identical set of four quantum numbers.

F

Fajans' Rules. Ionic compounds are most readily formed by :

1. ions with low charge (*e.g.*, Na^+ more than Al^{3+});
2. by large cations (Cs^+ more than Na^+);
3. by small anions (Cl^- more than I^-).

Fall-out (or radioactive fall-out)

1. Radioactive particles deposited from the atmosphere either from a nuclear explosion or from a nuclear accident. Local fall-out, within 250 km of an explosion, falls within a few hours of the explosion. Tropospheric fall-out consists of fine particles deposited all round the earth in the appproximate latitude of the explosion within about one week. Stratospheric fall-out may fall any where on earth over a period of years. The most dangerous radioactive isotopes in fall-out are the fission fragments iodine-131 and strontium-90. Both can be taken up by grazing animals and passed on to human populations in milk, milk products, and meat. Iodine 131 accumulates in the thyroid gland and stronium-90 accumulates in bones.

2. Chemical fall-out. Hazardous chemicals discharged into and subsequently released from the atmosphere, especially by factory chimneys.

F-block Elements. The block of elements in the periodic table consisting of the lanthanoid series (from cerium to lutetium) and the actinoid series (from thorium to lawrencium). They are characterized by having two s-electrons in their outer shell (n) and f-electrons in their inner (n-1) shell.

Feldspars. A group of aluminosilicates with framework structures with Al or Si in tetrahedral co-ordination. The most common constituents of igneous rocks. Subdivided into two groups depending on the detailed structural types. Used in the ceramic and enamelling industries.

F.C.C. See face-centred cubic crystral.

Feldspathoids. A group of alkali aluminosilicate minerals that are similar in chemical composion to the feldspars but are relatively deficient in silica and richer in alkalis. The structure consists of a framework of $(Si.Al)O_4$ tetrahedra with aluminium and silicon atoms at their centres. The feldspathoids occur chiefly with feldspars but do not coexist with free quartz (SiO_2) as they react with silica to yield feldspars. The chief varieties of feldspathaids are :

nepheline, $KNa_3(AlSiO_4)_4$;
leucite, $KAlSi_2O_6$;
analcime, $NaAlSi_2O_6.H_2O$;
cancrinite, $Na_8(AlSiO_4)_6(HCO_3)_2$;

and the sodalite subgroup comprising

sodalite, $3(NaAlSiO_4).NaCl$;
nosean, $3(NaAlSiO_4).Na_2SO_4$;
hauyne, $3(NaAlSiO_4).CaSO_4$;

and lazurite, $(Na, Ca)_8(Al, Si)_{12}O_{24}(S, SO_4)$.

Felspar. See feldspars.

Fermi. A unit of length equal to 10^{-15} metre. It was formerly used in atomic and nuclear physics.

Fermium. Symbol Fm. A radioactive metallic transsuranic element belonging to the actinoids; a.n. 100; mass number of the most stable isotope 257 (half-life 10 days). Ten isotopes are known. The element was first identified by A. Ghiorso and associates in debris from the first hydrogen-bomb explosion in 1952.

Feric Compounds. Compounds of iron in its +3 oxidation state; *e.g.*, ferric chloride is iron (III) chloride, $FeCl_3$.

Ferric Oxide. See iron (III) oxide.

Feric Sulphate. See iron (III) sulphate.

Ferricyanide. A compound having the complex ion $[Fe(CN)_6]^{-3}$, *i.e.*, the hexacyanoferrate (III) ion.

Ferrite. A member of a class of mixed oxides MO Fe_2O_3, where M is a metal such as cobalt, manganese, nickel, or zinc.

Ferroalloys. Alloys of iron with other elements made by smelting mixtures of iron ore and the metal ore; *e.g.*, ferrochromium, ferrovandium, ferromanganese, ferrosilicon, etc. They are used in making alloys steels.

Ferrocene. An orange-red crystalline solid, $Fe(C_5H_5)_2$; m.p. 173°C. It can be made by adding the ionic compound $Na^+C_5H_5$

(cyclopentadienyl sodium, made from sodium and cyclopentadiene) to iron (III) chloride. In ferrocene, the two rings are parallel, with the iron ion sandwiched between them (hence the name sandwich compound). The bonding is between pi orbitals on the rings and d-orbitals on the Fe^{2+} ion. The compound can undergo electrophilic substitution on the C_5H_5 rings (they have some aromatic character). It can also be oxidized to the blue ion $(C_5H_5)_2$ Fe^+. Ferrocene is the first of a class of similar complexes called metallocenes. Its systematic name is di-π cyclopentadienyl iron (III).

Ferrocyanide. A compound containing the complex ion [Fe $(CN_6]^{4-}$, *i e.*, the hexacyanoferrate (II) ion.

Ferromagnetism. Magnetic behaviour which does not follow the Curie-Weiss law and is field strength dependent. Has large positive magnetic susceptibility (χ) and the magnetic moment (M) does not fall to zero when the applied magnetic field is removed.

Ferro-manganese. Fe (14%), C (16.7%), Mn (80%) used for deoxidation and increasing C content in steel manufacture. A ferro-manganese containing 20% Mn is known as spiegeleisen.

Ferromolybdenum. Alloy of Fe and Mo (60-65%) used to introduce Mo into alloy steels and special cast irons. Obtained by reducing MoS_2 with Fe and C in an electric furnace.

Ferroprussiates. Cyanoferrates.

Ferrosilicon. Iron silicide, FeSi, obtained by reducing a silicon-containing iron-ore in an electric furnace; used in the manufacture of steels and to reduce the amount of iron carbide in cast iron.

Ferrous. Iran (II) compounds.

Ferrous, Nitroso Compounds. See iron nitrosyls.

Fertilizers. Although any substance which increases production when added to the soil could be called a fertilizer, usually in practice by fertilizers are meant compounds containing N, P or K. Ammonium sulphate is the most important nitrogenous fertilizer; sodium nitrate, ammonium nitrate, ammonia and urea are also used. A mixture of ammonium nitrate and calcium carbonate is sold under the name of Nitro-chalk. The most important fertilizers containing phosphorus are the superphosphates which are mineral tricalcium phosphates rendered more soluble by treatment with sulphuric acid and ammonium phosphate. Many mixed fertilizers are now sold, containing various proportions of nitrogen, phosphorus and potasium.

Fine Chemicals. Chemicals produced industrially in relatively small quantities and with a high purity, *e.g.*, dyes and drugs.

Fineness of Gold. A measure of the gold alloy, defined as the parts of gold in 1000 parts of the alloy by mass. Gold with a fineness of 750 contains 75% gold, *i.e.*, 18 carat gold.

Fine Structure. Closely spaced spectral lines arising from transitions between energy levels that are split by the vibrational or rotational motion of a molecule or by electron spin. They are visible only at high resolution. Hyperfine structure, visible only at very high resolution, results from the influence of the atomic nucleus on the allowed energy levels of the atom.

Fission-track Dating. A method of estimating the age of glass and other mineral objects by observing the tracks made in them by the fission fragments of the uranium nuclei that they contain. By irradiating the objects with neutrons to induce fission and comparing the density and number of the tracks before and after irradiation it is possible to estimate the time that has elapsed since the object solidified.

Fixation. The removal of unwanted silver halide from a photographic material by solubilization after development. The most widely used fixers are thiosulphates which form water-soluble complex salts with silver halides. Other inorganic fixers include thiocyanates, cyanides, ammonia and concentrated potassium iodide. Organic fixers include thiourea and allylthiourea (thiosinamine). Fixing baths may also include hardeners (chromium alum, potassium alum) to toughen the final gelatin layer.

Fare Stack. A chimney at the top of which unwanted gases are burnt.

Flint. A compact massive quartz mixed with opaline SiO_2 occurring mainly in chalk. After burning, quenching and grinding it is used in porcelain manufacture.

Fluoborite, $Mg_3(OH, F)_3BO_3$. A mineral with cp O^{-2}, OH^-, and F^- and octahedrally co-ordinated Mg.

Fluoridation. The introduction of small quantities of fluoride compounds into the water supply as a public-health measure to reduce the incidence of tooth decay in children.

Fluoride. See halide.

Fluorination. A chemical reaction in which a fluorine atom is introduced into a molecule.

Fluorine. Symbole F. A poisonous pale yellow gaseous element belonging to group VII of the periodic table (the halogens); a.n. 9; r.a.m. 18.9984; d. 1.7 g dm^{-3}; m.p. −219.62°C; b.p. −186.1°C. The main mineral sources are fluorite (CaF_2) and cryolite ($Na_3 AlF$). The element is obtained by electrolysis of a molten mixture of potassium fluoride and hydrogen fluoride. The element was identified by Scheele in 1771 and first isolated by Moissan in 1886.

Fluorite (fluorspar). A mineral form of calcium fluoride, CaF_2, crystallizing in the cubic system. Fluorite is used chiefly as a flux, material in the smelting of iron and steel.

Fluorite. Structure. See calcium fluoride structure.

Fluorspar. A naturally occuring form of calcium fluoride. CaF_2.

Flux

1. A substance used to keep metal surfaces free of oxide in soldering.
2. A substance used in smelting metals to react with silicates and other impurities and form a low-melting slag.

Fluxional Molecule. A molecule in which the constituent atoms change their relative positions so quickly at room temperature that the normal concept of structure is inadequate; *i.e.* no specific structure exists for longer than about 10^{-2} second and the relative positions become indistinguishable. For example ClF_3 at —60°C has a distinct 'T' shape but at room temperature the fluorine atoms are visualized as moving rapidly over the surface of the chlorine atom in a state of exchange and are effectively identical.

Forbidden Band. See energy bands.

Fossil Fuel. Coal, oil, and natural gas, the fuels used by man as a source of energy. They are formed from the remains of living organisms and all have a high carbon or hydrogen content. Their value as fuels relies on the exothermic oxidation of carbon to form carbon dioxide ($C + O_2 \rightarrow CO_2$) and the oxidation of hydrogen to form water ($H_2 + \frac{1}{2}O_2 \rightarrow H_2O$).

Frasch Process. A process for obtaining sulphur by passing superheated water down a shaft to liquefy sulphur which is blown to the surface with compressed air.

Francium Fr. At no. 87. Radioactive element which occurs as a member of the various natural decay series; m.p. approx 27°C. b.p. approx. 680°C. ^{223}Fr (half-life 21 minutes) is the most stable isotope. Fr is an alkali metal, electronic configuration, $7s^1$ and ils chemistry is known mainly from tracer studies. It forms a single series of Fr^+ compounds.

Franck-condon Principle. According to this principle the time required for an electronic transition in a molecule is very much less than the period of vibration of the constituent nuclei of the molecule. Consequently, it may be assumed that the during the electronic transition the nuclei do not change their positions or momenta. This principle is of great importance in discussing the energy changes and spectra of molecules.

Fraunhofer Lines. Dark lines in the solar spectrum that result from the absorption by elements in the solar chromosphere of some of the wavelengths of the visible radiation emitted by the hot interior of the sun.

Free Electron. See electron.

Froth Flotation. †An industrial technique for separating the required parts of ores from unwanted gangue. The mineral is pulverized and mixed with water, to which a frothing agent is added. The mixture is aerated and particles of one constituent are carried to the surface by bubbles of air, which adhere preferentially to one of the constituents. Various additives are also used to modify the surface properties of the particles (*e g.* to increase the adherence of air bubbles).

Fuel. A substance that is oxidized or otherwise changed in a furnace or heat engine to release useful heat or energy. For this purpose wood, vegetable oil, and animal products have largely been replaced by fossil fuels since the 18th century.

The limited supply of fossil fuels and the expense of extracting them from the earth has encouraged the development of nuclear fuels to produce electricity.

Fuller's Earth. A natural clay used as an absorbent and industrial catalyst.

Fulminate. See cyanic acid.

Fulminic Acid. See cyanic acid.

Fused. Describing a solid that has been melted and solidified into a single mass. Fused silica, for example, is produced by melting sand.

Fusible Alloys. Alloys that melt at low temperature (around 100°C). They have a number of uses, including constant-temperature baths. pipe bending, and automatic sprinkers to provide a spray of water to prevent fires from spreading. Fusible alloys are usually eutectic mixtures of bismuth, lead, tin, and cadmium. Wood's metal and Lipowitz's alloys are examples of alloys that melt at about 76°C.

Fusion

1. Melting,
2. See nuclear fusion.

G

Gadolinium. It is a ductile malleable silvery element of the lanthanoid series of metals. It is found in association with other lanthanoids. Gadolinium finds use in alloys, magnets, and in the electronics industry.

Symbol : Gd; m.p. 1313°C; b.p. 3266°C; r.d. 7.9: p.n. 64; r.a.m. 157.25.

Gadolinium Compounds. Gd forms a single series of typical lanthanide compounds in the +3 state Gd^{3+} (f^7 colourless) $Gd^{3+} \rightarrow Gd$(−2.40 volts in acid). GaI_2(GdI_3+Gd) has been reported but is probably Gd^{3+}, $2I^-e^-$. Gd_2Cl_3 and GdCl($GdCl_3$ plus Gd) have metal—metal bonding.

Galena, PbS. A common lead ore, black bluish-grey with a metallic lustre. It has the NaCl structure.

Gallates. Anionic Ga-containing specics formed by dissolving Ga. Ga_2O_3, $Ga(OH)_3$ in excess base.

Gallium, Ga. At. no. 31, at. wt. 69.72, m.p. 29.78°C, b.p. 2403°C, d 5.907. Found in traces in Zn and Al minerals and in germanite (0.7% Ga). The element, metallic in appearance, has a complex structure; it is obtained by electrolysis of Ga salts. It dissolves in acids (and alkalis) and becomes coated with a thin coating of oxide on heating in air. Used in doping semiconductors, light emitting dioxides (GaAs; GaP), electrical devices, phosphors.

Gallium Chemistry. Gallium is a group (III) element, electronic configuration $4s^2 4p^1$. Its main oxidation state is +3 (E° for $Ga^{3+} \rightarrow Ga$ −0.56 volts in acid) and there is an extensive aqueous chemistry of the Ga^{3+} ion (generally octahedrally co-ordinated). Halides, *e.g.* $(GaCl_3)_2$, are dimeric and predominantly covalent. Organic derivatives, GaR_3, are monomeric and $[R_2Ga]^+$ complexed species stable even in aqueous solution, *e.g.* $[Me_2Ga(H_2O)_2]^+$, $[Me_2GaOH]_4$. Complexes are readily formed. Gallium (I) compounds are relatively unstable and accur in $GaCl_2$, GaS, etc., contain Ga_2 units

Gallium Halide. Gallium (III) fluoride, GaF_3 is ionic and forms GaF_3,H_2O. Complexes, *e.g.* $(NH_4)_3GaF_6$ are formed in aqueous HF solution (decomposes to GaF_3). Gallium(III) chloride $[Cl_2Ga(\mu\text{-}Cl)_2GaCl_2]$ is covalent and gives GaOC with water. The bromide and iodide are similar to the chloride. $GaCl_3$ plus Ga gives $Ga^I[Ga^{III}Cl_4]$ (also Br and I) which are strong reducing agents.

Gallium Oxide, Ga_2O_3. Formed by heating the nitrate, sulphate, hydroxide, has various forms similar to those of Al_2O_3 Forms a hydroxide (NH_4OH on Ga^{3+} solution). Both oxide and hydroxide are amphoteric, dissolving in acids and bases (to give gallates).

Gallium Oxyacid Salts. These are all gallium(III) salts (Ga_2O_3 plus acid) which decompose on heating to Ga_2O_3. Known salts include $Ga(NO_3)_3$, $9H_2O$, $Ga(NO_3)_3,3H_2O$, $Ga(ClO_4)_3,6H_2O$, Ga_2, $(SO_4)_3,18H_2O$.

Gallium Sulphides. Ga_2S_3(Ga plus S), GaS(Ga_2S_3 plus H_2 contains Ga_2 units) and Ga_2S(Ga plus (H_2S) are known.

Galvanizing. A process which is used for coating steel with zinc, by dipping in a bath of molten zinc, or by electrodeposition. The zinc protects the steel from corrosion.

Gamma Radiation. A form of electromagnetic radiation emitted by changes in the nuclei of atoms. Gamma waves have very high frequency (short wavelength). Gamma radiation shows particle properties much more often than wave properties. The energy of a gamma photon, given by

$$W = hv$$

(where h is the Planck constant) can be very high A gamma photon of 10^{24} Hz has an energy of 6.6×10^{10} J. See also electromagnetic radiation.

γ-Rays have great penetrating power and are emitted in most nuclear disintegrations.

Gamma Rays. Streams of gamma radiation.

Gangue. The clay and other silicate material associated with ores.

Ganister. A hard sandstone used in refractory bricks for lining steel furnaces and convertors.

Garnet. Any of a group of silicate minerals that conform to the general formula $A_3B_2(SiO_4)_3$. The elements representing A may include magnesium, calcium, manganese, and iron(II); those representing B may include aluminium, ion (III), chromium, or titanium. Six varieties of garnet are generally recognized :

pyrope, $Mg_3Al_2Si_3O_{12}$;

almandine, $Fe2/3^+Al_2Si_3O_{12}$;

spessartite, $Mn_3Al_2Si_3O_{12}$;

grossularite, $Ca_3Al_2Si_3O_{12}$;

andradite, $Ca_3(Fe^{3+},Ti)_2Si_3O_{12}$;

and uvarovite, $Ca_3Cr_2Si_3O_{12}$.

Some garnets are used as gemstones with colours ranging from yellow-red to deep-red and also emerald green. Garnet sand is used as an abrasive.

Garnierite. A hydrated Mg—Ni silicate, an important nickel ore.

Gas Hydrates. Clathrates formed by water with atomic or small-molecular weight gases and some small organic molecules. The solid water forms open cages with the gases in holes. *e.g.* $Cl_2,7.3H_2O$.

Gas Mantle Oxide. 99% ThO_2, 1% CeO_2 made by ignition of a cellulose fabric impregnated with a mixture of the nitrates.

Gaylussite. A mineral consisting of a hydrated mixed carbonate of sodium and calcium, $Na_2CO_3CaCO_3.5H_2O$.

Geiger Counter. A device which is used to detect and measure amounsts of radioactivity. It consists of a cylindrical chamber which contains two electrodes, one of which is a fine wire along the axis of the cylinder. The chamber is filled with gas at a low pressure and the incident radiation causes ionization of the gas. If the chamber is polarized with an applied high voltage the ions produced migrate to the electrodes and the resulting current is detected and amplified.

It was first devised in 1908 by Hans Geiger 1882-1947). Geiger and W. Moller produced an improved design in 1928.

Gem. A prefix indicating that particular substituents are attached to the same atom, *e.g.* $F_3SiSiCl_2SiF_3$, gem-dicnlorohexafluoro-trisilane.

Gemstones. Materials which because of their appearance and variety are used in jewellery and decoration. Synthetic stones simulate natural gem-stones. x-Alumina (corundum) gives ruby (with Cr^{3-}) and sapphire (Fe plus Ti). Spinel, titania, $SrTiO_3$ are also used but are virtually all artificial. Diamonds can be synthesized but are mainly used for cutting.

Geochemistry. Refers to the scientific study of the chemical composition of the earth. It includes the study of the abundance of the earth's elements and their isotopes and the distribution of the elements in environments of the earth (lithosphere, atmosphere, biosphere and hydrosphere).

Germanates. Oxy componnds of germanium (IV) showing resemblances to simple silicates E g $SrGeO_3$ (cyclic$[Ge_3O_9]^{6-}$) and Mg_2GeO_4 (orthogermanate). Hydrated GeO_2 dissolve in excess base to give $[Ge(OH)_6]^{2-}$ and other oxide-hydroxide anions and cations.

Germanes, Germanium Hydrides. Monogermane, GeH_4, b.p. −88°C; Ge_2H_6, b.p. −29°C, and Ge_3H_8 are formea from GeO_2 and $LiAlH_4$ or $NaBH_4$. Higher germanes, to Ge_9H_{20}, are formed by electric discharge on GeH_4. Germanes are oxidized in air. An extensive chemistry of GeH_3—and related groups is known. Germanes have potentialities as sources of very pure Ge.

Germanite. A complex copper sulphide containing 6—10% Ge and 0.7% Ga.

Germanium. Symbol Ge. It is a lustrous hard, metalloid element belonging to group IV of the periodic table; a.n. 32; r.a.m. 72.59; r.d. 5.36; m.p. 937°C; b.p. 2830°C. It occurs in zinc sulphide and in certain other sulphide ores, and is mainly obtained as a by product of zinc smelting. It is also present in some coal (up to 1.6%). Small amounts are used in specilized alloys but the main use depends on its semiconductor properties. Predicted in 1871 by Mendeleev (eka-silicon), it was discovered by Winkler in 1886.

Germanium Chemistry. Germanium is a group IV element, electronic configuration $4s^2 4p^2$. The normal oxidation state is +4 in covalent $GeCl_4$, GeH_4 and ionic GeO_2, Complexes, *e.g.* $[GeF_6]^{2-}$,$GeCl_4$.2 pyridine, are known mainly with octahedral co-ordination. The +2 state is quite strongly reducing. Organo-germanium compounds of Ge(IV) are similar to the corresponding Si or Sn compounds.

Germanium Halides. Germanium fluorides, GeF_4, m.p.—15°C (heat on $BaGeF_6$, Ge plus F_2) is hydrolysed by water, forms $[GeF_6]^{2-}$ in aqueous HF. GeF_2 is polymeric (Ge plus GeF_4 at 100—300°C).

Germanium chlorides. $GeCl_4$, m.p.—49°C, b p. 86°C (Ge plus Cl_2) is hydrolysed by water. $GeCl_2$, colourless reducing agent ($GeCl_4$ plus Ge).

Bromides and iodides are similar to the chlorides.

Germanium (IV) Oxide (GeO_2) A compound which is obtained by strongly heating germanium in air or by hydrolysing germanium (IV) chloride. It occurs in two forms. One is slightly soluble in water forming an acidic solution; the other is insoluble in water. Germanium oxide dissolves in alkalis to form the anion GeO_3^{2-}.

Germanium Sulphides. GeS_2 and GeS are known.

German Silver (nickel silver). Refers to an alloy of copper, zinc, and nickel, often in the propositions 5:2:2. It resembles silver in appearance and finds use in cheap jewellery and cutlery and as a base for silver-plated wire.

Getter. A chemical for removing impurities. KF acts as a getter to remove HF. Li will remove O,S, etc. from Cu and Cu-alloys.

G. Factor. The proportionality factor in the relation between the magnetic moment μ and the number of unpaired electrons

$$\mu = g\sqrt{s(s+1)}$$

Sometimes called the gyromagnetic ratio. For a free electron g=2.003.

Gibbsite. Refers to a mineral form of hydrated aluminium hydroxide $Al(OH)_3$). It is named after US mineralogist George Gibbs (d. 1833).

Glass. A hard transparent material which is made by heating calcium oxide (lime), sodium carbonate, and sand (silicon (IV) oxide). This produces a calcium silicate—the normal type of glass, called soda glass. Special types of glass can be made by incorporating boron oxide in the glass (borosilicate glass, used for laboratory apparatus) or by including metals other than calcium, *e.g.* lead or barium.

Glass is an amorphous substance, in the sense that there has been no long-range ordering of the atoms on a lattice. It can be considered as a supercooled liquid, which has not crystallized. Solids with similar non-crystalline structures are also called glasses.

Glass Fibres Melted glass drawn into thin fibres some 0.005 mm−0.01 mm in diameter. The fibres may be spun into threads and woven into fabrics, which are then impregnated with resins to give a material that is both strong and corrosion resistant for use in car bodies and boat building.

Glauberite. A mineral consisting of a mixed sulphate of sodium and calcium, $Na_2SO_4.CaSO_4$.

Glauber's Salt. Sodium sulphate decahydrate, $Na_2SO_4.10H_2O$, used as a laxative. It it is named after J.R. Glauber 1604-68).

Glauconite. A green hydrated iron silicate. A constituent of the green sand associated with chalk.

Goethite. A yellow-brown mineral, FeO.OH, which is crystallizing in the orthorhombic system. It is formed as a result of the oxidation and hydration of iron minerals or as a direct precipitate from marine or fresh water (*e.g.* in swamps and bogs). Most limonite has been composed largely of cryptocrystalline goethite. Goethite is mined as an ore of iron.

Gold. A transition metal that is found native. It is unreactive and is very ductile and malleable. Gold finds use in jewellery, often alloyed with copper. Pure gold is 24 carat; 9 carat indicates that 9 parts in 24 consist of gold.

Symbol : Au; m.p. 1063°C; b.p. 2710°C; r.d. 19.3; p.n. 79; r.a.m. 196.97.

Gold (III) Chloride (gold trichloride, auric chloride, $AuCl_3$). A compound which is prepared by dissolving gold in aqua regia. The bright yelow crystals (chlorauric acid) produced on evaporation are heated to form dark red crystals of gold (III) chloride. The chloride decomposes easily (at 175°C) to give gold (I) chloride and chlorine; at higher temperatures it decomposes to give gold and chlorine, Gold (III) chloride is used in photography It exists as a diner. Au_2Cl_6.

Goldschmidt Process. A method of extracting metals by reducing the oxide with aluminium powder, *e.g.*

$$Cr_2O_3 + 2Al \rightarrow 2Cr + Al_2O_3$$

The reaction can also be used to produce molten iron. It was discovered by the German chemist Hans Goldschmidt (1861–1923).

Goldschmidt Reaction. The use of Al powder for the reduction of metal oxides.

Grain. A crystal in a metal that has been prevented from attaining its regular geometrical form.

Graphite, Plumbago, Black Lead. A form of carbon occurring and formed in crystalline and microcrystalline forms. The structure has parallel hexagonal sheets of carbon atoms. The C—C distance within the sheets is 1.42A, the distance found in aromatic rings. The C—C distance between sheets is 340A and there is high conductivity along sheets, easy cleavage and ready incorporation of other substances between the layers (graphite compounds, lamellar compounds, intercalation compounds) Used as a lubricant, refractory, pencils, etc, electrical components, paints, carbonizing steels. Very pure graphite coke from crude oil plus a pitch binder heated to 2800°C) is used as a and reflector in nuclear reactors.

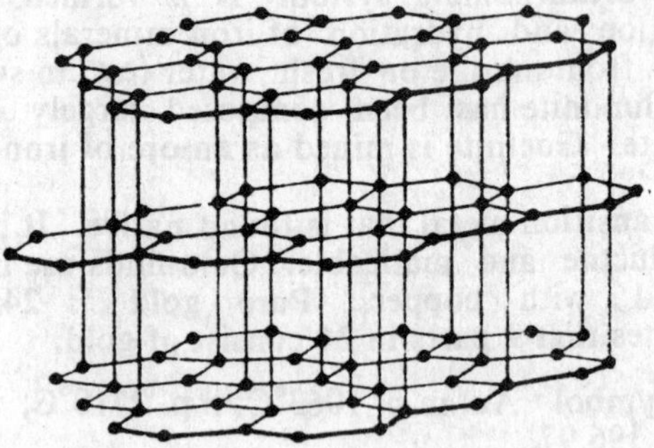

Graphitizer. An addition (*e.g.*, Ni, Cu, Si) to steel and cast iron which transforms Fe_3C to graphite plus Fe. Essential in cast irons but must be used in conjunction with carbide stabilizers in high speed steels.

Greenockite. A mineral form of cadmium sulphide, CdS.

Green Vitriol. $FeSO_4.IH_2O$.

Ground State. Refers to the lower stable energy state of a system, such as a molecule, atom, or nucleus.

Group. A vertical group of elements in the normal form of the periodic table. Within a group there are distinct chemical resemblances and gradations in properties.

Group 0 Elements. See noble gases.

Group I Elements. Refers to a group of elements in the periodic table which is divided into two subgroups group IA (the main group, alkali metals) and group IB. Group IB consists of the coinage metals, copper, silver, and gold, which have been usually classified with the transition elements.

Group II Elements. Refer to a group of elements in the periodic table, which is divided into two subgroups: group IIA (the main groups alkaline-earth metals) and group IIB. Group IIBincludes the three metals zinc (Zn), cadmium (Cd), and mercury (Hg, which have two s-electrons outside filled d-subshells. Moreover, none of their compounds have unfilled d-levels, and the metals are considered as nontransition elements. They are sometimes called the zinc group. Zinc and cadmium are relatively electropositive metals, forming compounds containing divalentions Zn^{2+} or Cd^{2+}. Mercury it more unreactive and also unusual in forming mercury (I) compounds, which contain the ion Hg_2^{2+}.

Group III Elements. A group of elements in the periodic table, which is divided into subgroup IIIB (the main group) and subgroup IIIA. The IIIA subgroup, consists of scandium (Sc), yttrium (Y), and lanthanum (La) which are generally classified with the lanthanoids, and actinium (Ac), generally classified with the, actinoids. The term group III elements usually refers to the main-group elements : boron (B) aluminium (Al), gallium (Ga), indium (In), and thallium (Tl), which, all have outer electronic configurations ns^2np^1 with no partly filled inner levels. They are the first members of the p-block. The group differs from the alkali metals and alkaline earth metals in displaying a considerable variation in properties as the group is descended.

As the group is descended, atomic radii increase and ionization energies are all lower than for boron. There is an increase in polar interactions and the formation of distinct M^{3+} ions. As the elements of group III have a vacant p-orbital they display many electron-acceptor properties. For example, many boron compounds form adducts with donors such as ammonia and organic amines (acting as Lewis acids). A large number of complexes of the type $[BF_4]^-$, $[AlCl_4]^-$,$[InCl_4]^-$, $[TlI_4]^-$ are known and the heavier members can expand their coordination numbers to six as in $[AlF_6]^{3-}$ and $[TlCl_6]^{3-}$. Another feature of group III is the increasing. The electron configuration ns^2np^1 suggests that only only one electron could be lost or shared in forming compounds. In fact, for the lighter members of the group the energy required to promote an electron from the s-subshell to a vacant p-subshell is small. It is more than compensated for by the resulting energy gain in forming there bonds rather than one. This energy gain is less important for the heavier members of the group. Thus, aluminium forms compounds of the type AlCl in the gas phase at high temperatures Gallium similarly forms such compounds and gallium(I) oxide (Ga_2O) can be isolated Indium has a number of known indium (I) compounds (*e g.* InCl, In_2O, In $^{1}[In^{111}Cl_6]$). Thallium has stable monovalent compounds. In aqueous solution, thallium(I) compounds are more stable than the corresponding thallium (III) compounds.

Group IV Elements. Refer to a group of elements in the periodic table, divided into subgroup IVB (the main group) and subgroup IVA The IVA subgroup includes titanium (Ti), zirconium (Zr), and hafnium (Hf), which are generally classified with the transition elements. The term group IV elements usually refers to the main-group elements : carbon (C). silicon (Si), germanium (Ge), tin (Sn), and lead (Pb), which all have outer electronic configurations ns^2np^2 with no partly filled inner levels.

The main valency of the elements has been 4, and the members of the group have been showing a variation from nonmetallic to metallic behaviour in moving down the group.

The general, the reactivity of the elements increases down the group form carbon to lead. All react with oxygen on heating. The first four from the dioxide; lead forms the monoxide (*i e.* lead (II) oxide, PbO). Similarly, all will react with chlorine to form the tetrachloride (in the case of the first four) or the dichloride (for lead). Carbon is the only one capable of reacting directly with hydrogen. The hydrides all exist from the stable methane (CH_4) to the unstable plumbane (PbH_4).

Group V Elements. Refers to a group of elements in the periodic table, divided into subgroup VB (the main group) and subgroup VA. The VA subgroup consists of vanadium (V), niobium (Nb), and tantalum (Ta), which are generally classified with the transition elements. The term group V elements usually refers to the main-group elements : nitrogen (N), phosphorus (P), arsenic (As), antimony (Sb), and bismuth (Bi), which all have outer electronic configurations ns^2np^3 with no partly filled inner levels.

Group VI Elements. A group of elements in the periodic table, divided into group VIB (the main group) and group VIA. The VIA subgroup consists of chromium (Cr), molybdenum (Mo), and tungsten (W), which are generally classified with the transition elements. The term group VI elements usually refers to the main-group elements : oxygen (O), sulphur (S), selenium (Se), tellurium (Te), and polonium (Po), which all have outer electronic configurations ns^2np^4 with no partly filled inner levels. They are also called the chalcogens.

Group VII Elements. Refer to a group of elements in the periodic table which are divided into two sub-group : group VIIA and group VIIB. Group VIIB is the main group (see halogens) and group VIIA consists of the elements manganese (Mn), technetium (Tc), and rhenium (Re), which are usually classified with the transition elements.

Group VIII Elements. Refer to a group of nine elements in the periodic table consisting of the platinum metals—cobalt (Cs), rhodium (Rh), iridium (Ir), nickel (Ni), palladium (Pd) and platinum (Pt)—together with iron (Fe), ruthenium (Ru), and osmium (Os). These metals have been now classified with the transition elements.

Gun Metal. A type of bronze usually having 88—90% copper, 8—10% tin, and 2—4% zinc. Formerly used for cannons, it is still used for bearings and other parts that require high resistance to wear and corrosion.

Gunpowder. A powdered mixture of sulphur, charcoal, and potassium nitrate, used as an explosive.

Gutzeit's Test. A test for arsenic involving formation of arsine, AsH_3 (solution plus Zn plus acid) which is reacted with $HgCl_2$ solution to give successively yellow, brown and black stains.

Gypsum. Mineral $CaSO_4,2H_2O$. Used in the production of plaster of paris, to control the rate of setting of Portland cement and as a filler.

It occurs in five varieties : rock gypsum, which is often red stained and granular; gypsite, an impure earthy form occurring as a surface deposit : alabaster, a pure fine-grained translucent form; stain spar, which is fibrous and silky; and selenite, which occurs as transparent crystals in muds and clays. It is used in the building industry and in the manufacture of cement, rubber, paper, and plaster of Paris.

H

Haber Process. An important industrial process which is used for the manufacture of ammonia, which is used for fertilizers and for making nitric acid. The reaction is the equilibrium :

$$N_2+3H_2 \rightleftharpoons 2NH_3$$

The nitrogen used is got by fractionation of liquid air and the hydrogen by the oxidation of hydrocarbons (from natural gas). The nitrogen and hydrogen are purified and mixed in the correct proportions. The equilibrium amount of ammonia gets favoured by low temperatures, but in practice the reaction would never reach equilibrium at normal temperatures. An optimum temperature of about 450°C is therefore used. High pressure also favours the reaction and a pressure of about 250 atmospheres is used The catalyst has been iron with small amounts of potassium and aluminium oxides present. The yield has been about 15%. Ammonia is condensed out of the gas mixture at -50°C.

Haematite, α-Fe_2O_3. An important iron ore, normally blood-red in colour but forms black crystals (specular iron). Has the corundum structure with c.p. O^{2-} and octahedrally co-ordinated Fe. Red ochre or reddle is a mixture of haematite, clay and some other impurities.

Hafnium Symbol Hf. Refers to a silvery lustrous metallic transition element; a.n. 72; r.a.m. 178.49; r.d. 13.3; m.p. 2230±20°C; b.p. 4602°C. The element occurs with zirconium and is extracted by formation of the chloride and reduction by the Kroll process. It finds use in tungsten alloys in filaments and electrodes and as a neutron absorber. The metal forms a passive oxide layer in air. Most of its compounds are hafnium(IV) complexes; less stable hafnium(III) complexes also exist. The element was first reported by Urbain in 1911, and its existence was finally established by D. Coster and G. C. de Hevesey in 1923.

Hahnium (Ekatantalum). (Symbol Ha; At No. : 105). This element has been reported by the Russian and also by American workers in April 1970. They obtained this new element by bombarding californium-246 with accelerated N-15 ions.

$$_{98}Cf^{246}+{}_7N^{15}\rightarrow{}_{105}Ha^{260}+{}_0n^1$$

Hahniu. 260 has a half-life of 1.6±0.3 second and decays to form lawrencium-256 and alpha particles of 91 MeV. Lawrencium rn, decays into mendelevium-252 and alpha particles

$$ {}_{105}Ha^{260} \rightarrow {}_{103}Lw^{256} + {}_{2}He^{4} $$

$$ {}_{103}Lw^{256} \rightarrow {}_{101}Md^{252} + {}_{2}He^{4} $$

The name hahnium and the symbol Ha have been proposed for the new element, 105 after Otto Hahn, the discoverer of nuclear fission.

Halide. A compound having a halogen. The inorganic halides of electropositive elements contain ions of the type Na^+Cl^- (sodium chloride) or K^+Br^- (potassium bromide). Transition metal halides often have some covalent bonding. Non-metal halides have been covalent compounds, which are usually volatile. Examples have been carbon tetrachloride (CCl_4) and silicon tetrachloride ($SiCl_4$). Halides are named as bromides, chlorides, fluorides, or iodides.

Halite (Rock Salt). Naturally occurring sodium chloride (common salt, NaCl), crystallizing in the cubic system. It is chiefly colourless or white (sometimes blue) when pure but the presence of impurities may colour it grey, pink, red, or brown. Halite is often found in association with anhydrite and gypsum.

Halogens. A group of elements (group VII of the periodic table) which includes fluorine, chlorine, bromine, iodine, and the short-lived element astatine. The halogens all have outer valence shells that are one electron short of a noblegas configuration. One to this the, halogens are characterized by high electron affinities and high electronegativities, fluorine being the most electronegative element known. The high electronegativities of the halogens have been favouring the formation of both uninegative ions, X, particularly in combination with electropositive elements (*e.g* NaCl, KBr, CsI, CaF_2), and the single covalent bond—X with elements of moderate to high electronegativity (*e.g.* HCl, SiF_4, SF_4, $BrCH_3$, Cl_2O). The halogens all form diatomic molecules, X_2, which have been characterized by their high reactivities with a wide range of other elements and compounds. As a group, the elements increase in both size bnd polarizability as the proton number increases and there is an attendant decrease in electronegativity and reactivity. This means that there is no chemistry of positive species apart from a few cationic iodine compounds. The heavier halogens Cl, Br, and I, all form oxo-species having formal positive oxidation numbers +1 and +5, chlorine and iodine also forming +3 and +7 species. (*e.g.* HOCl, $HBrO_3$, I_2O_5, HIO_4)

The elements decrease in oxidizing power in the order $F_2 > Cl_2 > Br_2 > I_2$ and the ions X^- may be arranged in order of increasing reducing power $F^- < Cl^- < Br^- < I^-$. Thus any halogen will displace the elements below it from their salts, for example.

$$Cl_2 + 2Br^- \rightarrow Br_2 + 2Cl^-$$

A wide range of organic halides gets formed in which the C-F bond is characteristically resistant to chemical attack; the C-Cl bond is also fairly stable, particularly in aryl compounds but the alkyl halogen compounds become increasingly susceptible to nucleophilic attack and generally more reactive.

Halogen-halogenoids. Halogens have a tendency to combine with pseudo halogens to form a new series of compounds known as halogen-halogenoids. A number of such compounds like fluorine azide (FN_3), chlorine azide (ClN_3), bromine azide (BrN_3), iodine azide (IN_3), cyanogen halides (CNCl, CNBr and ICN) and thiocyanogen and oxycyanogen halides are known.

Halogenoids. Walden and Audrieth defined halogenoid :

"Any univalent chemical aggregate composed of two or more electronegative atoms which shows in the free state certain characteristics of halogens and which combines with hydrogen to form an acid and with silver to form salt insoluble in water."

Hammer Mills. Devices which are used in the chemical industry for crushing and grinding solid materials at high speeds to a specified size. The impact between the particles, grinding plates, and grinding hammers pulverizes the particles. Hammer mills can be used for a greater variety of soft material than other types of grinding equipment.

Hard And Soft Acids And Bases. Bases may be arbitrarily divided into hard, non-deformable, *e.g.* F^-, and soft, polarinable, potentially π-bonded, *e.g.* PPn_3. Acids may be divided similarly into hard, non-polarizable, *e.g.* Na^+ and soft, polarizable, potentially π-bonded, *e.g.* Pt^{2+}. Hard and soft acid base theory (HSAB) postulates greatest stability for hard acid-hard base and soft acid-soft base pairs, *e.g* Si^{4+} forms its most stable halide complexes with F^-; Pt^{4+} with I^-.

Hardenability. A measure of the ease of hardening a steel. A steel of poor hardenability requires a drastic cold water quench and may crack. Addition of *e.g.* Mn, Ni, Cr improves hardenability so that oil quenching or even air cooling is sufficient for the hardening.

Hardness of Water. The property conferred on water by the presence of alkaline earth salts which prevent formation of a lather with soaps. Temporary hardness ($Ca(HCO_3)_2$) is removed by boiling; permanent hardness, other Ca and Mg salts, is removed by ion exchange (permutite process) or by detergents.

Temporary hardness is removed boiling.

$$Ca(HCO_3)_2(aq) \dashrightarrow (CaCO_3(s) + H_2O(l) + CO_2(g)$$

The precipitated calcium carbonate is the 'fur' (or 'scale') formed in kettles, boilers, pipes, etc. In some areas, hardness also results from dissolved calcium sulphate ($CaSO_4$), which cannot be removed by boiling (permanent hardness).

Hard water is a considerable problem in washing, reducing the efficiency of boilers, heating systems, etc.. and in certain industrial process. Various methods of water softening are used. In public supplies, the temporary hardness can be removed by adding lime (calcium hydroxide), which precipitates calcium carbonate

$$Ca(OH)_2(aq) + Ca(HCO_3)_2(aq) \dashrightarrow 2\,CaCO_3(s) + 2\,H_2O(l)$$

This is known as the Clark process (or as 'clarking'). It does not remove permanent hardness. Both temporary and permanent hardness can be treated by precipitating calcium carbonate by added sodium carbonate hence its use as a washing os da and in bath salts. Calcium (and other) ions can also beremoved from water by ion-exchange using zeolites (*e.g.*, Permutit). This method is used in small domestic water.softners. Another technique is not to remove the Ca^{2+} ions but to complex them and prevent them reacting further. For domestic use polyphosphates (containg the ion $P_5O_{18}^{6-}$, *e.g.* Calgon) are added. Other sequestering agents are also used for industrial water.

Hard Water. See hardness.

Heavy Hydrogen. See denterium.

Heavy Spar. A mineral form of barium sulphate, $BaSO_4$.

Heavy Water (deuterium oxide). Water in which hydrogen atoms, ^1H, are replaced by the heavier isotope deuterium, ^2H. It is a colourless liquid, which forms hexagonal crystals on freezing. Its physical properties differ from those of 'normal' water; r.d. 1.105; m.p. 3 8°C; b p 101.4°C. Deuterium oxide occurs to a small extent (about 0.003% by weight) in natural water, from which it can be separated by fractional distillation or by electrolysis. It is particularly useful in the nuclear industry because of its ability to reduce the energies of fast neutrons to thermal energies and because its quenching cross-section is lower than that of hydrogen and can sequently it does not appreciably reduce the neutron flux. In the laboratory it is used for labelling other molecules for studies of reaction mechanisms.

Heisenburg Uncertainty Principle. For small particles which possess both wave and particle properties, it is impossible to determine accurately both the position and momentum of the particle simultaneously. Mathematically the uncertainty in the position Δx and momentum Δp are related by the equation

$$\Delta p \times \Delta x = h/2\pi \text{ (h=Planck's constant)}$$

Heitler-london Theory. This theory is based upon the pairing and resultant neutralization of opposed electron spins. The fundamental idea about this theory is summarized below :

(i) In order for an atom tc enter into chemical combination, it must possess one or more unpaired electrons. The valency of a species is thus determined by the number of unpaired electrons which it possess.

(ii) Although participation in bond formation by paired electrons is formally ruled out, such electrons may be involved if they can be unpaired without the expenditure of excessive energy. For example, nitrogen with an outer configuration of $1s^1 2p_x^2\ 2p_y^1\ 2p_z^1$ shares its 2p electrons which flourine in the formation of nitrogen triflouride (NF_3) but cannot form a pentafluoride because no more orbitals in the n=2 level are available to accommodate one of the unpaired 2s electrons. With phosphorus, on the other hand, the configuration $s^2\ 3p_x^1\ 3p_y^1\ 3p_z\ 3p^0$ does permit the formation of a pentafluoride since 3d orbitals are presumably available.

(iii) A chemical bond results in when the two electrons are paired. For a stable bond formation, it is essential that the two electrons must have opposite spins.

(iv) Each atom in a molecule tends to assume a structure containing a closed shell of electrons.

(v) The pair of electrons is localised between the two bonded atoms.

Helium. It is a colourless monatomic gas. It is the first member of the noble gases (group 0 of the periodic table). Helium is having the electronic configuration $1s^2$ and consists of a nucleus of two protons and two neutrons (equivalent to an α-particle) with two extra-nuclear electrons. It has an extremely high ionization potential and is completely resistant to chemical attack of any sort. The gas accounts for only $5.2 \times 10^{-4}\%$ of the atmosphere; up to 70% occurs in some natural gas deposits.

Helium is the second most abundant element in the universe, the primary process on the Sun being nuclear fusion of hydrogen to give helium. Helium is recovered commercially from natural gas in both the USA and the USSR. Its application have been in fields in which inertness is required and where the cheaper alternatives, such as nitrogen, are too reactive; for example, high temperature metallurgy, powder technology, and welding. Helium is also favoured over nitrogen as an oxygen dilutent for deep-sea diving (lower solubility in blood) and as a pressurizer for liquefied gas fuels in rockets (total inertness). It is also used as an ideal gas for balloons (no fire risk) and for low-temperature physical research.

Hemihydrate. A crystalline hydrate containing two molecules of compound per molecule of water (*e.g.* $2CaSO_4.H_2O$).

Heptahydrate. A crystalline hydrate that has seven molecules of water per molecule of compound.

Hepataxodiphosphoric (V) acid. See phosphoric (V) acid.

Heptavalent (septavalent). Having a valence of seven.

Heteropoly Acids. These acids are obtained by the union of varying number of acid anhydride molecules (generally WO_3 MoO_2, V_2O_5) with some other acid which could furnish the central atom of central ion of the whole complex anion.

The ability to act as the central atom in heteropolyacids is found amongst the acid forming elements as well as among the metals of the transition series.

Group I	H, Cu
Group II	Be
Group III	B, Al
Group IV	C, Si, Ge, Sn, Ti, Zr, Ce, Th
Group V	N, P, As, Sb, V, Nb, Ta
Group VI	Cr, Mo, W, U, S, Se, Te
Group VII	Mn, I
Group VIII	Fe, Co, Ni, Rh, Os, Ir, Pt

Heusler Alloys. Ferromagnetic alloys containing no ferromagnetic elements. The original alloys contained copper, manganese, and tin and were first made by Conrad Heusler (19th-century mining engineer).

Hexagonal Close-packed Crystal. Refers to a crystal structure in which layers of close-packed atoms are stacked in an ABABAB arrangement. The second layer B fits into the holes of the first layer A, with the third layer over the first. The coordination number is 12. Zinc and magnesium have hexagonal close-packed structures.

The unit cell has been based on a hexagon with atoms occupying corner and mid positions.

High-speed Steel. Refers to a steel that will remain hard at dull red heat and can therefore be used in cutting tools for high-speed lathes It usually contains 12-22% tungsten, upto 5% chromium, and 0.4−0.7% carbon. It may also be having small amounts of vanadium, molybdenum, and other metals.

Holmium. A soft malleable silvery element of the lanthanoid series of metals. It occurs in association with other lanthanoids. It has few applications.

Symbol : Ho; m.p. 1474°C; b.p. 2695°C; r.d. 8.80; p.n. 67; r.a.m. 164.93.

Holium Compounds. Holmium forms a single series of typical lanthanide compounds in the +3 state

Ho^{3+} (f^{10} brown-yellow)→Ho

(−2.32 volts in acid).

Hornblende. Any of a group of common rock-forming minerals of the amphibole group with the generalized formula :

$$(Ca,Na)_2(Mg,Fe,Al)_5(Al,Si)_8O_{22}(OH,F)_2$$

Hornblendes consist mainly of calcium, iron, and magnesium silicate.

Hume-Rothery's Rule. The statement that the phase of many alloys is determined by the ratios of total valency electrons to the number of atoms in the empirical formula.

Hund's Rules. Rules which describe the electronic configuration of degenerate orbitals in the ground state. The electronic configuration will have the maximum number of unpaired electrons, *i e.* p^3. ↑ ↑ ↑ rather than ↑ ↓ ↑ . The ground state will have the maximum orbital angular momentum.

Hybridisation. The phenomenon of mixing of the atomic orbitals and formation of new orbitals of equal energy is known as hybridisation and the new orbitals formed are known as hybrid orbitals. The bonds formed by such orbitals are called hybrid bonds. The compounds which result in from these bonds are known as hybrids.

The four new bonds formed above are inclined to each other at an angle of 109° 28' and directed towards the four corners of a regular tetrahedron. The whole phenomenon can be diagrammatically represented in Fig. 1. The number of hybrid orbital produced is equal to the number of orbitals mixed.

Conditions for Hybridisation. Following are the rules for hybridisation.

Rule I. Only orbitals of similar energies belonging to the same atom or ion can undergo hybridisation.

Rule II. Number or hybrid orbitals produced is equal to the number of orbitals mixed Hybrid bonds are stronger than the single non-hybridised bonds of comparable energy.

Rule III. Most of the hybrid orbitals are similar but they are not necessarily identical in shape. They differ from one another in orientation in space.

Rule IV. For equivalent hybrids, the orientation in space is determined by :

(i) the number of orbitals mixed and consequently the number of hybrids obtained, and

(ii) which of x, y and z directions are preferred by the orbitals when pure.

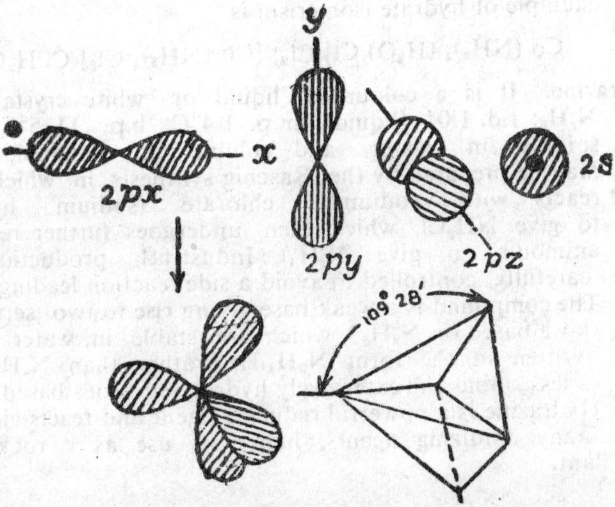

Rule V. The type of hybridisation indicates the geometry of molecules. From the type of hybridisation one can tell about the bond angles and structure-

Rule VI. Once an orbital has been used to build o hybrid it is then no longer available to hold electrons in its pure form.

Hybrid Orbitals. See orbital.

Hydracid. See binary acid.

Hydrate. A substance which is formed by combination of a compound with water.

Hydrate Isomerism. This type of isomerism arises due to different positions of a water molecule in a complex. For example, three isomers with the formula $CrCl_3,6H_2O$ are known.

(i)	$[Cr(H_2O)_6]Cl_3$	Violet	Hexa-aquochrominium (III) chloride
(ii)	$[Cr(H_2O)_5Cl]Cl_2[H_2O]$	Light green	Chloropentaquochronium (III) chloride)
(iii)	$[Cr(H_2O)_4Cl_2]Cl[2H_2O]$	Green	Dichlorotetraaquochromium (III) chloride

The violet coloured compound (i) does not lose any water molecule on dehydration over sulphuric acid and all the three Cl^- ions are precipitated by silver nitrate as silver chloride. The compound (ii), light green, loses one water molecule one dehydration over sulphuric acid and only two Cl^- ion are precipitated as silver chloride. The compound (iii), dark green, loses two water molecules on dehydration over sulphuric acid and only one chloride ion is precipitated as silver chloride. Another example of hydrate isomerism is

$$[Co(NH_3)_4(H_2O)Cl]Cl_2;\ [Co(NH_3)_4Cl_2]Cl.H_2O$$

Hydrazine. It is a colourless liquid or white crystalline solid, N_2H_4; r.d. 1.01 (liquid); m p. 1.4°C; b.p. 1135°C. It is very soluble in water and soluble in ethanol. Hydrazine may be prepared by the Raschig synthesis in which ammonia reacts with sodium(I) chlorate (sodium hypochlorite) to give NH_2Cl, which then undergoes further reaction with ammonia to give N_2H_4. Industrial production must be carefully controlled to avoid a side reaction leading to NH_4Cl. The compound is a weak base giving rise to two series of salts, those based on $N_2H_5^+$, which are stable in water (sometimes written in the form $N_2H_4.HCl$ rather than $N_2H_5^+Cl^-$), and a less stable and extensively hydrolysed series based on $N_2H_6^{2+}$. Hydrazine is a powerful reducing agent and reacts violently with many oxidizing agents, hence its use as a rocket propellant.

Hydrazoic Acid, Azoimide, N_3H, NNNH. M.p −80°C, b.p. 37°C. Prepared from sodium azide and acid or $[N_2H_5]^+$ plus nitrous acid, HNO_2. Heavy-metal salts, azides, are used as detonators, alkali metal salts are stable and azides are used synthetically in organic chemistry.

Hydrides. Hydrogen possesses a unique property by virtue of which it may combine with metals and non-metals by direct or indirect methods. These binary compounds of hydrogen are termed as hydrides.

The above definition of hydrides does not include hydrides of common occurrence. Examples and H_2O, HCl, H_2S, etc.

As the hydrides of a large number of elements are known, they are therefore characterised by a wide variety of properties and are thus best represented by different types of bondings.

There are several distinct types of hydrides.

(1) Salt-like. These are the hydrides of the most electropositive elements (*e g.* Na) and contain H^- ions.

(2) Covalent. Formed by most of the non-metals and transition metals. This class includes such diverse compounds as methane, CH_4 and iron carbonyl hydride, $H_2Fe(CO)_4$. In many compounds the hydrogen atoms act as bridges. Where there are more than one hydride sites there is often hydrogen exchange between the sites. Hydrogen may be inside metal clusters.

(3) Complexes. These derivatives contain complex anions which may be considered as derived from co-ordination of an H^- ion to a metal or non-metal. Examples are the BH_4^- and ReH_9^{2-} ions.

(4) Transition metal hydrides. These are formed by hydrogen uptake by the metal. The phases are often non-stoichi metric

Hydrides frequently function as hydrogenation catalysts.

Hydrobromic Acid (HBr). A colourless liquid which is obtained by adding hydrogen bromide to water. It shows the typical properties of a strong acid and it is a strong reducing agent. A convenient way of producing hydrobromic acid has been to bubble hydrogen sulphide through bromine water. Although it is not as strong as hydrochloric acid it dissociates extensively in water and is a good proton donor.

Hydrochloric Acid (HCl). A colourless fuming liquid which it obtained by adding hydrogen chloride to water :

$$HCl(g) + H_2O(l) \rightarrow H_3O^+(aq) + Cl(aq)$$

Dissociation into ions is extensive and hydrochloric acid shows the typical properties of a strong acid. It reacts with carbonates to give carbon dioxide and yields hydrogen when reacted with all but the most unreactive metals. Hydrochloric acid is used in the manufacture of dyes, drugs, and photographic materials. It is also used to pickle metals, *i e.* clean the surface prior to electroplanting. Hydrochloric acid donates protons with ease and has been the strongest of the hydrohalic acids. The concentrated acid gets oxidized to chlorine by such agents as potassium permanganate and manganese (IV) oxide.

Hydrofluoric Acid (HF). Refers to a colourless liquid which is produced by dissolving hydrogen fluoride in water. It is a weak acid, but will dissolve most silicates and hence can be used to etch glass. †As the interatomic distance in HF is relatively small, the H—F bond energy is very high, therefore—unable to donate protons easily—hydrofluoric acid is a weak acid.

Hydrogen Symbol H. A colourless odourless gaseous chemical element; a.n. 1; r.a.m. 1.0008; d. 0 0899 g dm^{-3} m.p.—259 14°C; b.p. —252.87°C. It has been the lightest element and the most abundant in the universe. It is bound in water and in all organic compounds. There are three isotopes: naturally occurring hydrogen consists of the two stable isotopes hydrogen—1 (99.985%) and deuterium. The radioactive tritium is made artificially. The gas is diatomic and has two forms: arthohydrogen, in which the nuclear spines are parallel, and parahydrogen, in which they are antiparallel. At normal temperatures the gas has been 25% parahydrogen. In the liquid it has been 99.8% parahydrogen. The main source of hydrogen is steam reforming of natural gas. It can also be made by the Bosch process and by electrolysis of water. The main use has been in the Haber process for making ammonia. Hydrogen has been also used in various other industrial processes, such as the reduction of oxide ores, the refining of petroleum, the production of hydrocarbons from coal, and the hydrogenation of vegetable oils. Considerable interest has also been shown in its potential use in a 'hydrogen fuel economy' in which primary energy sources not based on fossil fuels (*e.g*: nuclear, solar, or geothermal energy) are used to produce electricity, which is employed in electrolysing water. The hydrogen formed may be stored as liquid hydrogen or as metal hydrides. Chemically, hydrogen reacts with most elements. It was discovered by Henry Cavendish in 1776.

Hydrogen Azide. See hydrazoic acid.

Hydrogen Bond (H-bonding). Refers to an intermolecular bond between molecules in which hydrogen has been bound to a strongly electronegative element. Bond polarization by the electronegative element X leads to a positive charge on hydrogen $X^{\delta} \rightarrow H^{\delta}$; this hydrogen can then interact directly with electronegative elements of adjacent molecules. The H– bond is represented as a dotted line :

$$X^{\delta-}—H^{\delta+} \;\ldots\ldots\; X^{\delta-}—H^{\delta} \;\ldots$$

The length of a hydrogen bond has been characteristically 0.15 —0.2 nm.

Hydrogen bonding may lead to the formation of dimers (for example, in carboxylic acids) and has been used to explain the anomalously high boiling points of H_2O and HF.

Hydrogen Bromide (HBr). Refers to a colourless sharp-smelling gas that is very soluble in water. It is produced by direct combination of hydrogen and bromine in the presence of a platinum catalyst or by the reaction of phosphorus tribromide with water. It dissolves in water to give hydrobromic acid. Hydrogen bromide has been rather inactive chemically. It will not conduct electricity in the molten state, indicating that it is a molecular compound.

Hydrogencarbonate (bicarbonate). A salt of carbonic acid in which one hydrogen atom has been replaced; it thus contains the hydrogencarbonate ion HCO_3^-.

Hydrogen Chloride. A colourless fuming gas, HCl; m.p —114°C; b p.—85°C. It can be prepared in the laboratory by heating sodium chloride with concentrated sulphuric acid (hence the former name spirits of salt.) Industrially it is made directly from the elements at high temperature and used in the manufacture of PVC and other chloro compounds. It is a strong acid and dissociates fully in solution (hydrochloric acid).

Hydrogen Cyanide (hydrocyanic acid; prussic acid). A colourless liquid or gas, HCN, with a characteristic odour of almonds; r.d. 0 699 (liquid at 22°C); m.p.—14°C; b.p. 26°C. It is an extremely poisonous substance formed by the action of acids on metal cyanides. Industrially, it is made by catalytic oxidation of ammonia and methane with air and is used in producing acrylate plastics. Hydrogen cyanide is a weak acid (K_a=2 1 × 10^{-9} mol dm^{-3}). With organic carbonyl compounds it forms cyanohydrins.

Hydrogen Fluoride (HF). A colourless liquid produced by the reaction of concentrated sulphuric acid on calcium fluoride:

$$CaF_2(s)+H_2SO_4(aq)\rightarrow CaSO_4(aq)+2HF(l)$$

It produces toxic corrosive fumes and dissolves readily in water to give hydro-fluoric acid. †Hydrogen fluoride is a typical of the hydrogen halides as the individual H-F units are associated into much larger units, forming zigzag chains and rings. This is caused by hydrogen bonds that form between the hydrogen and the highly electronegative fluoride ions. Hydrogen fluoride is used extensively as a catalyst in the petroleum industry.

Hydrogen Iodide. A colourless gas, HI; m.p.--51°C; b.p.−36°C. It can be made by direct combination of the elements using a platinum catalyst. It is a strong acid dissociating extensively in solution (hydroxidic acid). It is also a reducing agent.

Hydrogen Peroxide (H_2O_2). A colourless syrupy liquid, usually used in solution in water. On contact with manganese (IV) oxide it gives off oxygen, the manganese (IV) oxide acting as a catalyst :

$$2H_2O_2\rightarrow 2H_2O+O_2$$

Hydrogen peroxide can act as an oxidizing agent, converting iron(II) ions to iron(III) ions, or as a reducing agent with potassium manganate (VII). It is used as a bleach and in rocket fuel. †The strength of solutions is usually given as volume strength—the volume of oxygen (dm^3) at STP given by decomposition of 1 dm^3 of the solution.

Hydrogen Spectrum. The atomic spectrum of hydrogen is characterized by lines corresponding to radiation quanta of sharply defined energy. A graph of the frequencies at which these lines occur against the ordinal number that characterizes their position in the series of lines, produces a smooth curve indicating that they obey a formal law. In 1885 J J. Balmer (1825−98) discovered the law having the form :

$$1/\lambda=R(1/n_1^2+1/n_2^2)$$

This law gives the so-called Balmer series of lines in the visible spectrum in which $n_1=2$ and $n_2=3, 4, 5..$ λ is the wavelength associated with the lines, and R is the Rydberg constant.

In the Lyman series, discovered by Theodore Lyman (1874−1954), $n_1=1$ and the lines fall in the ultraviolet. The Lyman series is the strongest feature of the solar spectrum as observed by rockets and satellites above the earth's atmosphere. In the Paschen series, discovered by F. Paschen (1865—1947). $n_1=3$ and the lines occur in the far infrared. The Brackett series also occur in the far infrared, with $n_1=4$.

Hydrogensulphate (bisulphate). A salt containing the ion HSO_4^- or an ester of the type $RHSO_4$, where R is an organic group. It was formerly called hydrosulphate.

Hydrogen Sulphide (sulphuretted hydrogen, H_2S). A colourless very poisonous gas having an odour of bad eggs. Hydrogen sulphide could be obtained by reacting hydrochloric acid with iron(II) sulphide. It is tested for by mixing with lead nitrate, with which it gives a black precipitate. Its aqueous solution is weakly acidic. Hydrogen sulphide reduces iron (III) chloride to iron (II) chloride, forming hydrochloric acid and a yellow precipitate of sulphur. Hydrogen sulphide precipitates insoluble sulphides, and is used in qualitative analysis. It burns with a blue flame in oxygen to form sulphur (IV) oxide and water. North Sea gas is having some hydrogen sulphide, which is removed before supply to the consumer.

Hydromagnesite. A mineral form of basic magnesium carbonate, $3MgCO_3.Mg(OH)_2.3H_2O$.

Hydroxide. A compound containing the ion OH^- or the group—OH.

Hydroscopic. Describing a substance that can take up water from the atmosphere.

Hyperons. Elementary particles of mass 2185 times that of the electron; intermediate between a proton and a deuteron. The best established hyperon is the lambda hyperon of zero charge but it is very likely that charged hyperons exist. Hyperons are found in cosmic radiation.

Hypo, $Na_2S_2O_3$. Sodium thiosulphate. Used in photography.

Hypochlorite. See chlorates.

Hypochlorous Acid. See chloric(I) acid.

Hypophosphorus Acid. See phosphinic acid.

I

Ideal Crystal. A single crystal with a perfectly regular lattice that contains no impurities, imperfections, or other defects.

Illinium. An obsolete name for promethium.

Ilmenite, $FeTiO_3$. A black Ti ore occurring in massive form and in sands. The structure has cp oxygen with Fe and Ti in octahedral co-ordination.

Imperfect Complexes. Imperfect complex compounds are those co-ordination compounds which remain as complexes either in solution state but not in the solid phase or which exist as complexes in the solid state but break up when dissolved in the solvent.

The imperfect complexes which exist only in solution are $K_2[Cd(CN)_4]$. $[Cu(NH_3)_2]Cl$, $K_2[CuCl_4]$, $K_2[Ni(CN)_4]$, $[Al(H_2O)_6]^{+3}$, etc.

The examples of imperfect complexes which exist only in the solid state are $K_2CoCl_4.Cu_2Cl_22CO,(NH_4)_2FeCl_4$ etc.

Indeterminacy. See uncertainty pr nciple.

Indium. In At. no. 49, at. wt. 114.82, m.p. 156.61°C, b.p. 2080°C, d 7.31. Occurs in some specimens of zinc blende; precipitated from solution electrolytically or with Zn. The metal is tetragonal with a distorted cp structure. The metal is oxidized on strong heating in air and dissolves in acids. Used in low melting alloys. In P is used in electric motors and other in compounds are used in electronic components and semi-conductors. Indium compounds are toxic.

Indium Chemistry. Indium is a group III element, electronic configuration $5s^25p^1$. It shows the normal group oxidation state of +3

($E°In^{3+} \rightarrow In$ in acid solution—3.34 volts) and a relatively stable +1 state.

($E°In^{+} \rightarrow In$ in acid solution—0.25 volts) which disproportionates in aqueous solution. The chemistry of the +3 state is similar to that of Al(III).

Inert Complex. A complex which participates in reactions, particularly ligand exchange reactions, either very slowly or sufficiently slowly that the rates of reaction can be readily followed. $[Co(NH_3)_6]^{3+}$ is an inert complex.

Inert Gases. A semi-obsolete name for the noble gases.

Inert Gas Formalism. The octet rule; the effective atomic number rule.

Inert Pair Effect. An apparent correlation between the existence of oxidation states n and (n—2) and the s^2 configuration for the latter for the heavy p block elements (n corresponds to the group-oxidation state). E.g. Pb shows oxidation states Pb(IV) and Pb(II).

Inert Pair Effect. Some of the heavier elements which are having three or four electrons in their outermost shell are expected to form trivalent or tetravalent ions. But some times they do not lose electrons and hence form cations with lower positive charge. Such electrons which do not participate in ion formation are called inert pairs. This effect is called inert pair effect When inert pair is also lost under suitable conditions, higher charged cations are formed.

Inner Sphere Mechanism. A mode of redox reaction in which the complexes form an intermediate with a shared ligand.

Inorganic Chemistry. The branch of chemistry concerned with compounds of elements other than carbon. Certain simple carbon compounds, such as CO, CO_2, CS_2, and carbonates and cyanides, are usually treated in inorganic chemistry.

Insertion Reaction. A common type of reaction in which an incoming molecule is inserted between previously bonded atoms, *e.g.* $[(OC)_5MnCH_3]$ plus CO gives $[(OC)_5MnC(O)CH_3]$ The inserted molecule contained a multiple bond or a lone pair.

Instability Constant. The reverse of a stability constant; the measure of dissociation.

Insoluble. Describing a compound that is having a very low solubility (in a specified solvent).

Interatomic Distances. The distances between the nuclei of atoms, bonded or non-bonded, in a molecule or crystal as determined by X-ray or electron diffraction, spectroscopic or other physical methods.

Intercalation Compounds. Derivatives, particularly of graphite, in which molecules are accommodated in holes or between layers in the lattice. The resulting compounds are easily handled and stored (*e g.* graphite-SbF_5, used as a fluorinating agent; graphite-$FeCl_3$ stable to water.

Interhalogen Compounds. The difference in electronegativity values among the elements of a halogen family gives rise to the formation of binary compounds. These are collectively acknowledged as interhalogen compounds. The structures of these compounds are known with the varying degree of accuracy.

The binary compounds formed by halogens amongst themselves are known as "Interhalogen Compounds." These have the general formula AX_n where n=1, 3, 5 and 7.

There are never more than two different halogens in any known interhalogen compound. It means that ternary compounds of halogens with one another are not known.

Interhalogenoids. Such compounds are formed by the combination of two pseudohalogens. These compounds are similar to the interhalogen compounds.

Cyanogen derivatives of other pseudo-halogens have been reported. Examples are cyanogen thiocyanate, (CNSCN), cyanogen azide (CNH_3) and cyanogen azido-di-thiocarbonate ($CNSCSN_3$).

Intermediate Bonding. A form of covalent bond that also has an ionic or electrovalent character.

Intermetallic Compound. A compound consisting of two or more metallic elements present in definite proportions in an alloy.

Intermolecular Forces. Weak forces occurring between molecules.

Interstitial Compounds. Compounds in which the structure may be considered to be derived by inserting small species (H, B, C, N, B) into the holes of a cp metallic lattice (particularly transition elements).

Invar. Refers to a tradename for an alloy of iron (63.8%), nickel 36%), and carbon (0.2%) that is having a very low expansivity over a a restricted temperature range. It is used in watches and other instruments to reduce their sensitivity to changes in temperature.

Iodic Acid. Refers to any of various oxoacids of iodine, such as iodic(V) acid and iodic(VII) acid. When used without an oxidation state specified, the term usually refers to iodic(V) acid (HO_3).

Iodic(V) Acid. A colourless of very pale yellow solid, HIO_3; r.d. 4.63; decomposes at 110°C. It is soluble in water but insoluble in pure ethanol and other organic solvents. The compound may be prepared by oxidizing iodine with concentrated nitric acid, hydrogen peroxide, or ozone. It is a strong acid and a powerful oxidizing agent.

Iodic(VII) Acid (Periodic acid). A hygroscopic white solid, H_5IO_6, which decomposes at 138°C and is very soluble in water, ethanol, and ethoxythane. Iodine(VII) acid is obtained by electrolytic oxidation of concentrated solutions of iodic(V) acid at low temperatures. It is a weak acid but a strong oxidizing agent.

Iodide. See halide.

Iodine Symbol I. It is a dark violet nonmetallic element belonging to group VII of the periodic table a.n. 53; r.a.m. 126.9045; r.d. 4.94; m.p. 113.5°C b p. 183.45C. The element is insoluble in water but soluble in ethanol and other organic solvents. When heated it gives a violet vapour that sublimes. Iodine is required as a trace element (see essential element) by living organisms, in animals it is concentrated in the thyroid gland as a constituent of thyroid hormones. The element occurs in sea and water and was formerly extracted from seaweed. It is now got from oil well brines (displacement by chlorine). There is one stable isotope, iodine—127, and fourteen radioactive isotopes. It finds use in medicine as a mild antiseptic (dissolved in ethanol as tincture of iodine), and in the manufacture of iodine compounds. Chemically, it has been less reactive than the other halogens and the most electropositive (metallic) halogen. It was discovered in 1812 by Courtoils.

Iodine Monochloride (ICl). A dark red liquid which is obtained by passing chlorine over iodine. It has properties similar to those of its constituent halogens, †Iodine monochloride is used as a nonaqueous solvent and as an iodating agent in organic reactions.

Iodine(V) Oxide (Iodine Pentoxide). A white solid, I_2O_5; r.d. 4.79; decomposes at 310°C. It dissolves in water to give iodic(V) acid and also acts as an oxidizing agent.

Iodine Pentoxide. See iodine(V) oxide.

Iodine Trichloride. Di iodine hexachloride.

Ionic Bond. See electrovalent bond.

Ionic Crystal. A crystal which is composed of ions of two or more elements. The positive and negative ions could be arranged in definite patterns and are held together by electrostatic attraction. Common examples have been sodium chloride and caesium chloride.

Ionic Radii. Ionic radius may be defined as the distance from the nucleus of an ion upto which it has influence on its electron cloud. The electron cloud, no doubt, may extent theoretically upto a very large distance. However, it is important to characterise each ion by some valve of ionic radii are very small and are generally expressed in Angstrom units.

Ionization Energy. See ionization potential.

Ionisation Isomerism. Compounds which have the same stoichiometric composition but yield different ions in solution are called ionisation isomers and the phenomenon is called Ionisation isomerism.

Such isomers are produced by an interchange of position of co-ordinating ions (ligands) inside and outside the complex. Ionisation isomers are easily detected by the quick determination of conductance in solution and by chemical tests. For example, the violet bromopentammine sulphate of cobalt with a formula of $[Co(NH_3)_5Br]SO_4$ is an ionisation isomer of the red sulphatopentammine bromide with a formula of $[Co[NH_3)_5SO_4]Br$. Both of these compounds have the similar empirical formula, but the first complex reacts with $BaCl_2$ to precipitate immediately $BaSO_4$ and no precipitate with $AgNO_3$. On the other hand, sulphatopentammine cobalt gives a precipitate of AgBr with $AgNO_3$ and no precipitate with $BaCl_2$. Other illustrations of ionisation isomerism are :

$$[Co(NH_3)_4Cl_2]\ NO_2 \text{ and } [Co(NH_3)_4Cl(NO_2)]\ Cl$$
$$[Pt(NH_3)_4(OH)_2]\ SO_4 \text{ and } [Pt(NH_3)_4SO_4]\ (OH)_2$$
$$[Pt(NH_3)_4Cl_2]\ Br_2 \text{ and } [Pt(NH_3)_4\ Br_2]\ Cl_2$$

Ionization Potential (I P) Symbol : I The energy required to remove an electron from an atom (or molecule or group) in the gas phase, *i.e.* the energy required for the process:

$$M \rightarrow M^+ + e^-$$

It gives a measure of the ability of metals to form positive ions. The second ionization potential is the energy required to remove two electrons and form a doubly charged ion :

$$M \rightarrow M^{2+} + e^-$$

Ionization potentials stated in this way are positive; often they are given in electronvolts Ionization energy referers to the energy required to ionize one mole of the substance, and is usually stated in kilojoules per mole (kJ mol^{-1}).

In chemistry, the terms 'second', 'third', etc., ionization potentials are usually used for the formation of doubly, triply, etc., charged ions. However, in spectroscopy and physics, they are often used with a different meaning The second ionization potential is the energy to remove the second least strongly bound electron in forming a single charged ion. For lithium ($1s^2\ 2s^1$) it would refer to removal of a 1s electron to produce an excited ion with the configuration $1s^1\ 2s^7$ Note also that ionization potentials are now stated as energies. Originally they were the potential through which an electron had to be accelerated to cause ionization by electron impact :

$$M + e^- \rightarrow M^{2+} + 2e^-$$

They were thus stated in volts.

Ion Pair. A pair of oppositely charged ions which are produced as a result of a single ionization; *e.g.*

$$HCl \rightarrow H^+ + Cl^-$$

Sometimes a positive ion and an electron are referred to as an ion pair, as in

$$A \rightarrow A^+ + e^-$$

Iridium Symbol Ir. A silvery metallic transition element. a.n. 77; r a.m. 192.20; r.d. 22.42; m p. 2410°C; b.p. 4130°C. It occurs with platinum and is mainly used in alloys with platinum and osmium. The element is known to form a range of iridium(III) and iridium(IV) complexes. It was discovered in 1804 by Tennant.

Iron. A transition element occurring in many ores, especially the oxides (haematite and magnetite) and carbonate. It could be extracted in a blast furnace using coke, limestone, and hot air. The coke and air form carbon monoxide, which then reduces the iron ore to iron. The limestone removes acidic impurities and forms a layer of slag above the molten iron at the base of the furnace. Steel is formed by reducing the carbon content to between 0.1 and 1.5%. Iron is used as a catalyst in the Haber process for ammonia production. It corrodes in air and moisture to hydrated iron(III) oxide (rust).

The most stable oxidation state has been + 3, which is yellow, but +2 (green) and + 6 (easily reduced) also exist. Solutions of iron(II) ions yield a green precipitate with sodium hydroxide solution, whereas iron(III) ions give a brown precipitate.

Symbol : Fe; m.p. 1539°C; b p. 2887°C; r.d. 7.9; p.n. 26; r.a m. 55.85.

Iron(II) Chloride. A green-yellow deliquescent compound, $FeCl_2$; hexagonal; r.d. 3.16; m.p. 670°C. It also exists in hydrated forms: $FeCl_2.2H_2O$ (green monoclinic; r.d.2 36) and $FeCl_2.4H_2O$ (blue-green monoclinic deliquescent; r.d. 1.93). Anhydrous iron (II) chloride can be prepared by passing a stream of dry hydrogen chloride over the heated meta he hydrated forms can be made using dilute hydrochloric acid or by recrystallizing with water. It is converted into iron(III) chloride by the action of chlorine.

Iron(III) Chloride (ferric chloride, $FeCl_3$). A compound prepared in the anhydrous state as dark red crystals by passing dry chlorine over heated iron. The product sublimes and is collected in a cooled receiver. The hydrated salt ($FeCl_3.6H_2O$) is prepared by adding excess iron(III) oxide to concentrated hydrochloric acid. On crystallization yellow-brown crystals of the hexahydrate are formed. Iron(III) chloride is very soluble in water and undergoes salt hydrolysis. †At temperatures bellow 400°C the anhydrous salt exists as a dimer, Fe_2Cl_6.

Iron(III) Oxide (ferrous oxide, FeO). A black powder which is obtained by the careful reduction of iron(III) oxide using either carbon monoxide of hydrogen. It can also be prepared by heating iron(II) oxalate in the absence of air. It is only really stable at high temperatures and disproportionates slowly on cooling to give iron(III) oxide and iron. Iron(II) oxide can be reduced by heating in a stream of hydrogen. When exposed to air, it is oxidized to iron(III) oxide. It is a basic oxide, dissolving readily in dilute acids to form iron(II) salt solutions. †If heated to a high temperature in an inert atmosphere, iron(II) oxide disproportionates to give iron and triiron tetroxide.

Iron(III) Oxide. A red-brown to black insoluble solid, Fe_2O_3; trigonal; r.d. 5.24; m.p. 1565°C. There is also a hydrated form. $Fe_2O_3 \times H_2O$, which is red-brown powder; r.d. 2.44—3.60.

Iron(III) oxide is found naturally as haematite and can be prepared by heating iron(III) hydroxide or iron(II) sulphate. It could be readily reduced on heating with carbon or in a stream of carbon monoxide, hydrogen, or coal gas :

$$F_2O_3 + 3C \rightarrow 2Fe + 3CO$$

Iron(II) Sulphate. An off-white solid, $FeSO_4.H_2O$; monoclinic r.d. 2.970. There is also a heptahydrate. $FeSO_4.7H_2O$; blue-green monoclinic; r.d. 1.898; m.p. 64°C. The heptahydrate has been the best known iron(II) salt and is sometimes called green vitriol or copperas. It is obtained by the action of dilute sulphuric acid on iron in a reducing atmosphere. The anhydrous compound is very hygroscopic. It decomposes at red heat to give iron(III) oxide, sulphur trioxide, and sulphur dioxide. A solution of iron(II) sulphate is gradually oxidized on exposure to air, a basic iron(III) sulphate being deposited.

Iron(III) Sulphate. A yellow hygroscopic compound, $Fe_2(SO_4)_3$; rhombic; r.d. 3,097; decomposes above 480°C. It is prepared by heating in aqueous acidified solution of iron(II) sulphate with hydrogen peroxide :

$$2FeSO_4 + H_2SO_4 + H_2O_2 \rightarrow Fe_2(SO_4)_3 + 2H_2O$$

On crystallizing, the hydrate $(Fe_2(SO_4)_3 \, 9H_2O$ is formed. The acid sulphate $Fe_2(SO_4)_3.H_2SO_4.8H_2O$ gets deposited from solutions having a sufficient excess of sulphuric acid.

Isobars. Two or more nuclides that have the same nucleon numbers but different proton numbers.

Isoelectronic. Describing compounds that are having the same number of electrons. For example, carbon monoxide (CO) and nitrogen (N_2) are isoelectronic.

Isometric

1. (in crystallography) Denoting a system in which the axes are perpendicular to each other, as in cubic crystals.
2. Denoting a line on a graph illustrating the way in which temperature and pressure have been interrelated at constant volume.

Isomorphism. Having the same crystal structure (isostructural). Isomorphous substances will often induce crystallization of each other and may even crystallize on each other (*e.g.* violet chromium alum will crystallize on colourless potassium alum).

Isomorphism, Law of. Sec Mitscherlich's law.

Isopoly Acids. These acids are formed if the oxygen atoms in any oxyacid are replaced by radicals of the same acid. For example, the introduction of one or more CrO_4^{2-} radicals in H_2CrO_4 acid gives rise to vatious isopolyacids like $H_2CrO_3 \, CrO_4$, $H_2CrO_2.2CrO_4 \; H_2CrO.3CrO_4$.

Isopolyacids may also be defined in another way in which these acids are derived from a single acid by the elimination of water from two or more molecules of the same acid.

$$2H_2CrO_4 \xrightarrow{-H_2O} H_2Cr_2O_7$$

$$3H_2CrO_4 \xrightarrow{-2H_2O} H_2Cr_3O_{10}$$

$$4H_2CrO_4 \xrightarrow{-3H_2O} H_2Cr_4O_{13}$$

The important isopolyacids are isopolymolybdates, isopolytungstates and isopolyvanadates. These acids are discussed one by one.

Isopolymorphism. Substances which have more than one crystalline form isomorphous with the crystalline forms of another substance are said to show isopolymorphism. Thus, the octahedral and rhombic forms of As_2O_3 are isomorphous with the corresponding forms of antimony trioxide. These oxides are isodimorphous.

Isotope. One of two or more atoms of the same element that are having the same number of protons in their nucleus but different numbers of neutrons. Hydrogen (1 proton, no neutrons), deuterium (proton 1 neutron), and tritium (1 proton, 2 neutrons) are isotopes of hydrogen. Most elements in nature consist of a mixture of isotopes.

Isotope Separation. The separation of the isotopes of an element from each other on the basis of slight differences in their physical properties. For laboratory quantities the most suitable device is often the mass spectrometer. On a larger scale the methods used include gaseous diffusion (widely used for separating isotopes of uranium in the form of the gas uranium hexafluoride), distillation (formerly used to produce heavy water), electrolysis (requiring cheap electrical power), thermal diffusion (formerly used to separate uranium isotopes, but now considered uneconomic), centrifuging (a method in which there is renewed interest), and laser methods (involving the excitation of one isotope and its subsequent separation by electromagnetic means).

Isotopic Dilution Methods. It is an example of radiometric method of trace analysis. It is used to determine the quantity of constituent (radioactive or non-radioactive) in a mixture of closely related compounds which are difficult to separate and to estimate quantitatively by the usual conventional methods.

Isotopic Mass (isotopic weight). The mass number of a given isotope of an element.

Isotopic Number (neutron excess). The difference between the number of neutrons in an isotope and the number of protons.

J

Jade. Refers to a hard semiprecious stone consisting either of jadeite nephrite. Jadeite, the most valued of the two, is a sodium aluminium pyroxene, $NaAlSi_2O_6$. It is prized for its intense translucent green colour but white, green and white, brown, and orange varieties also occur. The only important source of jadeite has been in the Mogaung region of upper Burma. Nephrite has been one of the amphibole group of rock-forming minerals. It occurs in a variety of colours, including green, yellow, white, and black. Important sources include the Soviet Union, New Zealand, Alaska, China, and W USA.

Jade. A semi-precious stone, which consists of two distinct minerals: jadeite, $NaAl(SiO_3)_2$, a pyroxene; and nephrite, $Ca_2Mg_5(Si_4O_{11})_2(OH)_2$, an amphibole. Both are tough due to interlocking fine fibres. Jade is white to green, greasy and translucent when polished.

Jahn—Teller Effect. The Jahn—Teller theorem states that, when any degenerate electronic state contains a number of electrons such that the degenerate orbitals are not completely filled, the geometry of the species will change so as to produce non-degenerate orbitals. Particularly applied to transition metal compounds where the d^9 state is d^6d^3. Cu(II) generally has the distorted octahedral co-ordination typical of the Jahn—Teller effect. Where a species is not distorted the vibrational spectra frequently show anomalous intensities if subject to the Jahn—Teller theorem.

The effect is observed in inorganic complexes. For example, the ion $[Cu(H_2O)_6]^{2+}$ is octahedral and the six ligands might be expected to occupy equidistant position at the corners of a regular octahedron. In fact, the octahedron is distorted, with four ligands in a square and two opposite ligands further away. If the 'original' structure has a centre of symmetry, the distorted structure must also have a centre of symmetry. The effect was predicted theoretically by H.A. Jahn and Edward Teller in 1937.

Jasper. An impure variety of chalcedony. It is associated with iron ores and as a result contains iron oxide impurities that give the mineral its characteristic red or reddish-brown colour. Jasper is used as a gemstone.

Jeweller's Rouge. Red powdered haematite, iron(III) oxide, Fe_2O_3. It is a mild abrasive used in metal cleaners and polishers.

K

K. Potassium.

Kainite, $MgSO_4,KCl,3H_2O$. Occurs in the Stassfurt salt deposits as beds. Used as a source of K salts and as fertilizer.

Kalinite. A mineral form of aluminium potassium sulphate $(Al_2(SO_4)_3,K_2SO_4\,24H_2O)$.

Kaodin (china clay). Refers to a soft white clay that is composed chiefly of the mineral kaolinite (see clay minerals). It gets formed during the weathering and hydrothermal alteration of other clays or feldspar. Besides its vital importance in the ceramics industry it is also used extensively as a filler in the manufacture of rubber, paper, paint, and textiles and as a constituent of medicine.

Keesom Forces One dipolar molecule at effective intermolecular distance orients the other dipolar molecule in such a way that the positive end of one attracts the negative end of the other and vice-versa, thereby giving rise to dipole-dipolar interactions. These interactions are present at low or moderate temperatures between closely spaced molecules having permanent dipole moments and represented as shown in Fig.

The effect was first studied by Keesom in 1912; it is now referred to as the orientation effect.

Kieselguhr. A soft fine-grained deposit which consists of the siliceous skeletal remains of diatoms, formed in lakes and ponds. Kieselguhr finds use as an absorbent, filtering material, filler, and insulator.

Kieserite. A mineral form of magnesium sulphate monohydrate, $MgSO_4\,H_2O$.

Killed Spirits. Zinc(II) chloride solution used as a flux for solder, so called because it is made by adding zinc to hydrochloric acid ('spirits of salt').

Kimberlite. Refers to a rare igneous rock that often contains diamonds. It occurs as narrow pipe instrusions but is often altered and fragmented. It consists of olivine and phlogopite mica, usually with calcite, serpentine, and other minerals. The chief occurrences of kimberlite have been in South Africa. especially at Kimberley (after which the rock is named), and in the Yakutia area of Siberia.

Kovar. A tradename for an alloy of iron, cobalt, and nickel with an expansivity similar to that of glass. It is therefore used in making glass-to-metal seals, especlally in circumstances in which a temperature variation can be expected.

Kroll Process A method of getting certain metals by reducing the metal chloride with magnesium. Titanium can be obtained in this way.

$$TiCl_4+2Mg \rightarrow 2MgCl_2+Ti$$

Krypton, Kr. At. no. 36, at. wt. 83.80, m.p.—156.6°C, b.p.—152.3°C. One of the noble gases (% abundance 1.1 ×10.5% in air) obtained pure by fractional distillation of liquid air. Used (with Ar) for fluorescent lights, and in certain photograhic flash lights. Electronic configuration $4s^2 4p^6$. Kr forms a limited range of chemical compounds and some clathrates. Linear KrF_2(Kr plus F_2 in presence of an electrical discharge) forms adducts with Lewis acid fluorides, *e.g.* $KrF_2.2SbF_5$.

Kurchatovium (Eka-Hafnium) or Rutherfordium (At. No: 104: Symbol Ku). In August 1965, Flerov and his associates at the laboratory Dubna, U S.S.R., reported that they had detected this element having atomic number 104 by bombarding plutonium-242 with accelerared neon-22 ions.

$$_{94}Pu^{242}+_{10}Ne^{22} \rightarrow _{104}Ku^{260}+4_0n^1$$

The fission tracks of the reaction, when examined under the microscope, revealed the existence of an atomic species which decayed by spontaneous fission with a half-life of 0.3 sec.

The Russians proposed the name Kurchatovium and the symbol Ku for the element-144 after I.V. Kurchatov, the first director of the Russian atomic energy project.

In April 1969, American workers of Albert Ghiorso's group confirmed the formation of Kurchatovium by another process. However, the American workers named this element rutherfordium.

$$_{98}Cl^{249}+_6C^{12} \rightarrow _{104}Ku^{260}+_0n^1$$

Krypton. A colourless odourless monatomic element of the rare-gas group. It is now known to form compounds. It occurs in minute quantities (0.001% by volume) in air. Krypton finds use in fluorescent lights.

Symbol: Kr: m.p.—157.3°C; b.p.—153.4°C; d. 3.73 kg m^{-3}; p.n. 36; r.a.m. 83.80.

L

La. Lanthanum.

Label. A stable or radioactive nuclide which is used to investigate some process, such as a chemical reaction.

Labelling. Refers to the process of replacing a stable atom in a compound with a radioisotope of the same element to enable its path through a biological or mechanical system to be traced by the radiation it emits. In some cases a different stable isotope has been used and the path is detected by means of a mass spectrometer. A compound having either a radioactive or stable isotope is called a labelled compound. If a hydrogen atom in each molecule of the compound is replaced by a tritium atom, the compound is called a tritiated compound. A radioactive labelled compound will behave chemically and physically in the same way as an otherwise identical stable compound, and its presence can easily be detected using a Geiger counter. This process of radioactive tracing has been widely used in chemistry, biology, medicine, and engineering. For example, it can be used to follow the course of the reaction of a carboxylic acid with an alcohol to give an ester.

$$CH_3COOH+C_2H_5OH \rightarrow C_2H_5COOCH_3+H_2O$$

To determine whether the noncarbonyl oxygen in the ester comes from the acid or the alcohol, the reaction is performed with the labelled compound $CH_3CO^{18}OH$, in which the oxygen in the hydroxyl group of the acid has been 'labelled' by using the ^{18}O isotope. It is then found that the water product is $H_2\,^{18}O$ *i.e.* the oxygen in the ester comes from the alcohol, not the acid.

Labile Complex. A complex which participates in very fast reactions, particularly ligand exchange reactions, generally within the time of mixing. E.g. $[Fe(H_2O)_6]^{3+}$. $[Co(H_2O)n^3]^{2+}$. Compare inert complex.

Liability. A measure of the ease of replacement of attached groups in complexes and molecules. Cr^{3+},Co^{3+} complexes are inert, many other complexes are relatively liable.

Labradorite. A plagioclase feldspar consisting of albite and anorthite in proportions ranging from 1:1 to 1:3. A primary constituent of some basic igneous rocks, weathers to kaolin.

Lake. A pigment which is made by combining an organic dyestuff with an inorganic compound (*e.g.* an oxide).

Lamellar Compound. Refers to a compound with a crystal structure composed of thin plates or layers. Silicates form many compounds with distinct layers. Examples include talc ($Mg_3(OM)_2$ Si_4O_{10}) and pyrophyllite ($Al_2(OH)_2Si_4O_{10}$).

Lamellar Solids. Refer to solid substances in which the crystal structure has distinct layers (*i.e.*, has a layer lattice). The micas are an example of this type of compound. Intercalation compounds have been lamellar compounds formed by interposition of atoms, ions, etc., between the layers of an existing element or compound. For example, graphite has been a lamellar solid. With strong oxidizing agents (*e.g.*, a mixture of concentrated sulphuric and nitric acid) it forms a nonstoichiometric 'graphitic oxide', which has been an intercalation compound having oxygen atoms between the layers of carbon atoms.

Lamp Black. A black pigment; a finely divided form of carbon which is formed by incomplete combustion of an organic compound.

Lansfordite. A mineral form of magnesium carbonate pentahydrate, $MgCO_3.5H_2O$.

Lanthanides. The elements from lanthanum to lutetium, in which the 4f orbitals are being filled. The elements lanthanum, cerium, prasesdymium. neodymium, promethium, samarium, europium, gadolinium, terbium, dysprosium, holmium, erbium, thulium, ytterbium and lutetium. Except for Pm the elements are not uncommon the normal source is monazite which contains 90% La, Ce, Pr, Nd in its lanthanide content. The elements are separated by chromatography or ion exchange. The metals are prepared by MF_3+Ca at 1450—1700°C but for Sm, Eu and Yb the reaction $La+M_2O_3$ with volatilization of M is used. The metals are shiny and are attacked by water and acids and burn in oxygen or air to M_2O_3 or MO_2 and react in hydrogen to MH_2 and MH_3.

The outer electron configurations are as follows :

57 lanthanum (La) $5d^16s^2$
58 cerium (Ce) $4f5d^16s^2$ (or $4f^26s^2$)
59 praseodymium (Pr) $4f^36s^2$
60 neodymium (Nd) $4f^46s^2$
61 promethium (Pm) $4f^56s^2$
62 samarium (Sm) $4f^66s^2$
63 europium (Eu) $4f^76s^2$
64 gadolinium (Gd) $4f^75d^16s^2$
65 terbium (Tb) $4f^96s^2$
66 dysprosium (Dy) $4f^{10}6s^2$
67 holmium (Ho) $4f^{11}6s^2$
68 erbium (Er) $4f^{12}6s^2$
69 thulium (Tm) $4f^{13}6s^2$
70 ytterbium (Yb) $4f^{14}6s^2$
71 luteium (Lu) $4f^{14}5d^16s^2$

Note that lanthanum itself does not have a 4f electron but it is generally classified with the lanthanoids because of its chemical similarities, as are yttrium (Yt) and scandium (Sc). Scandium, yttrium, and lanthanum are d-block elements; the lanthanodis and actioids make up the f-block.

Lanthanide Contraction. There is a regular decrease in ionic and atomic radii across the lanthanides. This means that radii of elements after the lanthanides are very similar to radii of corresponding elements before the series. Thus properties depending on ionic size, *e.g.*, structure, solubility, of elements before and after the lanthanides are very similar, *e.g.*, Mo and W.

Lanthanum, La. At no. 57, at wt. 138.9055, m.p. 920°C, b.p. 3454°C d 6.14. The first lanthanide (4f) element. The metal is hcp (to 310°C) (ccp to 868°C) and bcc (to m.p.), it is a superconductor below 4.9 K (hcp) or 5.9K (ccp). The metal is used in steels and in Mg and Al alloys. La_2O_3 is used in glasses, as a glass polish, as a high temperature refractory and as a host matrix for fluorescent phosphors.

Lanthanum Compounds. Lanthanum forms a single series of typical lanthanide compounds in the +3 state La^{3+} (f^0 colourless) → La (—2.52 volts in acid). $La(NO_3)_3.6H_2O$ is used as an analytical reagent for ethanoate ions. LaI_2 is probably La^{3+}, $2I^-$, e^-

Lapis Lazuli. Refers to a blue rock that is widely used as a semi-precious stone and for ornamental purposes. It has been composed chiefly of the deep blue mineral lazurite embedded in a matrix of white calcite and usually also contains small spocks of pyrite. It is found in only a few places in crystalline limestones as a contact metamorphic mineral. The cheif source has been Afghanistan; lapis lazuli also occurs near Lake Baikal in Siberia and in Chile. It was formerly used to make the artists' pigment ultramarine.

Lattice. The regular arrangement of atoms, ions, or molecules in a crystalline solid.

Lattice Energy. The lattice energy of an ionic solid is defined as the energy released when the correct number of gaseous cations and gaseous anions are brought together from infinite distance to form one mole of an ionic solid.

The formation of a mole of ionic solid $C^+ a^-$ from the constituent gaseous ions may be represented as

$$C^+ (g) + a^- (g) \rightarrow C^+ a^- (s) + \text{Energy released}$$

The energy released in the above equation is known as lattice energy. It is usually denoted by U_{oa}. According to the conventions of the 'First Law of Thermodynamic' the energy released *i.e.*, the lattice energy, has a negative sign.

Laughing Gas. See dinitrogen oxide.

Law of Constant Composition. See chemical combination.

Law of Definite Proportions. See chemical combination.

Law of Multiple Proportions. See chemical combination.

Law of Octaves (newlands' law). Refers to an attempt at classifying elements made by J. A. R. Newlands (1837-98) in 1863. He arranged 56 elements in order of increasing atomic mass in groups of eight, pointing out that each element resembed the element eight places from it in the list. He drew an analogy with the notes of a musical scale. Newlands' octaves were groups of similar elements distinguished in this way: *e.g.*, oxygen and sulphur; nitrogen and phosphorus; and fluorine, chlorine, bromine, and iodine. In some cases it was necessary to put two elements in the same position. The proposal got rejected at the time.

Law of Reciprocal Proportions. See chemical combination.

Lawrencium. A radioactive transuranic element of the actinoid series which is not found naturally on Earth. Several very shortlived isotopes have been synthesized.

Symbol : Lr; p.n. 103.

Layer Lattice. Refers to a crystal structure in which the atoms are chemically bonded in plane layers, with relatively weak forces between atoms in adjacent layers. Graphite and micas have been examples of substances having layer lattices *i.e.*, they are lamellar solids).

L-D Process. Basic-oxygen process.

Leaching. Refers to the washing out of a soluble material from an insoluble solid using a solvent. This is often done in batch tanks or by dispersing the crushed solid in a liquid.

Lead Symbol Pb. A heavy dull grey soft ductile metallic element which belongs to group IV of the periodic table; a.n. 82; r.a.m. 207.19; r.d. 11.35; m.p. 327.5°C; b.p. 1740°C. The main ore has been the sulphide galena (PbS); other minor sources include anglesite ($PbSO_4$), cerussite ($PbCO_3$), and litharge (PbO). The metal is extracted by roasting the ore to give the oxide, followed by reduction with carbon. Silver is also recovered from the ores. Lead finds a variety of uses including construction, lead-plate accumulators, bullets, and shot, and is a constituent of such alloys as solder, pewter, bearing metals, type metals, and fusible alloys. Chemically, it is known to form compounds with the + 2 and + 4 oxidation states, the lead (II) state being the more stable.

Lead (II) Acetate. See lead (II) ethanoate.

Lead (II) Carbonate. A white solid, $PbCO_3$ insoluble in water; rhombic r.d. 6.6. It occurs as the mineral cerussite, which has been isomorphous with aragonite and may be prepared in the laboratory by the addition of cold ammonium carbonate solution to a cold solution of a lead (II) salt (acetate or nitrate). It decomposes at 315°C to yield lead(II) oxide and carbon dioxide.

Lead (II) Carbonate Hydroxide (white lead, $2PbCO_3Pb(OH)_2$). It is the most important basic lead carbonate. It can be manufactured electrolytically. It finds use as a pigment in white and coloured paints but it has the considerable drawback that it is poisonous.

Lead-chamber Process. A former process which is used for the manufacture of sulphuric acid by oxidizing sulphur (IV) oxide with nitrogen monoxide. The reaction, which was carried out in large lead chambers, has now been replaced by the contact process.

Lead Dioxide. See lead (IV) oxide.

Lead (II) Ethanoate (lead (II) acetate, $Pb(CH_3CO_2)_2$). A compound usually occurring as the hydrate $Pb(CH_3CO_2)_2\,3H_2O$, forming monoclinic crystals. At 100°C it loses ethanoic acid and water forming a basic lead (II) ethanoate. Its solubility in water has been 50 g per 100 g of water at 25°C. Lead (II) ethanoate can be prepared as an anhydrous salt. Its chief asset has been that it is one of the few common lead (II) salts that are soluble in water.

Lead (IV) Ethanoate (Lead Tetra-acetate). A colourless solid, $Pb(CH_3COO)_4$, which decomposes in water and is soluble in pure ethanoic acid; monoclinic; r.d. 2,228; m.p. 175°C. It could be obtained by dissolving dilead (II) lead (IV) oxide in warm ethanoic acid. In solution it behaves essentially as a covalent compound (no measurable conductivity) in contrast to lead (II) salt, which is a weak electrolyte.

Lead (IV) Hydride. See plumbane.

Lead Monoxide. See lead (II) oxide.

Lead (II) Oxide (Lead Monoxide). A solid yellow compound, PbO, which is insoluble in water; m.p. 886°C. It is known to exist in two crystalline forms: litharge (tetrahedral; r d. 9.53) and massicot (Thombic; r.d. 8.0). It can be obtained by heating the nitrate, and is manufactured by heating molten lead in air. If the temperature used is lower than the melting point of the oxide, the product is massicot; above this, litharge gets formed. Variations in the temperature and in the rate of cooling give rise to crystal vacancies and red, orange, and brown forms of litharge can be produced. The oxide has been amphotoric, dissolving in acids to yield lead (II) salts and in alkalis to yield plumbates.

Lead (IV) Oxide (lead dioxide, PbO_2). A dark-brown solid which is farming hexagonal crystals. On heating to 310°C it decomposes into lead (II) oxide and oxygen. It can be obtained electrolytically or by reacting lead (II) oxide with potassium chlorate. Lead (IV) oxide finds use in the manufacture of matches.

Lead (II) Sulphate ($PbSO_4$). A white crystalline solid that is found naturally as the mineral anglesite. It is almost insoluble in water and can get precipitated by reacting an aqueous solution containing sulphate ions with a solution of a soluble lead (II) salt (*e.g.* lead (II) ethanoate). It forms basic lead (II) sulphates when shaken together with lead (II) hydroxide and water. Basic lead (II) sulphates have been useful as pigments.

Lead (II) Sulphide. A black crystalline solid, PbS, which is insoluble in water; r.d. 7.5; m.p. 1 114°C. It is found naturally as the metallic-looking mineral galena (the principal ore of lead). It may be prepared in the laboratory by the reaction of hydrogen sulphide with soluble lead (II) salts. Lead (II) sulphide finds use as an electrical rectifier.

Lead Tetra-acetate. See lead (IV) ethanoate.

Lead Tetraethyl (tetraethyl lead $Pb(CH_2CH_3)_4$). A poisonous liquid that is insoluble in water but soluble in orgaaic solvents. It may be manufactured by the reaction of an alloy of sodium and lead with 1-chloroethane. The product is obtained by steam distillation. Lead tetraethyl finds use as an additive in internal-combustion engine fuel to increase its octane number.

Leblane Process. Refers to an obsolete process for manufacturing sodium carbonate. The raw materials were sodium chloride, sulphuric acid, coke, and limestone (calcium carbonate), and the process involved two stages. First the sodium chloride was heated with sulphuric acid to yield sodium sulphate :

$$2NaCl(s) + H_2SO_4(l) \rightarrow Na_2SO_4(s) + 2HCl(g)$$

The sodium sulphate was then heated with coke and limestone :

$$Na_2SO_4 + 2C + CaCO_3 \rightarrow Na_2CO_3CaS + 2CO_2$$

Calcium sulphide was a by-product, the sodium carbonate being extracted by crystallization. The process, invented in 1783 by the French chemist Nicolas Leblane (1742-1806), was the first for producing sodium carbonate synthetically (earlier methods were from wood ash and other vegetable sources). By the end of the 19th century it was largely replaced by the Solvay process.

Lechatelierite. A mineral form of silicon (IV) oxide, SiO_2.

Levelling Solvents. The Bronsted-Lowry theory can be extended to acid-base reactions in non-equeous solvents. This study is of immense use in differentiating the acid strength of a particular acid and in titrations of weak bases. In water solvent, mineral acids appear to be equally strong because of their complete ionization. Water is called here a levelling solvent because it levels all the acids to the same strength. If instead of water solvent, we take the mineral acids in pure acetic acid solvent (which is a poor proton acceptor compared to water), it is found acids become weak and can be differentiated. The following equilibrium may be written for HCl in acetic acid medium;

$$\underset{\text{Acid}}{HCl} + \underset{\text{Base}}{CH_3COOH} \rightleftharpoons \underset{\text{Base}}{Cl^-} + \underset{\text{Acid}}{CH_3COOH_2}$$

Acetic acid and Cl^- ions both complete for protons and the former being a poor proton acceptor does it much less effectively than water. Thus, HCl in acetic acid solvent appears to be a much weaker acid than that in water. With the help of the conductivity measurements of the mineral acid solutions in acetic acid solvent, it can be shown that their strengths follow the order :

$$HNO_3 < HCl < H_2SO_4 < HBr < HClO_4$$

A weak base like acetamide or acetanilide in aqueous medium cannot be titrated with acids. If however, the weak base is taken in glacial acetic acid solvent, the former behaves as a strong base and can be titrated. This is because acetic acid (which acts as a better proton donor) exerts a levelling effect on the base.

Lewis Acid. Refers to a substance that can accept an electron pair to form a coordinate bond: a Lewis base has been an electron-pair donor. In this model, the neutralization reaction is seen as the acquisition of a stable octet by the acid, for example:

$$Cl_3B + :NH_3 \rightarrow Cl_3B:NH_3$$

Metal ions in coordination compounds have been also electron-pair acceptors and therefore Lewis acids. The definition includes the 'traditional' Bronsted acids since H^+ is an electron acceptor, but in common usage the terms Lewis acid and Lewis base have been reserved for systems without acidic hydrogen atoms.

Lewis Base. See Lewis acid.

Ligand. Any atom, ion or molecule which is capable of donating a pair of electrons to the metal atoms is called a co-ordinating group or ligand. In a ligand, the particular atom which actually donates the electron pair is called the donor atom. For example, in the complex potassium ferrocyanide $K_4[Fe\ CN_6]$, the six $(CN)^-$ ions are ligands and the nitrogen in $(CN)^-$ is the donor atom. Ligands may be classified as unidenate and polydenate ligands.

The ligands have been found to be arranged around the central metal ion inside the first sphere of attraction in preferred geometries. The common geometries found in complexes are linear, equilateral, triangular, tetrahedral, square planar, trigonal bipyramidal, square pyramidal and octahedral.

Ligand Field Theory. Recently based on crystal field theory of Bethe (1929) and Van Vleck (1931-1935), a new approach known as ligand field theory was postulated.

In the crystal field theory, attraction between central metal and ligand in a complex is regarded as purely electrostatic, *i.e.*, the bonding between the central metal ion and ligands is purely ionic. Thus, in crystal field theory covalent character of bond between metal and ligand is not taken into account. However, the enough evidence is available which suggests there is some measure of covalent bonding in complexes.

In ligand field theory, it is assumed that molecular orbitals are formed by the overlap of orbitals from the ligand with the atomic orbitals of the central atom. (The procedure is similar as adopted in case of simple molecules). The molecular orbitals thus formed may be of a bonding, antibonding or a non-bonding character. The antibonding orbitals are similar to the bonding orbitals except that the antibonding orbitals lie higher in energy and have nodes or regions of low electron density between the central atom and the ligands. The antibonding orbitals are of interest because it is these into which electrons may be excited from the t_{2g} orbitals by absorptions of energy. The non-bonding orbitals in octahedral complexes are simply the dxy, dxz and dyz orbitals of the central atom.

As the number of molecular orbitals formed is always equal to the number of atomic orbitals taking part in the overlappings and this number is quite large for complexation processes, the MO energy level diagrams for metal complexes are highly complicated. The fact that σ as well as π (both types of overlaps are involved in the coordination process) and more complication to these diagrams.

The three common types of complexes are octahedral, tetrahedral and square planar.

Ligation or Complexation Energy (E_{lig}). It is the energy released when a gaseous cation combines with a group of gaseous ligands (anions or neutral molecules) to form a gaseous complex as shown by the general equation.

$$M^{n+}(g) + x\,L^{n-}(g) \rightarrow [MLx]^{x-n}(g)$$

Lignite. See coal.

Lime. See calcium oxide.

Limestone. Refers to a sedimentary rock that has been composed largely of carbonate minerals, especially carbonates of calcium and magnesium. Calcite and aragonite have been the chief minerals; dolomite is also present in the dolomitic limestones. There are many varieties of limestones but most are deposited in shallow water. Organic limestones (*e.g* chalk) are formed from the calcareous skeletons of organisms; precipitated limestones include oolite, which is composed of ooliths—spherical bodies formed by the precipitation of carbonate around a nucleus; and clastic limestones have been derived from fragments of pre-existing calcareous rocks.

Lime Sulphur, eau grison. Solution containing calcium polysulphides and thiosulphate made by dissolving sulphur in a suspension of $Ca(OH)_2$. Used as a fungicide and insecticide (sheep dips). Can decompose to liberate H_2S.

Linde Process. A high-pressure process for the production of liquid oxygen and nitrogen by compression to about 200 bar ($20 MN/m^2$) followed by refrigeration and fractionation in a double column. The Linde-Frankl process operates at a much lower pressure.

Linz-Donnewitz Process. Basic oxygen process.

Liquation. Readily fusible metals, *e.g.*, Sn, Pb and Bi, are refined by this method. Impure metal in the form of ingots (blocks) is placed in the upper part of a sloping hearth (usually of a reverberatory furnace) maintained at a temperature slightly above the melting point of the metal. The impurities remain behind as dross while the pure metal melts and flows down into a well at the bottom of the incline. (Fig.).

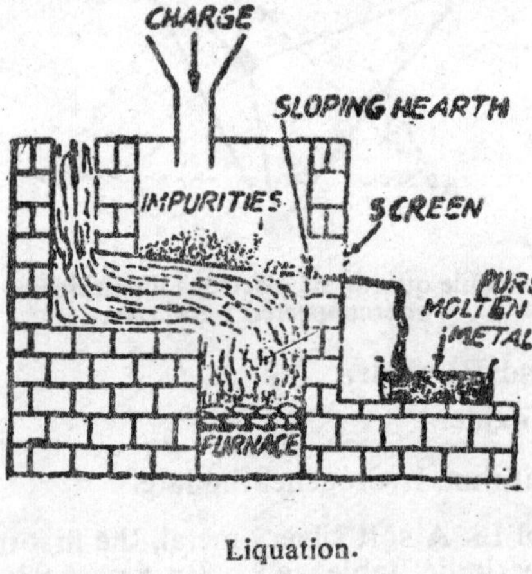

Liquation.

Liquid Air. Refers to a pale blue liquid. It is a mixture of liquid oxygen (boiling at -182 9°C) and liquid nitrogen (boiling at -195.7°C).

Linnett Double Quarter Theory (Ldqr) : In 1961. Linnett proposed an interesting modification of Lewis-Langmuir octet theory. According to him.

(i) The eight valency electrons present around atoms of first short period in their molecules should not be treated as four pairs but should be treated as two groups of four electrons. Each group of four electrons is termed as quartet. As eight electrons are assumed to be made up of two quartets in this theory, it is named as Linnett double quartet theory (LDQT).

(ii) Linnet further proposed that a set of four electrons (quartet) be considered to have the same spin, ↑, another quartet to have the opposite spin, ↓. In a quartet,! the four electrons will naturally tend to arrange themselves in such a way that their mutual repulsion effects are minimised. This is possible in the tetrahedral configuration. The second quartet will also take up the tetrahedral configuration. Thus, the eight electrons arrange themselves in two tetrahedra. The two tetrahedra take up relative positions that keep the interelectronic repulsion at a minimum. Thus, the vertices of one tetrahedron are expected to lies at the centres of the faces of the other as shown in Fig.

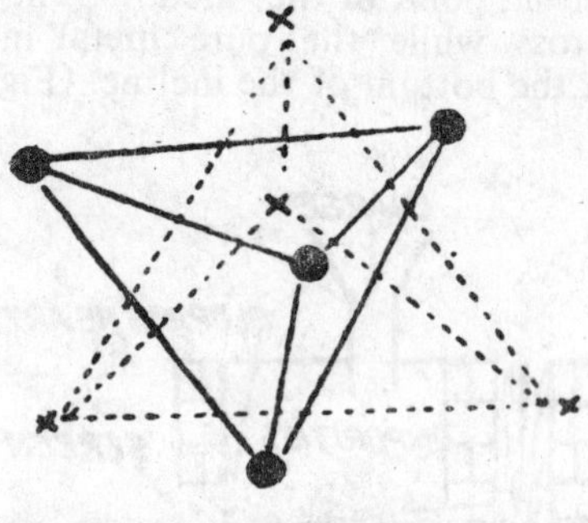

The double quartet of electrons illustrated by two interconnected tetrahedra

Litharge. See lead(II) oxide.

Lithia. Lithium oxide.

Lithia Water. Lithium hydrogencarbonate.

Lithium. Symbol Li. A soft silvery metal, the first member of group I of the periodic table; a.n. 3; r.a.m. 6.936; r.d. 0.534; m.p. 180.54°C; b.p. 1340°C. It is a rare element found in spodumene ($LiAlSi_2O_6$), petalite ($LiAlSi_4O_{10}$), the mica lepidolite, and certain brines. It could be usually extracted by treatment with sulphuric acid to give the sulphate, which is converted to the chloride. This is mixed with a small amount of potassium chloride, melted, and electrolysed The stable isotopes have been lithium-6 and lithium-7. Lithium-5 and lithium-8 are shortlived.

radioisotopes. The metal is used to remove oxygen in metallurgy and as a constituent of some Al and Mg alloys. It finds also use in batteries and is a potential tritium source for fusion research. Lithium salts have been used in psychomedicine. The element reacts with oxygen and water, on heating it also reacts with nitrogen and hydrogen. Its chemistry differs somewhat from that of the other group I elements because of the small size of the Li^+ ion.

Lithium Aluminium Hydride. See lithium tetrahydroaluminate(III).

Lithium Carbonate (Li_2CO_3). A white solid which is obtained by the addition of excess sodium carbonate solution to a solution of a lithium salt. It is soluble in the presence of excess carbon dioxide, forming lithium hydrogencarbonate. It heated (to 780°C) in a stream of hydrogen it undergoes thermal decomposition to yield lithium oxide and carbon dioxide.

Lithium Deuteride. See lithium hydride.

Lithium Hydride (LiH). A white crystalline solid which is (produced by direct combination of the elements at a temperature above 500°C. Electrolysis of the fused salt yields hydrogen at the anode. Lithium hydride reacts violently with water to give lithium hydroxide and hydrogen; the reaction is highly exothermic. The compound finds use as a reducing agent and in the preparation of other hydrides. Lithium hydride is having a cubic structure and is more stable than the group I hydrides.

Lithium Hydrogencarbonate. A compound, $LiHCO_3$ formed which by the reaction of carbon dioxide with aqueous is lithium carbonate and known only in solution. It has found medicinal uses similar to those of lithium carbonate and is sometimes included in proprietary mineral waters.

Lithium Hydroxide (LiOH). A white solid which is prepared industrially as the monohydrate ($LiOH.H_2O$) by reacting lime with a lithium ore or with a salt made from the ore. Lithium hydroxide is having a closer resemblance to the group II hydroxides than to the group I hydroxides.

Lithium Oxide (Lithia). A white crystalline compound, Li_2O; cubic; r.d. 2.01; m.p. 1700°C. It can be obtained from a number of lithium ores; the main uses have been in lubricating greases, ceramics, glass and refractories, alkaline storage batteries, and as a flux in brazing and welding.

Lithium Sulphate (Li_2SO_4). A white solid which is obtained by the addition of excess lithium oxide or carbonate to a solution of sulphuric acid. It is readily soluble in water from which it crystallizes as the monohydrate.

Lithium Tetrahydroaluminate(III) (Lithium Aluminium Hydride; LAH). A white or light grey powder, $LiAlH_4$; r.d. 0.917; decomposes at 125°C. It may be obtained by the reaction of excess lithium hydride with aluminium chloride. The compound is soluble in ethoxyethane, reacts violently with water to release hydrogen, and is widely used as a powerful reducing agent in organic chemistry. It should always be treated as a serious fire risk in storage.

Localized Bond. Refers to a bond in which the electrons contributing to the bond remain between the two atoms concerned, *i.e.* the bonding orbital is localized. The vast majority of bonds have been of this type.

Lone Pair. A pair of electrons having opposite spin in an orbital of an atom. For instance, in ammonia the nitrogen atom has five electrons, three of which are used in forming single bonds with hydrogen atoms. The other two occupy a filled atomic orbital and constitute a lone pair. The orbital containing these electrons is equivalent to a single bond (sigma orbital) in spatial orientation, accounting for the pyramidal shape of the molecule. In the water molecule, there are two lone pairs on the oxygen a.o.m. In considering the shapes of molecules, repulsions between bonds and lone pairs can be taken into account :

Lone pair lone pair→lone pair-bond→bond-bond.

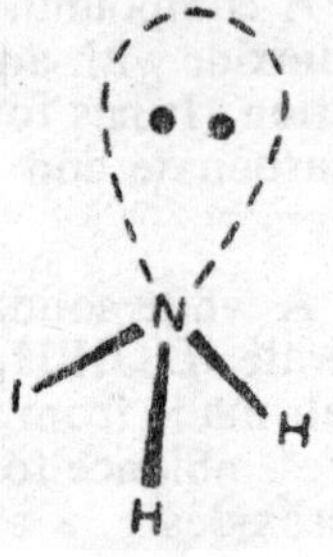

Lone pair of electrons in ammonia

Lodestone. See magnetite.

London Forces : With non-polar substances, such as iodine and the inert gases, there are no initial dipoles; the molecules are electrically symmetrical. Any slight relative displacement of the nuclei or the electrons in a molecule will, however, give rise to an electrical dipole and it is thought that, momentarily, such displacements are constantly occurring within molecules or atoms.

The electrical displacement necessary to establish a dipole is temporary and random, so that a molecule, as a whole, over a period of time, will have no observable dipole. But if it has a dipole at any one instant, it will induce another in a neighbouring molecule and an attractive force will be established between the two. Such interaction must be wholly responsible for van der Waal's bonding between non-polar atoms and molecules; its functioning was first explained by London (1930) and it is now referred to as the dispersion effect.

The dispersion effect makes a big contribution even to the bonding between polar molecules. Of the 29.7 K J. required to break down the crystal structure of solid ammonia, for example, it is estimated that 1.33 KJ. are needed for the orientation effect, 1.55 KJ. for the induction effect and 14.85 KJ. for the dispersion effect.

Long Period. See periodic table.

Lowry-bronsted Theory. See acid.

Lutetium, Lu. At.no. 71, at.wt. 174.967, m p. 1656°C, b.p. 3315°C, d 9 84. The last of the lanthanide (4f) elements. The metal is hcp. After neutron irradiation Lu nuclides emit pure β radiation which can be used in cracking, polymerization and other catalytic process.

Lutetium Compounds. Lutetium forms a single series of typical lanthanide compounds in the +3 state

Lu^{3+}(f^{14} colourless)→Lu (−2.25 volts in acid).

Lux-Flood Concept. The proton plays a vital role in explaining the acid-base behaviour in the Bronsted Lowry concept. It was observed by Lux (1939) that the acid-base reactions are also feasible in oxide systems without the aid of protons. This approach was extended by Flood (1947) and applied to non-protonic systems which were not covered by the Bronsted-Lowry concept. According to the Lux-Flood definition, a base (like CaO, BaO or Na_2O) is an oxide ion donor and an acid (like SiO_2, CO_2 or P_4O_{10}) is an oxide ion acceptor. Some acid-base reactions, in terms of the Lux-Flood concept, may be written as :

Base		Acid		
CaO	+	SiO_2	→	$CaSiO_3$
BaO	+	CO_2	→	$BaCO_3$
$6Na_2O$	+	P_4O_{10}	→	$4Na_3PO_4$

The concept is particularly applicable to reactions which take place at high temperature, *i.e.*, in metallurgical operations or during the manufacture of ceramics and glass. The approach can be extended to include other negative ion systems (like halides, sulphide or carbanion).

Lyman Series. A series of lines in the ultraviolet spectrum emitted by excited hydrogen atoms. They correspond to the atomic electrons falling into the lowest energy level and emitting energy as radiation. The wavelength (λ) of the radiation in the Lyman series is given by

$$1/\lambda = R(1/1^2 - 1/n^2)$$

where n is an integer and R is the Rydberg constant.

M

Magnadur. Tradename for a ceramic material which is used to make permanent magnets. It consists of sintered iron oxide and barium oxide.

Magnalium. Tradename for an aluminium-based alloy of high reflectivity for light and ultraviolet radiation that is having 1—2% of copper and between 5% and 30% of magnesium. Strong and light, these alloys also sometimes have other elements, such as tin, lead, and nickel.

Magnesia. See magnesium oxide.

Magnesite. Refers to a white, colourless, or grey mineral form of magnesium carbonate, $MgCO_3$, crystallizing in the trigonal system. It could be formed as a replacement mineral of magnesium-rich rocks when carbon dioxide is available. Magnesite is mined both as a source of magnesium carbonate.

Magnesium Symbol Mg. It is a silvery metallic element belonging to group II of the periodic table (see alkaline-earth metals); a.n. 12; r.a.m. 24.321; r.d. 1.74; m.p. 651°C; b.p. 1107°C. The element occurs in a number of minerals, including magnesite ($MgCO_3$), dolomite ($MgCO_3.CaCO_3$), and camallite ($MgCl_2$. KCl.$6H_2O$). It also occurs in sea water, and it is an essential element for living organisms. Extraction may be carried out by electrolysis of the fused chloride. The element is used in a number of light alloys (*e.g.* for aircraft). Chemically, it is very reactive. In air it forms a protective oxide coating but when ignited it burns with an intense white flame. It also reacts with the halogens, sulphur, and nitrogen. Magnesium was first isolated by Bussy in 1828.

Magnesium Bicarbonate. See magnesium hydrogencarbonate.

Magnesium Carbonate ($MgCO_3$). A white solid that is found naturally as the mineral magnesite and in association with calcium carbonate in dolomite. Magnesium carbonate is sparingly soluble in water but reacts with dilute acids to give the salt, carbon dioxide, and water. Because of this it is used in medicine as a mild antacid. Magnesium carbonate also decomposes easily to yield magnesium oxide, which is an important refractory material. It is also mixed with asbestos and used to cover hot pipes.

Magnesium Chloride ($MgCl_2$). A solid that exists in a variety of hydrated forms, most commonly as the hexahydrate, $MgCl_2.6H_2O$. When heated, this hydrated form is hydrolysed by its water of crystallization, with the evolution of hydrogen chloride and the formation of the oxide :

$$MgCl_2 + H_2O \rightarrow MgO + 2HCl$$

The anhydrous chloride must therefore be obtained by evaporating an aqueous solution in an atmosphere of hydrogen chloride. It finds use in the preparation of cotton fabrics.

Magnesium Hydrogencarbonate (magnesium bicarbonate, $Mg(HCO_3)_2$) The solid hydrogencarbonate is unknown at room temperature. The compound could be formed in solution when water containing carbon dioxide dissolves magnesium carbonate :

$$MgCO_3 + H_2O + CO_2 \rightarrow Mg(HCO_3)_2$$

It is a cause of temporary hardness in water.

Magnesium Hydroxide. A white solid compound. $Mg(OH)_2$; trigonal r.d. 2.36; decomposes at 350°C. Magnesium hydroxide occurs naturally as the mineral brucite and can be prepared by reacting magnesium sulphate or chloride with sodium hydroxide solution. It finds use in the refining of sugar and in the processing of uranium. Medicinally it is important as an antacid (milk of magnesia) and as a laxative.

Magnesium Oxide (magnesia). A white compound, MgO; cubic; r.d. 3.58; m.p.−2800°C. It is found naturally as the mineral periclase and is prepared commercially by thermally decomposing the mineral magnesite :

$$MgCO_3(s) \rightarrow MgO(s) + CO_2(g)$$

It is having a wide range of uses, including reflective coatings on optical instruments and aircraft windscreens and in semi-conductors. Its high melting point makes it useful as a refractory lining in metal and glass furnaces.

Magnesium Peroxide (Mg_2O_2) A white insoluble solid which is prepared by reacting sodium or barium peroxide with a concentrated solution of a magnesium salt. Magnesium peroxide finds use as a bleach for dyestuffs and silks.

Magnesium Sulphate. A white soluble compound, $MgSO_4$, existing as the anhydrous compound (rhombic; r.d. 2.66; decomposes at 1124°C) and in hydrated crystalline forms. The mono-hydrate $MgSO_4.H_2O$ (monoclinic; r.d. 2.45) is found naturally as the mineral kieserite. The commonest hydrate has been the heptahydrate, $MgSO_4.7H_2O$ (rhombic; r.d. 1.68), which is called Epsom salt(s), and occurs naturally as the mineral epsomite. This is a white powder with a bitter saline taste, which losses $6H_2O$ at 150°C and $7H_2O$ at 200°C. It finds use in sizing and fireproofing cotton and silk, in tanning leather, and in the manufacture of fertilizers, explosives, and matches. In medicine, it is used as a laxative. It finds also use in veterinary medicine for treatment of local inflammations and infected wounds.

Magnetic Moment. Refers to the ratio between the maxium torque (T_{max}) exerted on a magnet, current-carrying coil, or moving charge situated in a magnetic field and the strength of that field. It has been thus a measure of the strength of a magnet or current-carrying coil. In the Sommerfield approach this quantity (also called electromagnetic moment or magnetic area moment) is the ratio $T_{max}B$. In the Kennelly approach the quantity (also called magnetic dipole moment) is $T_{max}H$

An orbital electron is having an orbital magnetic moment IA, where I is the equivalent current as the electron moves round its orbit. It is given by $I=q\omega/2\pi$, where q is the electronic charge and ω is its angular velocity. The orbital magnetic moment has been therefore $IA=q\omega A/2\pi$, where A is the orbital area. If the electron has been spinning there is also a spin magnetic moment (see spin); atomic nuclei also have magnetic moments.

Magnetic Quantum Number. This quantum number was introduced to describe the Zeeman effect (*i e.*, splitting spectral lines under the influence of an applied magnetic field).

We know that each sub-shell is further sub-divided into various orbitals; the number of these orbitals, being given by n^2, thus when n=2, there are 4 possible orbitals. Each orbital contains two electrons.

When an atom is placed in a magnetic field (Zeeman effect), due to the presence of negatively charged electrons in it, the orbitals orient themselves relative to the magnetic field and their orientation in the magnetic field is determined by the magnetic quantum number (m).

This number is usually designated as 'm' and can have values from +l to −l through zero, *i.e.*,

$$m=+l, (l-1), (l-2) \ldots\ldots 0, -1, -2 \ldots\ldots -(l-1), -l$$

In other words, the number of m values for a given l value is 2l+1, *i e*,

$$m=(2l+1) \text{ values through zero.}$$

Magnetite. Refers to a black mineral form of iron oxide which is crystallizing in the cubic system. It is a mixed iron(II)-iron (III) oxide. Fe_3O_4, and is one of the major ores of iron. It is strongly magnetic and some varieties, known as lodestone, are natural magnets; these were used as compasses in the ancient world. Magnetic is widely distributed and occurs as an accessory mineral in almost all igneous and metamorphic rocks. The largest deposits of the mineral are found in N Sweden.

Malachite. A green mineral consisting of copper(II) carbonate and hydroxide ($CuCO_3.Cu(OH)_2$). It is used as an ore and a pigment.

Malleability. The property which enables a body—particularly a metal—to be extended in all directions by hammering or rolling. The degree of malleability is gauged by the thinness of leaf or foil which it is possible to produce.

Manganate (VI). A salt having the ion MnO_4^{2-}. Manganate (VI) ions are dark green; they are produced by manganate (VII) ions in basic solution.

Manganate (VII) (permanganate). Refers to a salt having the ion MnO_4. Manganate (VII) salts have been purple, and strong oxidizing agents. In basic solutions the dark green manganate (VI) ion gets formed.

Manganese. A transition metal which is found naturally as oxides, *e.g.* pyrolusite (MnO_2). Its main use has been in alloy steels which are made by adding pyrolusite to iron ore in an electric furnace. Manganese decomposes cold water and dilute acids to yield hydrogen and reacts with oxygen and nitrogen when heated. The oxidation states are +7, +6, +4, and (the most stable) +2. Manganese(II) salts are pale pink and with alkali the solutions precipitate manganese(II) hydroxide, which rapidly oxidizes in air to brown manganese(III) oxide.

Symbol: Mn; m p. 1244°C; b.p. 2040°C; r.d. 7.4; p.n. 25; r.a.m. 54.94.

Manganese Dioxide. See manganese (IV) oxide.

Manganese(II) Oxide (manganous oxide, MnO). A green powder which is obtained by heating manganese(II) carbonate or oxalate in the absence of air. Alternatively it may be obtained by heating the higher manganese(II) oxide is a basic oxide and is almost insoluble in water. At high temperatures it gets reduced by hydrogen to manganese. On exposure to air, manganese(II) oxide rapidly oxidizes. Manganese(II) oxide is having crystal lattice similar to that of sodium(I) chloride.

Manganese (VI) Oxide (manganese dioxide). A black oxide which is obtained by heating manganese(II) nitrate. The compound also occurs naturally as pyrolusite. It is a strong oxidizing agent, and finds use as a depolarizing agent in violtic cells.

Manganese (III) Oxide (manganic oxide, manganese sesquioxide, Mn_2O_3). A black powder which is prepared by igniting manganese (IV) oxide or a manganese (II) salt in air at 800°C. It reacts slowly with cold dilute acids to form manganese (III) salts. Manganese (III) oxide is bound in nature as braunite ($3M_2nO_3$ SiO_3), and the monohydrate ($Mn_2O_3H_2O$) is found in the mineral manganite. †The manganese (III) oxide crystal lattice is having Mn^{3+} and O^{2-} ions. With concentrated alkalis, it undergoes disproportionation to give manganese (II) and manganese (IV) ions.

Manganese(IV) Oxide (manganese dioxide, MnO_2). A black powder which is obtained by the action of heat on manganese (II) nitrate. Hydrated forms are found naturally as the minerals pyrolusite and wad (a less pure form). Manganese (IV) oxide is insoluble in water. It acts as a powerful oxidizing agent : it reacts with hot concentrated hydrochloric acid to produce chlorine and with warm sulphuric acid to give oxygen. The dihydrate ($MnO_2.2H_2O$) gets formed when potassium permanganate is reduced in alkaline solution It is used as a catalyst in the laboratory preparation of chlorine, as a depolarizer in electric dry cells, and in the glass industry. At 500–600°C, manganese (IV) oxide decomposes to yield manganese (III) oxide and trimanganese tetroxide. It is having a good electrical conductivity.

Manganese Sesquioxide. See manganese (III) oxide.

Manganic Compounds. Compounds of manganese in its +3 oxidation state; *e.g.* manganese oxide is manganese (III) oxide, Mn_2O_3.

Manganin. A copper alloy containing ∫13—18% of manganese and 1—4% of nickel. It has a high electrical resistance, which is relatively insensitive to temperature changes. It is therefore suitable for use in resistance wire.

Manganic Oxide. See manganese (III) oxide.

Manganous Compounds. Compounds of manganese in its +2 oxidation state; *e.g.* manganous oxide is manganese (II) oxide, MnO.

Manganous Oxide. See manganese (II) oxide.

Marble. A dense mineral form of calcium carbonate ($CaCO_3$) which is formed from lime stone.

Marsh's Test for Arsenic. The arsenic-containing specimen is converted to volatile AsH_3 which is decomposed to a brown stain on heating. Estimation is by comparison of stoins. Sb reacts similarly but the Sb stain is not soluble in NaOCl.

Martensite. Refers to a solid solution of carbon in alpha-iron (see iron) formed when steel gets cooled too rapidly for pearlite to form from austenite. It has been responsible for the hardness of quenched steel.

Mascagnite. A mineral form of ammonium sulphate, $(NH_4)_2 SO_4$.

Masking. An analytical technique involving addition of a masking agent which stops interfering species participating in a reaction. Demasking is the reverse process *e.g.* breakdown of a cyanide complex with acid.

Massicot. See lead (II) oxide.

Mass Number. See nucleon number.

Masurium. A former name for technetium.

Matte. Refers to a mixture of iron copper sulphides obtained at an intermediate stage in smelting copper ores.

Matte. The indefinite mixture of artificial sulphides produced by a smelting operation.

Mendeleev's Law. See periodic law.

Mendelevium, Md. At. no. 101. ^{256}Md (1.5 hours) and ^{258}Md (2 months) have been formed by bombardment of Es with ^{4}He particles in a cyclotron or linear accelerator. The Md^{2+} ion is fairly stable in aqueous solution

$$Md^{3+} \xrightarrow{-0.2v} Md^{2+}$$

No solid compounds are known.

Mercuric Chloride. See mercury (II) chloride.

Mercuric Compounds. Compounds of mercury in its + 2 oxidation state; *e.g.* mercuric chloride has been mercury (II) chloride $HgCl_2$.

Mercuric Oxide. See mercury (II) oxide.

Mercuric Sulphide. See mercury (II) sulphide.

Mercurous Compounds. Compounds of mercury in its + 2 oxidation state; *e.g.* mercury chloride has been mercurous chloride, $HgCl_2$.

Mercurous Chloride. See mercury (I) chloride.

Mercurous Oxide. See mercury (I) oxide.

Mercurous Sulphide. See mercury (I) sulphide.

Mercury Symbol Hg. Refers to a heavy silver liquid metallic element which belong to the zinc group; a.n. 80; r.a.m. 200.59; r.d. 13.55; m.p. −38.87°C: b.p. 356 58°C. The main ore is the sulphide cinnabar (HgS), which can be decomposed to the elements. Mercury finds use in thermometers, barometers, and other scientific apparatus, and in dental amalgams. The element has been less reactive zinc and calcium and will not displace hydrogen from acids. It has been also unusal in forming mercury (I) compounds containing the Hg_2^{2+} ion, as well as mercury (II) compounds containing Hg^{2+} ions. It also forms a number of complexes and organomercury compounds (*e.g.* the Grignard reagents).

Mercury Amalgams. Intermetallic compounds of mercury. Some have definite compounds, *e.g.* $NaHg_2$. and are effectively alloys. Hg rich amalgams are liquid. Fe does not form amalgams and is used for containers for mercury. Used in dental stoppings, cadmium amalgam is the cathode in the Weston standard cell.

Mercury (I) Chloride. A white salt, Hg_2Cl_2; r.d. 70; m.p. 302°C; b.p. 384°C. It is obtained by heating mercury (II) chloride with mercury and finds use in calomel cells (so called because the salt was formerly called calomel) and as a fungicide.

Mercury (II) Chloride (mercuric chloride, $HgCl_2$). A colourless crystalline compound which is prepared by direct combination of mercury (II) chloride is soluble in water; it also dissolves in concentrated hydrochloric acid due to the formation of complex ions, $HgCl_4^{2-}$ and $HgCl_3^-$. On heating, mercury (II) chloride sublimes yielding a white translucent mass. It has been extremely poisonous, but in dilute solution (1 : 1000) it finds use as an antiseptic.

Mercury (I) Oxide (mercurous oxide, Hg_2O). There has been some doubt as to the existence of this compound.) The black precipitate formed on addition of sodium hydroxide solution to a solution of mercury (I) nitrate by some to be mercury (I) oxide; others, who doubt its existence, think that the blackness of the precipitate is due to some free mercury. X-ray examination of this black compound has shown it to be an intimate mixture of mercury (II) oxide and mercury.

Mercury (II) Oxide. Refers to a yellow or red oxide of mercury, HgO. The red form is obtained by heating mercury in oxygen at 350°C; the yellow form, which differs from the red in particle site gets precipitated when sodium hydroxide solution is added to a solution of mercury (II) nitrate. Both forms decompose to the elements at high temperature. The black precipitate formed when sodium hydroxide gets added to mercury (I) nitrate solution is sometimes referred to as mercury (I) oxide (Hg_2O) but has been probably a mixture of HgO and free mercury.

Mercury (I) Sulphide (mercurods sulphide, Hg_2S). There is some doubt as to the existence of this compound). The brownish black precipitate formed when mercury has been treated with cold concentrated sulphuric acid from a long time is thought to be mercury(I) sulphide. Alternatively it may be obtained by the action of hydrogen sulphide or an alkaline sulphide on a mercury (I) salt solution. As soon as the mercury (I) sulphide gets formed it disproportionates to yield mercury (II) sulphide and mercury.

Mercury (II Sulphide. Refers to a red or black compound, HgS, occurring naturally as the minerals cinnabar (red) and metacinabar (black). It can be prepared as a black precipitate by bubbling hydrogen sulphide through a solution of mercury (II) nitrate. The red form is prepared by sublimation. The compound is also called vermilion (used as a pigment).

Meson, Mesotron. Mesons are subatomic particles with either a positive, negative or zero charge, which can most conveniently be detected by means of photographic emulsions. Mesons found the atmosphere at sea level have a mass 210 times that of an electron and a mean lifetime of 2×10^{-6} seconds. These are μ mesons, in cosmic ray studies. πMesons, mass 276 times that of electrons, lifetime 2.6×10^{-8} seconds, have been found at greater heights. They are currently considered to be part of the bonding mechanism in the atomic nucleus. Other mesons (τ, θ, etc.) have been described; they are all found in cosmic radiation and result from high energy nuclear collisions.

Mesothorium. Names used for Ra and Ac isotopes produced in natural decay series.

Metaboric Acid. See boric acid.

Metal. Refers to any of a class of chemical elements that are typically lustrous solids that have been good conductors of heat and electricity. Not all metals have all these properties (*e.g.* mercury is a liquid). In chemistry, metals fall into two distinct types. Those of the *s*- and *p*-blocks (*e.g.* sodium and aluminium) have been generally soft silvery reactive elements. They tend to form positive ions and so on have been described as electropositive. This is contracted with typical nonmetallic behaviour of forming negative ions. The transition elements (*e.g.* iron and copper) have been harder substances and generally less reactive. They form coordination complexes. All metals have oxides that are basic.

Metal Carbonyls. Co-ordination derivatives between CO and a metal in which the carbon is generally bonded to the metal. Most transition metal and actinide elements form carbonyls, $M(CO)_x$ ($x=1$, 2, 3, etc.) using matrix isolation techniques involving metal atoms and CO. These derivatives are only stable in matrices at low temperature. Most of the transition metals form, stable carbonyls (MCl_n, Macac$n$, etc, plus reducing agent plus CO under pressure). The carbonyl groups can be bonded.

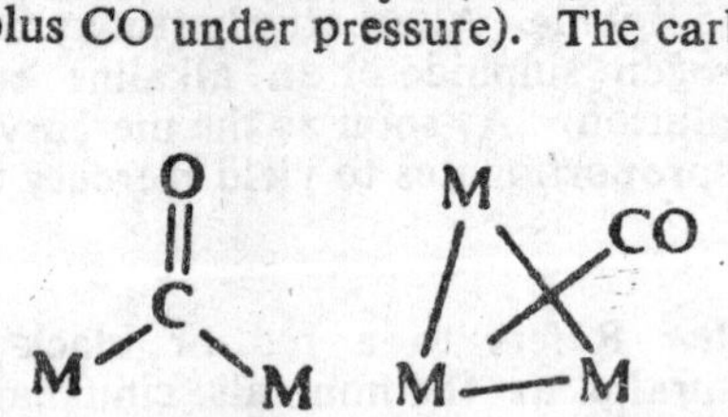

Simple stable carbonyls, except $V(CO)_6$, have an electronic configuration corresponding to the next noble gas. Carbonyl groups can be substituted by other unchanged ligands (*e.g.* PPh_3, pyridine) organic groups (olefines, aromatic derivatives) and also from other derivatives, *e.g.* halides, hydrides, sulphides.

Metal Cluster Compounds. Compounds containing clusters of metal atoms linked together by covalent for co-ordinate bonds.

Metallic Bond. The geometric arrangement of atoms in a metal crystal is the most compact one. The forces that keep the atoms so closely bound together in a metallic crystal constitute what is generally known as the metallic bond. The various characteristics of the metallic bond are :

(i) The metallic bond is essentially is covalent bond, without saturation, allowing a large number of atoms to be held together by a mutual sharing of valency electrons.

(ii) The density of electrons between the atoms in much lower than the allowed by the Pauli's exclusion principle. This allows electrons to move freely from point without a significant increase in their energy.

However, the metallic, bond differs from the covalent bond in the following respects :

(i) The valency electrons in a covalent bond are localized and, therefore, a covalent bond has a directional character while in a metallic bond the valency electrons are spread all over the crystal more or less uniformly. The metallic bond therefore is non-directional in character.

(ii) The valency electrons in a covalent bond are attracted strongly towards the nuclei of the atoms while the valency electrons in a metallic bond, being mobile, do not experience such strong forces of attraction towards the nuclei. Metallic bonds are, therefore, weaker than the covalent bonds. Due to this reason, the energy required to convert a mole of a metal (say, copper) into free atoms is smaller than the energy required to vaporise a mole of a covalent substance into free atoms (say, graphite).

Metallic Crystal. A crystal which is formed by metal atoms in the solid state. Each atom contributes its valence (outer) electrods to a 'sea' of electrons, which have been free to migrate through the solid, and the ions remaining are arranged in a lattice. The ability of electrons to move through the lattice accounts for the electrical and thermal conductivity of metals.

Metallic Hydrides or Interstitial Hydrides. The transition elements and rare-earth metals combine with hydrogen to produce hydrides which are metallic in appearance but do not possess definite composition. These types of hydrides are known as metallic hydrides.

In these hydrides, the hydrogen atoms seem to occupy interstitial positions in the metallic lattice and so form a solid solution. Because of this reason, metallic hydrides are also interstitial hydrides.

The metallic hydrides are related to other industrial compounds such as borides, carbides and nitrides in the same way that the lattices of the parent metal undergo no change.

Some of the metallic hydrides are given below :

$LaH_{2.76}$, $Ce\ H_{2.69}$ $PrH_{2.85}$ $TiH_{1.78}$
$ZrH_{1.93}$, $TaH_{0.73}$, $Pd\ H_{0.6}$ etc.

Apart from these simpler hydrides, transition metals form complex hydrides but their chemistry is of recent origin.

Structure of meiallic Hydrides. The exact state in which the hydrogen gas exists in the metal lattice has been a subject of much investigation. In many cases, adsorption of hydrogen results in no fundamental changes in the metal lattice, although density and X-ray determination show the lattices to be expand-

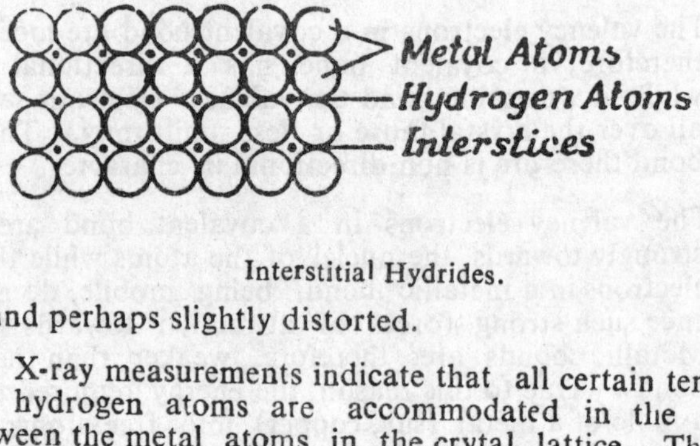

Interstitial Hydrides.

ed and perhaps slightly distorted.

X-ray measurements indicate that all certain temperatures, the hydrogen atoms are accommodated in the interestices between the metal atoms in the crytal lattice. The hydrogen is given out on raising the temperature of the metallic hydrides.

Metallic Radii. The metallic radius may be defined as one-half of the distance between the atoms of a metal in the metallic close-packed crystal lattice in which the metal exhibits a co-ordination number of twelve.

It is generally observed that metallic radii are about 10 to 15% higher than the single covalent bond radii. For example, the internuclear distance between two adjacent sodium atoms in a metallic crystal of sodium metal is 3 72 A°, so that the metallic or atomic radius of sodium is 3.72/2 or 1 oA°.

Assuming the atoms to be spheres, knowing the density of the metal, the nature of the unit cell, *i.e.*, the arrangement of other atoms around the central atom, the radius of an atom in a metallic crystal can be calculated.

Metallocenes. The bis-(h^5-cyclopentadienyl-metal) compounds $(C_5H_5)_2M$, formed by reaction between the cyclopentadienyl anion and suitable derivatives of the transition metals, *e.g.* M=Fe, ferrocene: M=Co, cobaltocene: M=Ni, nickelocene. The neutral uncharged complexes of the transition elements Mg & Ti to Ni are iso-structural, m.p. 174°C. They vary in colour, Ti green: Cr, red: Fe, orange; Co, purple: Os, yellow. Only those metallocenes of the iron group (Fe, Ru, Os) are diamagnetic, the others exhibiting varying degrees of paramagnetism, *e g.* Cp_2Co, one unpaired electron: Cp_2Mn, five unpaired elements. Only the iron group metallocenes exhibit aromatic behaviour (see ferrocene) and are not particularly sensitive to oxidation. Many metallocenes may be oxidized to cations, $[(C_5H_5)_2Fe^+]$, ferricinium: $[(C_5H_5)_2Co^+]$, cobalticinium. Metallocene derivatives are readily formed.

Metallography. Refers to the microscopic study of the structure of metals and their alloys. Both optical microscopes and electron microscopes have been used in this work.

Metalloids. Elements which have properties between the obvious metals and obvious non-metals. The classification is arbitrary and often based on the structure and properties of the free element. E g. As, Sb, Bi.

Metallurgy. The science of the extraction, working and use of metals and their alloys.

Process metallurgy is concerned with the extraction and production of metals, while physical metallurgy concerns the mechanical behaviour of metals.

Metals. Refer to any of a class of chemical elements with certain characteristic properties. In everyday usage, metals have been elements (and alloys) such as iron, aluminium, and copper, which have been lustrous malleable solids—usually good conductors of heat and electricity. Note that this is not a strict definition—some metals are poor conductors, mercury is a liquid, etc.

In chemistry, metals have been distinguished by their chemical properties, and there are two main groups. Reactive metals, such as the alkali metals and alkaline-earth metals, are electropositive elements. They have been high in the electrochemical series and tend to form compounds by losing electrons to give positive ions. They have basic oxides and hydroxides. This typical metallic behaviour decreases across the periodic table and increases down a group in the table.

The other type of metals have been the transition elements, which are less reactive, have variable valences, and tend to form complexes. In the solid and liquid states metals are having metallic bonds, formed by positive ions with free electrons.

Metamict State. Refers to the amorphous state of a substance that has lost its crystalline structure because of the radioactivity of uranium or thorium. Metamict minerals are minerals whose structure has been disrupted by this process. The metamictization is caused by alpha-particles and the recoil nuclei from radioactive disintegration.

Metaphosphoric Acid. See phosphoric (VI) acid.

Metaplumbate. See plumbate.

Metastable Species. Refers to an excited state of an atom, ion, or molecule, that has a relatively long lifetime before reverting to the ground state. Metastable species have been intermediates in some reactions.

Metastannate. See stannate.

Micas. Aluminosilicates with layers of linked (Si, Al)O_4 tetrahedra similar to those of clays or talc. Readily split into thin insulating sheets; used as fillers and to impart lustre to wall-papers and paints. The most important micas are phlogopite, $KMg_3(OH)_2Si_3AlO_{10}$, and muscovite, $KAl_2(OH)_2Si_3AlO_{10}$. If there are hydrated ions between the layers (instead of K) 'hydrated micas' montmorillonite, vermiculite, normally classified as clays are found. Fluorophlogopite, $K_2Mg_6Al_2Si_6O_{20}F_4$ with other micas is widely used in the electrical industry.

Mineral. Refers to a naturally occurring substance that has a characteristic chemical composition and, in general, a crystalline structure. The term is also often applied generally to organic substances that are obtained by mining (*e.g.* coal, petroleum, and natural gas) but strictly speaking these are not minerals, being complex mixtures without definite chemical formulas. Rocks are composed of mixtures of minerals. Minerals may be identified by the properties of their crystal system, hardness (measured on the Mohs' scale) relative density, lustre, colour, cleavage, and fracture. Many names of minerals end in ite.

Mineral Acid. The term used for a common inorganic acid, such as hydrochloric acid, sulphuric acid, or nitric acid.

Mirabilite. A mineral form of sodium sulphate, $Na_2SO_4.10H_2O$.

Mischmetal. An alloy of approximate composition La (25%) Ce (50%), Nd (20%) with some iron, prepared by electrolysis of fused lanthanide chlorides. Used to improve the malleability of cast iron and to improve the mechanical properties of Mg alloys. Small particles of the alloy are pyrophoric and ignite in air. Used in lighters to ignite fuels and on tracer bullets.

Mitcherlich's Law (law of isomorphism). The law which states that substances that crystallize in isomorphous forms (*i.e.* have identical crystalline forms and form mixed crystals) have similar chemical compositions. The law can be used for indicating the formulae of compounds. For instance, the fact that chromium oxide is isomorphous with Fe_2O_3 and Al_2O_3 implies that its formula is Cr_2O_3.

Mole Symbol mol. The SI unit of amount of substance. It may be defined as equal to the amount of substance that contains as many elementary units as there are atoms in 0.012 kg of carbon—12. The elementary units may be atoms, molecules, ions, radicals, electrons, etc. and must be specified. 1 mole of a compound has a mass equal to its relative molecular mass expressed in grams.

Molecular Crystal. Refers to a crystal in which molecules, as opposed to atoms, occupy lattice points. Examples include iodine and solid carbon dioxide (dry ice). As the forces holding the molecules together are weak, molecular crystals have low melting points. When the molecules have been small, the crystal structure approximates to a close-packed arrangement.

Molecular Orbitals. The orbitals which belong to a group of atoms forming a molecule. The orbitals may be bonding, anti-bonding or non-bonding according to whether the presence of an electron tends to hold the molecule together, to cause disruption of the molecule or to have no bonding effect on the molecule. The total number of molecular orbitals must be equal to the total number of atomic orbitals used in the bonding. In general only outer electrons need be considered for molecular orbitals.

Molecular Orbital Theory. According to the valence bond approach, a bond is formed between two atoms when they are brought together close enough to ensure maximum overlap of the atomic orbitals. When the bond is formed only one electron from each bonded atom loses its identity and moves in the outer atomic orbitals of both bonded atoms. The inner atomic orbitals from each atom forming the bond are, however, undisturbed, *i.e.*, each atom retains its own identity.

According to the molecular orbital theory, developed by Mullikan, Huckel and Hund, the orbitals are associated with the molecule as a whole. The quantized molecular orbitals of varying energy levels surround both the nuclei of the bonded atoms. These molecular orbitals are formed by the coalescence of the individual atomic orbitals when the atoms to be bonded come together. The orbitals of the bonded atoms lose their individual identity. The electrons are considered moving over the whole nuclear system instead of moving around a particular nucleus. The configuration of the molecule is obtained by placing electrons into these quantized molecular orbitals in an aufbcu order analogous to that employed in the filling of atomic orbitals. Corresponding to the practice of indicating atomic orbitals by the letters s, p and d molecular orbitals are assigned similar Greek letter designations ρ (sigma), π pi), δ delta).

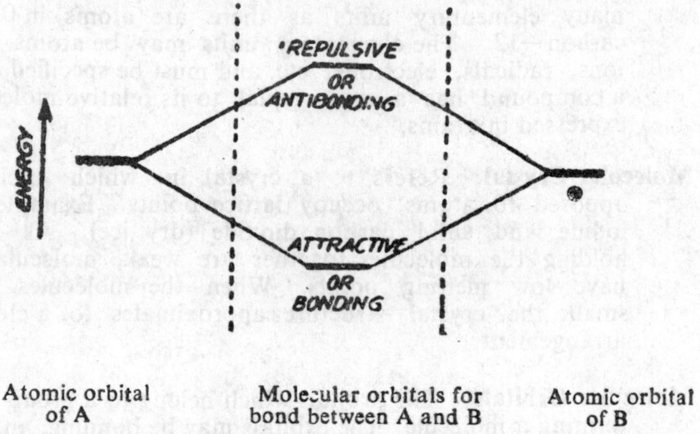

The formation of two molecular orbitals from the coalescence of two atomic orbitals in atoms A and B.

Molecular Spectrum. Refers to the absorption or emission spectrum that is characteristic of a molecule. Molecular spectra have been usually band spectra.

Molecule. Refers to one of the fundamental units forming a chemical compound; the smallest part of a chemical compound that can take part in a chemical reaction. In most covalent compounds, molecules are consisting of groups of atoms held together by covalent or coordinate bonds. Covalent substances that form macromolecular crystals have no discrete molecules (in a sense the whole crystal is a molecule). Similarly, ionic compounds do not have single molecules, being collections of oppositely charged ions.

Molybdenum Blue. A blue hydrated hydrous oxide formed by mild reduction of molybdenum (VI) solutions.

Molybdenum. A transition element that is found naturally in molybdenite (MoS_2).. It is used in alloy steels and in lamp bulbs. The compound ammonium molybdate, dissolved in nitric acid, finds use as a test for phosphates (V).

Symbol : Mo; m.p. 2620°C; b.p. 4800°C; r.d. 10.2: p.n. 42; r a m. 95.94.

Mond Process. Refers to a method of obtaining pure nickel by heating the impure metal in a stream of carbon monoxide at 50—60°C. Volatile nickel carbonyl ($Ni(CO)_4$) is formed, and this can be decomposed at higher temperatures (180°C) to give pure nickel. The method was invented by the German—British chemist Ludwig Mond 1839—1909).

Monel Metal. Refers to an alloy of nickel (60—70%), copper (25—35%), and small quantities of iron, manganese, silicon, and carbon. It is used to make acid-resisting equipment in the chemical industry.

Monodentate Ligand. A ligand with only one co-ordinately active lone pair of electrons which interact through a single atom, *e.g.* NH_3

Monotropy. Refers to the existence of a single allotrope of an element that is always more stable than the other(s) regardless of temperature; phosphorus, for example, exhibits monotropy. The phase diagram for a monotropic element reveals that one allotrope always has a lower vapour pressure than the other(s) at all temperature; this has been the stable allotrope.

Monovalent (univalent). Having a valency of one.

Mosaic Gold. See tin (IV) sulphide.

Moseley's Law. The frequencies of the lines in the X-ray spectra of the elements are related to the atomic numbers of the elements. If the square roots of the frequencies of corresponding lines of a set of elements are plotted against the atomic numbers a straight line is obtained. The law was discovered by H.G. Moselay (1887-1915).

Muffle Furnaces. A gas-heated furnace designed so that the charge does not come into contact with the hot combustion gases. This is effected by putting the charge in a retort or 'box with hot gases heating the outside walls. Electrically-heated 'muffles' are also available.

Multicentre Bond. A bond formed by a bonding molecular orbital between three or more atoms containing a single pair of electrons. Such bonds occur in the boron hydrides.

Multidentate Ligand, Polydentate ligand. A ligand which has two or more sites at which it can co-ordinate. Such a ligand can form a bridge or can form a chelate compound. An example is $H_2NCH_2CH_2NH_2$.

Multiple Bonding. Bonding between atoms in which there is overlap of orbitals in more than one position in space. The electrons may come equally from each atom or be derived from only one of the atoms (co-ordinate bond, back bonding). Generally the primary bond is a σ-bond with extra interaction by π-or σ bonding.

Multiplet

1. A spectral line formed by more than two (see doublet) closely spaced lines
2. A group of elementary particles that are identical in all respects except that of electric charge.

Mumetal (Trademark) A magnetic alloy containing about 75% nickel, the remainder being iron, copper, and chromium and sometimes molybdenum. It is easily magnetized and demagnetized, has a high permeability, and is used in transformer cores and electro-mechanical equipment.

Muntz Metal. A form of brass containing 60% copper, 39% zinc, and small amounts of lead and iron. Stronger than alpha-brass, it is used for hot forgings, brazing rods, and large nuts and bolts. It is named after G.F. Muntz (1794-1857).

muon. A μ-meson. See meson.

Muonium. A electron plus a muon, formally analogous to hydrogen.

Muriate. An obsolete name for a chloride; *e.g.* muriate of potash (KCl).

Muscovite (white mica; Potash Mica). A mineral form a potassium aluminosilicate, $K_2Al_4(Si_6Al_2)O_{20}(OH,F)_4$; one of the most important members of the mica group of minerals. It is chemically complex and has a sheetlike crystal structure. It is usually silvery-grey in colour, sometimes tinted with green, brown, or pink. Muscovite is a common constituent of certain granites and pegmatites. It is also common in metamorphic and sedimentary rocks. It is widely used in industry, for example in the manufacture of electrical equipment and as a filler in roofing materials, wallpapers, and paint.

N

Nano-symbol. n A prefix denoting 10^{-9}. For example, 1 nanometre (nm) $= 10^{-9}$ metre (m).

Nascent Hydrogen. Refers to a reactive form of hydrogen which is generated in situ in the reaction mixture (*e.g.* by the action of acid on zinc). Nascent hydrogen can be able to reduce elements and compounds that do not readily react with 'normal' hydrogen. It was once thought that the hydrogen was present as atoms, but this is not the case. Probably hydrogen molecules get formed in an excited state and react before they revert to the ground, state.

Natrolite. $Na_2(Al_2Si_3O_{10}),2H_2O$.

Natron. A mineral form of hydrated sodium carbonate, $Na_2CO_3\,H_2O$.

Natural Abundance. See abundance.

Neodymium. Symbol Nd. A soft silvery metallic element whcih belongs to the lanthanoids; a.n. 60; r.a.m. 144.24; r.d. 7.040 (20°); m.p. 1016°C; b.p. 3068°C. It occurs in bastnasite and monazite, from which it is recovered by an ion-exchange process. There are seven naturally occurring isotopes, all of which have been stable, except neodymium-144, which is slightly radioactive (half-life $10^{10}-10^{15}$ years). Seven artificical radioisotopes have been produced. The metal has been used to colour glass violet-purple and to make it dichroic. It is also used in mischmetal (18% neodymium). It was discovered by C.A. von Wels-bach in 1885.

Neodymium Compounds. Stable Nd compounds are typical lanthanide compounds in the +3 state

$$Nd^{3+}\ (f^3\ \text{red}) \rightarrow Nd\ (-2.44\ \text{volts in acid})$$

A +4 fluoride, Cs_3NdF_7, has been prepared by the action of F_2 on $CsCl+NdCl_3$. $NdCl_2(NdCl_3+Nd)$ and NdI_2 are known but it is not certain that these compounds contain Nd^{2+} ions.

Neon. An inert colourless odourless monatomic element of the rare-gas group. It is found in minute quantities (0.0018% by volume) in air. It is used mainly in neon signs and lights.

Symbol : Ne; m.p -248.67°C; b.p. -246.05°C; d. 0.9 kg m^{-3}; p.n. 10; r.a.m. 20.18.

Neptunium. Symbol Np. A radioactive metallic transuranic element which belongs to the actinoids; a.n. 93; r.a.m. 237.0482. The most stable isotope, neptunium-237, is having a half-life of $2.2+10^6$ years and is produced in small quantities as a by-product by nuclear reactors Other isotopes are having mass numbers 229-236 and 238-241. The only other relatively long-lived isotope is neptunium-236 (half-life 5×10^3 years). The element was first produced by McMillan and Abelson in 1940.

Neptunium Series. See radioactive series.

Nesquehonite. A mineral form of magnesium carbonate trihydrate, $MgCO_3.3H_2O$.

Nessler's Reagent. A solution of mercury (II) iodide (HgI_2) in potassium iodide and potassium hydroxide. It finds use in testing for ammonia, with which it forms a brown coloration or precipitate.

Neutrino. An elementary particle without charge and with spin $\frac{1}{2}$, postulated by Pauli to account for the conservation of angular momentum in nuclear transformations. Neutrinos appear to have variable finite mass and to oscillate from one type to another.

Neutron. Discovered by Chadwick in 1932, the neutron is one of the fundamental particles in an atom. It has almost the same mass as a proton but is electrically neutron. A free neutron, obtained from an atomic pile, has a mean lifetime of 750s; it decays to yield a proton and an electron. A neutron has magnetic moment 1.91 B.M : and interacts with atoms with unpaired electrons. Neutrons have been used for the determination of molecular structure by the diffraction which occurs when a substance is subjected to a neutron flux.

Neutron Number. Symbol : N The number of neutrons in the nucleus of an atom; *i.e.* the nucleon number (A) minus the proton number (Z).

Newland's Law (law of octaves). The observation that, when the elements are arranged in order of increasing atomic weight, there has been a similarity between members that are eight elements apart (inclusive). For example, lithium is having similarities to sodium, beryllium to magnesium, and so on. Newlands discovered this relationship in 1863 before the announcement of Mendeleev's famous periodic law (1869). It is now recognized that Newland's law arises from there being eight members in each short period.

Nichrome. Tradename for a group of nickel chromium alloys which are used for wire in heating elements as they possess good resistance to oxidation and have a high resistivity. Typical is Nichrome V having 80% nickel and 19.5% chromium, the balance consisting of manganese, silicon, and carbon.

Nickel. Symbol Ni. A malleable ductile silvery metallic transition element: a.n. 28; r.a.m. 5870; r.d. 8.9; m p. 1450°C; b.p. 2840°C. It occurs in the minerals penitlandite (NiS), pyrrhoite ((Fe,Ni)S), and garnierite ($(Ni,Mg)_6(OH)_6Si_4O_{11}.H_2O$). Nickel also occurs in certain iron meteorites (up to 20%). The metal could be extracted by roasting the ore to give the oxide, followed by reduction with carbon monoxide and purification by the Mond process. Alternatively electolysis is used. Nickel metal finds use in special steels, in Invar, and being ferromagnetic, in magnetic alloys, such as Mumetal. It is also an effective catalyst, particularly for hydrogenation reactions (see also Rancy nickel). The main compounds have been formed with nickel in the +2 oxidation state; the +3 state also exists (*e.g.* the black oxide, Ni_2O_3). Nickel was discovered by A. F. Cronstedt in 1751.

Nickel Carbonyl ($Ni(CO)_4$). A colourless liquid with a highly toxic vapour. It may be prepared by passing carbon monoxide over finely divided nickel at a temperature of approximately 60°C. The product may be purified by fractional distillation. Nickel carbonyl decomposes on heating to give pure nickel. It finds use as a catalyst and in the preparation of nickel (Mond process).

Nickelic Oxide. See nickel(III) oxide.

Nickelous Oxide. See nickel(II) oxide.

Nickel (II) Oxide. A green powder, NiO; r.d. 6.6 It can be prepared by heating nickel (II) nitrate or carbonate with air excluded.

Nickel (III) Oxide (Nickel Peroxide; Nickel Sesquioxide). A black or grey powder. Ni_2O_3; r.d. 4.8. It is prepared by heating nickel (II) oxide in air and used in nickel-iron accumulators.

Nickel silver. (German silver). Refers to any of a group of alloys having nickel, copper, and zinc in various proportions (but no silver). They are white or silvery in colour, can be highly polished, and have good corrosion-resistance. They find use in silver plating, chromium plating, and enamelling.

Nicrosilal. Austenitic alloy cast iron containing about 18% Ni, and 4–6% Si. Has high corrosion, heat and oxidation resistance; used in high-temperature equipment.

Nido. Applied to open structures, *e.g.* $B_4C_2H_8$ (nest-like).

Niobium Symbol Nb. A soft ductile grey—blue metallic transition element; a.d. 41; r.a.m. 92.91; r.d. 8.57; 2468°C; b p. 4742°C. It is found in several minerals, including niobite ($Fe(NbO_3)_2$), and may be extracted by several methods including reduction of the complex fluoride K_2NbF_7 using sodium. It finds use in special steels and in welded joints (to increase strength). Niobium—zirconium alloys are used in superconductors. Chemically, the element reacts with the halogens and oxidizes in air at 200°C. It forms a number of compounds and complexes with the metal in oxidation states 2, 3, or 5. The element was discovered by Charles Hatchett in 1801 and first isolated by Blomstrand in 1864. Formerly, it was called columbium.

Nitrate. A salt or ester of nitric acid.

Nitre (Saltpetre). Commercial potassium nitrate; the name was formerly applied to natural crustlike efforescences, occurring in some arid regions.

Nitre Cake. See sodium hydrogensulphate.

Nitric Acid (HNO_3). Refers to a colourless fuming corrosive liquid that is a strong acid. Nitrate acid can be prepared in a laboratory by the distillation of a mixture of an alkali metal nitrate and concentrated sulphuric acid. Commercially it is obtained by the catalytic oxidation of ammonia and is supplied as concentrated nitric acid, which contains 68% of the acid and is often coloured yellow by dissolved oxides of nitrogen. Nitric acid is a strong oxidizing agent. Most metals get converted to their nitrates with the evolution of oxides of nitrogen (the composition of the mixture of the oxides concentration of the nitric acid used).

Nitric Oxide. See nitrogen monoxide.

Nitrides. Compounds of nitrogen with a more electropositive element. Boron nitride is a covalent compound having macromolecular crystals. Certain electropositive elements, such as lithium, megnesium and calcium, react directly with nitrogen to form ionic nitrides containing the N^{3-} ion. Transtion elements form a range of interstitial nitrides (*e.g.* Mn_4 N, W_2 N), which can be formed by heating the metal in ammonia.

Nitriding. A method of surface hardening steel by treating it in an atmosphere of NH_3 which gives a surface layer of nitride. The steel must contain specific alloying elements, *e.g.* Al, Cr. Mo.

Nitrido Complexes. Complexes containing the N^{3-} ligand, *e.g.* $[OsO_3N]^-$, the osmiamate ion.

Nitrification. Refers to a chemical process in which nitrogen (mostly in the form of ammonia, in plant and animal wastes and dead remains gets oxidized at first to nitrites and then to nitrates. These reactions have been effected mainly by the bacteria Nitrosomans and Nitrobacter respectively. Unlike ammonia, nitrates could be readily taken up by plant roots; nitrification is therefore a crucial part of the nitrogen cycle Nitrogen-containing compounds are usually applied to soils deficient in this element, as fertilizer. Compare denitrification.

Nitrite. A salt or ester of nitrous acid; the nitride ion, NO^-_2 is having a bond angle of 115°.

Nitrogen Symbol N. A colourless gaseous element which belongs to group V of the periodic table; a n. 7; r.a.m 14.0067; d. 1.2506 g dm^{-3}, m.p. —209.86°C; b.p. —195.8°C. It is found in air (about 78% by volume) and is an essential constituent of proteins and nucleic acids in living organisms (see nitrogen cycle). Nitrogen is prepared for industrial purposes by fractional distillation of liquid air. Pure nitrogen can be obtained in the laboratory by heating a metal azide. There have been two natural isotopes: nitrogen—14 and nitrogen—15 (about 38). The element finds use in the Haber process for making ammonia and is also used to provide an inert atmosphere in welding and metallurgy. The gas is diatomic and relatively inert—it reacts with hydrogen at high temperatures and with oxygen in electric discharges. It also forms nitrides with certain metals. Nitrogen was discovered in 1772 by D. Rutherford.

Nitrogen Cycle. Refers to one of the major cycles of chemical elements in the environment. Nitrates in the soil would be taken up by plant roots and may then pass along food chains into animals. Decomposing bacteria are able to convert nitrogen-containing compounds (especially ammonia) in plant and animal wastes and dead remains back into nitrates, which get released into the soil can again be taken up by plants (see nitrification). Though nitrogen has been essential to all forms of life, the huge amount present in the atmosphere is not directly available to most organisms (compare carbon cycle). It can, however, get assimilated by some specialized bacteria and algae (see nitrogen fixation) and is therefore made available to other organisms indirectly. Lightning flashes also made some nitrogen available to plants by causing the combination of atmosphoric nitrogen and oxygen to form oxides of nitrogen, which enter the soil and form nitrates. Some nitrogen, gets returned from the soil to the atmosphere by dentrifying bacteria (see denification).

Nitrogen Dioxide (NO_2). A brown gas which is produced by the dissociation of dinitrogen tetroxide, the dissociation being com-

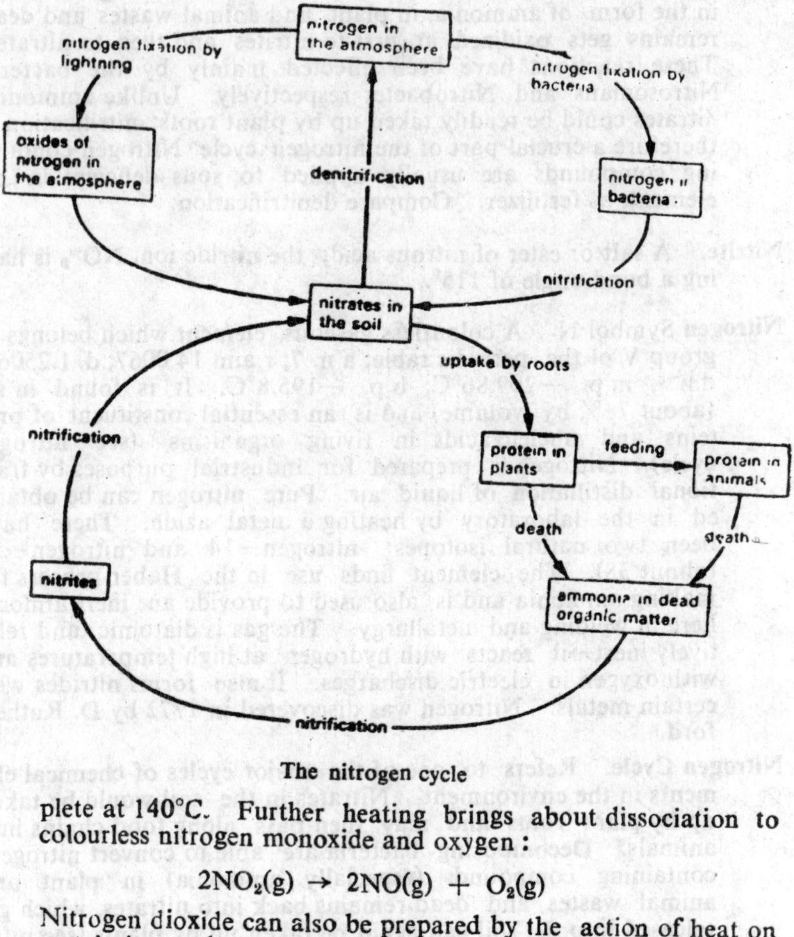

The nitrogen cycle

plete at 140°C. Further heating brings about dissociation to colourless nitrogen monoxide and oxygen :

$$2NO_2(g) \rightarrow 2NO(g) + O_2(g)$$

Nitrogen dioxide can also be prepared by the action of heat on metal nitrates (not the nitrates of the alkali metals or some of the alkaline-earth metals).

Nitrogen Fixation. Refers to a chemical process in which atmospheric nitrogen gets assimilated into organic compounds in living organisms and hence into the nitrogen cycle. The ability to fix nitrogen has been limited to certain bacteria (*e.g.* Azotobacter) and blue-green algae (*e.g.* Anabaena). Rhizobium bacteria are able to fix nitrogen in association with cells in the roots of leguminous plants such as peas and beans, in which they are known to form characteristic root nodules; cultivation of legumes is therefore one way of increasing soil nitrogen. Various chemical processes are used to fix atmospheric nitrogen in the manufacture of fertilizers. These include the Birkeland—

Eyde process, the cyanamide process (see calcium dicarbide), and the Haber process.

Nitrogen Monoxide (nitric oxide, NO). A colourless gas which is insoluble in water but dissolves in a solution containing iron (II) ions owing of the formation of the complex ion $(FeNO)^{2+}$; the nitrogen monoxide can get released by heating. Nitrogen monoxide is prepared by the action of nitric acid on copper turnings; the impure product can be purified by using a solution of iron (II) ions to absorb the product. Commercially, nitrogen monoxide may be obtained by the catalytic oxidation of ammonia or by the direct union of nitrogen and oxygen in an electric arc. Nitrogen monoxide is the most heat-stable of the oxides of nitrogen, only decomposing above 1000°C. At ordinary temperatures it reacts immediately with oxygen to yield nitrogen dioxide :

$$2NO(g) + O_2(g) \rightarrow 2NO_2(g)$$

Nitrogenous Base. A basic compound having nitrogen. The term is used especially of organic ring compounds such as adenine, guanine, cytosine, and thymine, which have been constituents of nucleic acids.

Nitrosonium salts. Salts containing the NO^+ ion which result from the reaction of nitrosyl chloride with metallic chlorides, *e.g.* $[NO]^+[AlCl_4]^-$, or from reactions in BrF_3,

Nitrosyls. Compounds containing NO bonded to metals through the nitrogen. In general similar to carbonyls but the M—N bonds are more stable to substitution than M—C bonds. For electron-counting assumed to have 3-electrons donated to the metal (*i.e.* NO^+ complexes). Formed direct by direct action of NO or indirectly by the action of nitrosyl halides. The iron group forms very stable nitrosyls, *e.g.* $Fe(CO)_2(NO)_2$, $[Fe(CN)_5NO]^{2-}$ (nitroprusside ion), $[Ru[NO)Cl_5]^{2}$ Thionitrosyls, containing MNS groups are also known.

Nitrosyl Ion. The ion NO^+. See nitrogen monoxide.

Nitrous Acid A weak acid, HNO_2 which is known only in solution and in the gas phase. It may be prepared by the action of acids upon nitrites, preferably using a combination that removes the salt as an insoluble precipitate (*e.g.* $Ba(NO_2)_2$ and H_2SO_4). The solutions have been unstable and decompose on heating to give nitric acid and nitrogen monoxide. Nitrous acid can function both as an oxidizing agent (forms NO) with I^- and Fe^{2+} or as a reducing agent (forms NO_3^-) with, for example, Cu^{2+}; the latter is most common. It is widely used (prepared in situ) for the preparation of diazonium compounds in organic chemistry. The full systematic name has been dioxonitrict (III) acid..

Nitrous Oxide. See dinitrogen oxide.

Nitryl Ion (Nitrouium Ion). The ion NO_2^+, found in mixtures of nitric acid and sulphuric acid and solutions of nitrogen oxides in nitric acid. Nitryl salts, such as $NO_2^+ClO_4^-$, can be isolated but are extremely reactive. Nitryl ions generated in situ are used for nitration in organic chemistry.

Nobelium.. (Symbol: No; At No. 102).

In July 1957, a group of scientists from the United States, the United Kingdom and Sweden (working under the auspices of the Nobel Institute for Physics, Stockholm) announced that they were successful in preparing a new element 102 by the bombardment of curium 244 with carbon-13 nuclei. The isotope was reported to possess the half-life of 10 minutes and to decay with emission of 8.5 MeV alpha-particle. These investigators suggested the name for the new element as Nobelium and the symbol No. This name was proposed in honour of Alfred Nobel. This name was later accepted by the commission on atomic weights of the International Union of Pure and Applied chemistry.

Soon after the report of the discovery of the new element, attempts to duplicate the experiments were made both in Berkeley, U.S.A. and in the U.S.S.R., but without success.

In April 1258, however, A Ghiorso, T. Sikkeland J.R. Walton and G.T. Seaborg at Berkeley provided positive proof of the formation of element with a mass number of 254. They synthesized it by bombarding curium-246 with a beam of carbon-12 nuclei from the newly installed heavy ion accelerator.

$$_{96}Cm^{246} + {}_6Cl^2 \rightarrow {}_{102}NO^{254} + 4{}_0n^1$$

The collected the nobelium ions (positively charged) which were found to be an alpha emitter

$$_{102}NO^{254} \rightarrow {}_{100}Fm^{250} + {}_2He^4$$

In 1956, Russian reported that they obtained nobelium-254 by bombarding uranium-238 with neon-22 and by bombarding americium—243 with nitrogen-15.

$$_{92}U^{238} + {}_{10}Ne^{22} \rightarrow {}_{102}NO^{254} + 6{}_0n^1$$

$$_{95}Am^{248} + {}_7N^{15} \rightarrow {}_{102}Na^{254} + 4{}_0n^1$$

cium-243. By the using of double recoil technique, the product of two stages alpha-particle emission and electron capture-namely fermium 252, was collected and identified chemically.

As the discovery of lawrencium is based entirely on nuclear evidence and since the quantities obtainable are extremely minute, it is not, therefore, possible to study the chemical or physical properties of the element or even to obtain the chemical identification.

Properties : It is the past member of the 14 member actinide series, A has been show *i.e.* to exihibit a stable (III) oxidation state in aqueous solution.

Noble Gases (rare gases, inert gases, group 0 elements). Refer to a group of monatomic, low-boiling gases which may be classified as belonging to group 0 of the periodic table and occupying a position between the highly electronegative group VII elements and the highly electropositive group I elements. They have been : helium (He), neon (Ne), argon (Ar), krypton (Kr), xenon (Xe), and radon (Rn). This complete filling of the s and p levels yields a closed-shell structure and is associated with a general lack of chemical reactivity and very high ionization energies. It has been the acquisition of a stable noble-gas configuration that is asssociated with determining the valence of many covalent and ionic compounds. Prior to 1962, the noble gases were frequently called inert gases as no chemical compounds were known (there were a few clathrates and 'hydrates'), but the realization that the ionization potential of xenon was sufficiently low to be accessible to chemical reaction led to the preparation of several fluorides, oxides, oxyfluorides, and a hexafluoroplatinate of xenon. Several unstable krypton fluoride species have also been reported.

Argon forms about 1% of the atmosphere but the other gases occur in only minute traces (excluding radon). They are get by fractionation of liquid air or of natural gas. Considerable amounts of argon are obtained as a by-product from ammonia-plant tail gas. Their applications depending heavily on their inertness; for example, welding (Ar), filling light bulbs (Ar), gas-stirring in high temperature metallurgy (Ar), powder technology (Ar), discharge tubes (Ne), and chemical research as inert carriers (He, Ar). Neon has been proposed as an oxygen diluent for deep-sea diving because of its lower solubility in blood. Liquid helium is extensively used in low-temperature work and randon is used therapeutically as a source of α-particles.

Node. A region of very low electron density in an atom or molecule as predicted from wave equations.

Nomag. Austenitic alloy cast iron, non-magnetic with high electrical resistance. Contains 10-12% Ni, 5–6% Mn.

Non-aqueous Solution. A solution in which the solvent is not water but may be organic (*e.g.* Et_2O) or inorganic (*e.g.* NH_3), protonic (*e.g.* EtOH), or non-protonic (*e.g.* BrF_3). Non-aqueous solvents may be self-ionized, *e.g.* $2NH_3 \rightleftharpoons NH_4^- + NH_2^-$.

Non aqueous Solvents. See non-aqueous solution.

Nonmetal. Refers to any of a class of chemical elements that are typically poor conductors of heat and electricity and that do not form positive ions. Nonmetals have been electronegative elements, like carbon, nitrogen, oxygen, phosphorus, sulphur, and the halogens. They form compounds that contain negative ions or covalent bonds. Their oxides have been either neutral or acidic.

Nonpolar Compound. A compound that has covalent molecules having no permanent dipole moment. Examples of nonpolar compounds have been methane and benzene.

Non-polar Solvent. Refers to a solvent that does not have a permanent electric dipole moment and consequently has no drive towards intermolecular association with polar species. Non-polar solvents have been usually good solvents for non-polar solutes. For example, carbon tetrachloride, hexane, and benzene are good solvents for iodine.

†Solvation energies are usually low and the driving force for dissolution by non-polar solvents has been the increased disorder of the solution.

Non-relativistic Quantum Theory. See quantum theory.

Non-stoicheiometric Compounds. Compounds in which the ratios of the numbers of atoms present do not follow simple integers. Many transition metal oxides, silicates, etc. are non-stoicheiometric and the phenomenon is common.

Nontronite. A montmorillonite mineral, a constituent of some clays.

Nordhausen Sulphuric Acid. See disulphuric (VI) acid.

Nucleon. A collective name for particles of mass number 1, *i.e.* protons and neutrons.

Nucleon Number (mass number) Symbol : A. The number of nucleons (protons plus neutrons) in an atomic nucleus.

Nucear Fission. In 1939, Hahn and Stratsman discovered that when uranium nucleus was bombarded with neutrons it splits up into two radioactive nuclei, *i.e.*, barium (Z=56) and krypton (Z=36). The new process of splitting of a heavy nucleus into two lighter nuclei of nearly comparable masses with release of energy was called fission by Friesh and Meitner in 1939.

A nuclear fission is generally accompanied by the release of a small number of neutrons also.

Immediately after the annoucement of the discovery of nuclear fission, physicists in many laboratories repeated and confirmed these experiments. Within next two years same process was confirmed for thorium and protactinium.

In addition to the two fragments, on the average 3 neutrons and about 200 Mev energy were released per fission The energy was much larger than that released in any of the exoergic nuclear reactions.

Nuclear Fusion. When lighter nuclei moving at a high speed are fused together to form a heavy nucleus, the process is called nuclear fusion. In other words, it is the process of fusing light nuclei together to form a single heavy nucleus, for example, the fusion of hydrogen to form deuterium or helium, lithium and hydrogen to form helium. All these reactions take place in the presence of very high temperatures at about 10^8K, so they are called thermonuclear reactions. All the reactions are exothermic and release huge energies. The energy comes from conversion of mass to energy by the relation E mc^2. The mass of the final single nucleus formed as a result of fusion is always less than that of the individual lighter nuclei. Some of the examples of such reactions along with the energies released by them are detailed below :

$$ {}_1H^1 + {}_2H^2 \rightarrow {}_2He_3 + \gamma + 5.5 \text{ MeV} $$
$$ {}_1H^2 + {}_2H^2 \rightarrow {}_2He^4 + \gamma + 3.2 \text{ MeV} $$
$$ {}_1H^3 + {}_1H^1 \rightarrow {}_2He^4 + \gamma + 19.7 \text{ MeV} $$
$$ {}_1H^3 + {}_1H^2 \rightarrow {}_2He^4 + {}_0n^1 + 17.6 \text{ MeV} $$
$$ {}_2Li^6 + {}_1H^2 \rightarrow {}_2He^4 + {}_2He^4 + 22.4 \text{ MeV} $$
$$ {}_3Li^7 + {}_1H^1 \rightarrow {}_2He^4 + {}_2He^4 + 17.3 \text{ MeV} $$

It may appear that it is probably easier to bring about fusion process than that of fission. But fusion reactions not only require highest kind of skill, but huge temperature conditions for their initiation and thus require suitable triggering devices.

Nuclear Reactors. A device to produce the nuclear energy in a controlled way to be used for peaceful purposes. It works on the same principle as the atomic bomb but under controlled conditions. All the neutrons produced are not allowed to carry out the chain reactions.

In India, the first nuclear reactor was put into operation at Trombay in the year 1956 with the help of Canadian

Government. In order to have an effective control on the fission process, cadmium rods are suitably inserted in the fissionable material. Cadmium has the property of absorbing neutrons of all energies—cadmium rods are called arrestors.

Necleus. Refers to the compact positively charged centre of an atom made up of one or more nucleons (protons and neutrons) around which is a cloud of electrons. The density of nuclei has been about 10^{15} kg m^{-3}. The number of protons in the nucleus defines the element, being its proton number (or atomic number). The nucleon number, or atomic mass number has been the sum of the protons and neutrons. The simplest nucleus is that of a hydrogen atom, ^{1}H, being simply one proton (mass 1.67×10^{-27} kg). The most massive naturally occurring nucleus is ^{238}U of 92 protons and 146 protons (mass 4×10^{-25} kg, radius $9\ 54 \times 10^{-15}$ m). Only certain combinations of protons and neutrons form stable nuclei. Others undergo spontaneous decay.

O

Occlusion

1. Refers to the process in which small amounts of one substance are trapped in the crystals of another; for example, pockets of liquid occluded crystallization from a solution.
2. Refers to the absorption of a gas by a solid; for example, the occlusion of hydrogen by palladium.

Ochre. Earthy pigments yellow to brownish red in colour. The yellow varieties occur as the mineral limonite mixed with clay and certain 15– 65% Fe_2O_3. The red varieties contain haematite. Siennas are similar. Synthetic yellow ochres are hydrated Fe_2O_3, reds are obtained by heating the yellow ochres.

Octahedral Co-ordination. Co-ordination of six ligands which are disposed at the corners of an octahedron. SF_6 has octohedral co-ordination about sulphur. An octahedron has eight triangular faces.

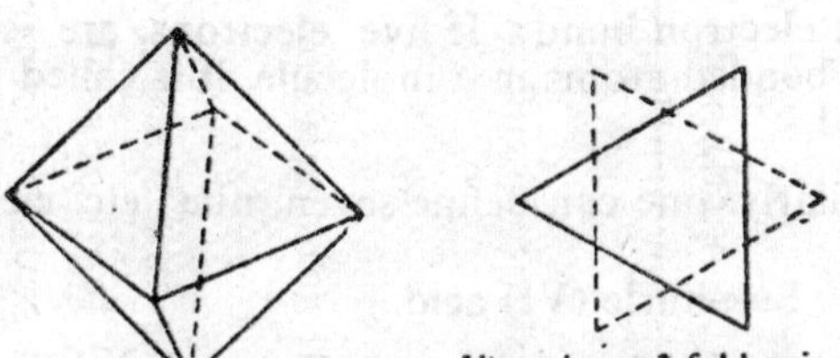

Octahydrate. A crystalline hydrate which is having eight molecules of water per molecule of compound.

Octavalent. Having a valency of eight.

Octave. See law of octaves.

Octaves, Law of. Newlands (1863-1864) found that when the elements were arranged in order of increasing atomic weights, the first element was similar to the eighth, the second to the ninth, the third to the tenth and so on. He termed this relationship the 'law of octaves'. See periodic law. The law is a natural consequence of the periodic table.

Octet. Refers to a stable shell of eight electrons in an atom. The completion of the octet gives rise to particular stability and this forms the basis of the Lewis octet theory, thus :

(1) the noble gases are having complete octets and are chemically inert.

(2) the bonding in small covalent molecules is frequently achieved by the central atom completing its octet by sharing electrons with surrounding atoms, *e.g.* CH_4,H_2O,

(3) the ions formed by electropositive and electronegative elements are generally those with a complete octet, *e.g.* Na^+,Ca^{2+},O^{2-},Cl^-.

Odd Electron Bonds. In 1916, W.F. Luder postulated that there are a number of stable molecules in which the bonds are formed by sharing of an odd number of electrons (*i.e.*, one, three, five, etc) between the two bonded atoms. The bonds of this type are called odd electrons bonds.

Odd electron bonds are of following types :

(a) One electron bond : If a bond is formed by sharing of one electron between the two bonded atoms, it is known as a one electron bond.

(b) Three electron bond : When a bond is formed by sharing of three electrons between the two bonded atoms, it is termed as a three electron bond.

(c) Five electron bond : If five electrons are shared between two bonded atoms in a molecule, it is called a five electron bond.

Similarly, one can define seven, nine, etc. electron bonds.

Oil of Vitriol. Sulphuric (VI) acid.

Oil Sand (tar **sand : bituminous sand**). A sandstone or porous carbonate rock which gets impregnated with hydrocarbons. The largest deposit of oil sand is found in Alberta, Canada (the Athabasca tar sands); there have been also deposits in the Orinoco Basin of Venezuela, the USSR, USA, Madagascar, Albania, Trinidad, and Romania.

Olefine Complexes. Co-ordination compounds in which an olefine is bonded to a metal by overlap of the π-bonding orbitals of the olefin with a suitable orbital of the metal and by back-bonding from a suitable orbital of the metal into a π-anti-bonding orbital of the olefin. The olefin (or acetylene) is thus bonded sideways on to the metal. Olefin complexes are formed by many transition and non-transition metals and are intermediates during many reactions of olefins in the presence of metals. E.g.

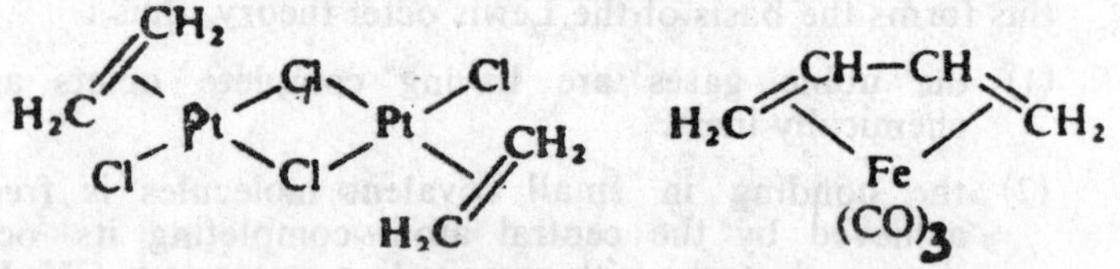

Oleum (**pyrosulphuric acid $H_2S_2O_7$**). A colourless fuming liquid which is obtained dissolving sulphur (VI) oxide in concentrated sulphuric acid. It gives sulphuric acid on diluting with water.

Olivine. Refers to an important group of rock-forming silicate minerals which are crystallizing in the orthorhombic system. Olivine conforms to the general formula $(MgFe)_2SiO_4$ and comprises a complete series from pure magnesium silicate (forsterite. Mg_2SiO_4) to pure iron silicate (fayalite, Fe_2SiO_4). It has been, brown-green, or yellow-green in colour.

Onium Compounds. Groups of compounds containing cations of the type R_xA^+ which are analogous to ammonium compounds. Ions include phosphonium R_4S^+, oxonium R_3O^+, sulphonium R_3S^+, iodonium R_2T^+.

Onium Ion. An ion which is formed by adding a proton to a neutral molecule, *e.g.* the hydroxonium ion (H_3O^+) or the ammonium ion (NH_4^+).

Opal. Refers to a hydrous amorphous form of silica. Many varieties of opal occur, some being prized as gemstones Common opal has been generally milk white but the presence of impurities may colour it yellow, green, or red. Precious opals, which are used as gemstones, display the property of opalescence—a characteristic internal play of colours resulting from the interference of light rays within the stone. Black opal is having a black background against which the colours get displayed. Geysenite has been a variety deposited by geysers or hot springs. Another variety, diatomite, has been made up of the skeletons of diatoms.

Open-hearth Process Refers to a method of making steel by heating pig iron in an open-hearth furnace by burning a mixture of gas and air. Oxygen gets injected through pipes to oxidize impurities and burn off carbon, and lime has been used to form a slag.

Open Shell Compound. A species for which the total number of valency electrons does not correspond to a noble gas configuration or the octet rule. E.g BF_3 (6 outer electrons) SF_6 (12 outer electrons). In the transition series applied to compounds in which there is a partially filled d shell.

Orbit. The path of an electron as it travels round the nucleus of an atom.

Orbital. According to wave-mechanics, an electron is treated as a three dimensional wave system which is represented by a wave function ψ. The Magnitude of ψ will give the amplitudes of the wave system at various points surrounding the nucleus. The volume containing all points at which ψ has an appreciable magnitude is called the orbital of an electron. The orbital of an electron in a free atom is described as an atomic orbital.

There is another way of explaining orbitals. There is probability of finding the electron at all finite distances from the nucleus. We draw a surface that connects points of equal probability and encloses a volume in which there is a high probability. This region in space around the nucleus where there is high probability of finding the electron, is termed as an orbital.

An orbital is the most probable space in which the electron spends most of its time while in constant motion. In simple words, it is the spatial description of the motion of an electron corresponding to a particular energy level. The energy of electron in an atomic orbital is always the same.

Interpretation of Atomic Orbital

(i) Consider the electron as a charge cloud. The density of the cloud at any instant is proportional to ψ^2 The density of the charged cloud decreases with increasing distance. It means that the total density of the charge must be equal to the charge of a single electron. But at different distances from the nucleus the effective charge will have only fractional values (0.8, 0.4, 0.2, 0.1) of the total electronic charge. Hence once again it is possible to draw a three-dimensional region around the nucleus where almost all the electron charge cloud is found. This region of density of charge cloud is known as atomic orbital.

(ii) Another interpretation is to represent the electron by means of dots. The density of these dots at any instant is proportional of ψ^2. This density decreases with increasing distance from the nucleus Again, one can draw three dimensional region around the nucleus to contain all the dots. This region of density of dots is known as atomic orbital.

(iii) If only atom is considered and is observed as many times as possible, each time the electron would have changed its position with respect to the nucleus. If each small region of space is shaded in proportion, how often the electron is found, again this figure is obtained.

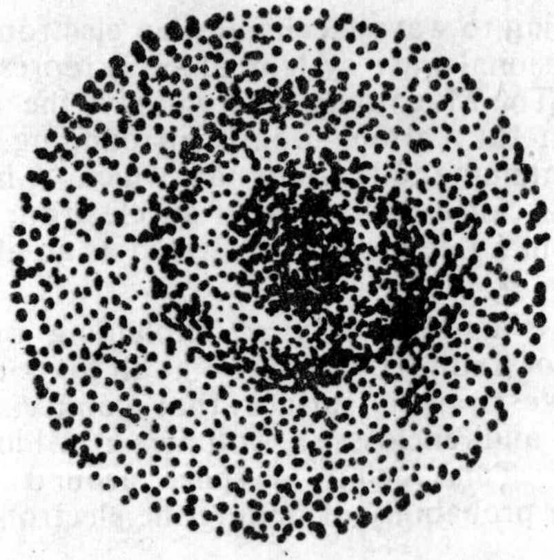

(iv) Suppose an ideal camera is available to photograph the position of an electron at a given instant; the three dimensional photography would enable its position to be specified exactly. The electron would be in continual collisions with photons from the light source and would, therefore, be in constant motion. Consequently, photographs taken at small time intervals would reveal the electron in a series of new positions. Thousands of such photographs would produce a picture resembling a cloud of varying density. Fig. represents a section through such a picture, and is apparent that the cloud will appear most dense where the points are most crowded together, *i.e.*, in the regions where the electron is most probably to be found. It is then possible to construct a boundary surface enclosing the regions of highest electron density. Fig represents such a surface, but it will be seen later the surfaces of other shapes are also of great importance.

Orbital Quantum Number. See atom.

Orbital Splitting. Refers to the raising of the degeneracy of a series of orbitals of equivalent energy, *e.g.* by placing in a crystal field.

Ore. A naturally occurring mineral from which a metal could be extracted, usually on a commercial basis. The metal may be present in the ore as the native metal, but more commonly it is found in a combined form as an oxide, sulphide, sulphate, silicate, etc.

Ore Dressing. See beneficiation.

Orford Process. An old process for Ni extraction involving treatment of mixed sulphides with Na_2SO_4 and coke when NiS produces a lower layer on cooling.

Organometallic Compounds. The term organometallic is usually restricted to those compounds in which a metal atom (including non-metallic elements like boron, silicon, phosphorus, arsenic etc, which are less electronegative than carbon) is bonded directly to a carbon atom of atleast one organic group radical or molecule). Organometallic compounds would include various hydrocarbon derivatives, carbonyls, carbides, certain chelate complexes, but donot include carbonates of metals complexes of the metals with organic amines and metallic salts of organic acids.

Let us illustrate the above by taking two compounds namely $B(CH_3)_3$ and $B(OCH_3)_3$. Trimethyl boron, $B(CH_3)_3$ is called as organometallic compound whereas trimethyl orthoborate, $B(OCH_3)_3$ is not, because there is a metal to carbon bond in the first compound and no such bond in second compound Examples of a few organometallic compounds are $(\pi\ C_5H_5)_2Fe$, $(C_2H_5)_2Zn$, $(CH_3)_4Si$, $(C_5H_5)_4Ce$, $(C_9H_7)_4Ce$, etc.

The transition elements form organometallic compounds with π bonding ligands, *i e.*, the organo-metallic compounds of transition elements must have the π-bonding ligands as the binary organometallic compounds of transition elements are not stable at room temperature. Thus, the presence of π-bonding ligand in organometallic compounds transition elements increases the stability of the σ-bond and thus makes the compound stable.

Orgel Diagrams. Simple graphs showing the relation between the energies of various electronic states and the crystal field splitting.

Orpiment. Refers to a natural yellow mineral form of arsenic(III) sulphide, As_2S_3. The name has been also used for the synthetic compound, which finds use as a pigment.

Ortho

1. Designating the form of a diatomic molecule in which both nuclei have the same spin direction; *e.g.* orthohydrogen, orthodeuterium.
2. Certain acids, considered as formed from an anhydride and water, were named ortho acids to distinguish them from the less hydrated meta acids. For example, H_4SiO_4 (from SiO_2+2H_2O) has been orthosilicic acid; H_2SIO_3 (SiO_2+H_2O) is metasilicic acid.

Ortho-and Para-hydrogen. The hydrogen atom consists of a proton around which one electron revolves. Both the proton and the electron spin about an axis like a top. When two hydrogen atoms combine to form a molecule, the spins of the electrons are always in the opposite directions as otherwise the molecule would not be stable according to Pauli exclusion principle. The spins of the protons may, however, be either in the same direction or in the opposite directions. Corresponding to these differences in the spins of the protons there are two allotropes of hydrogen. When the spins are in the same direction, we get ortho-hydrogen and when they are in opposite directions, we get para-hydrogen. The two forms are similar in chemical properties but differ in thermal conductives and specific heats.

Ortho-Hydrogen Para-Hydrogen

Orthoboric Acid. See boric acid.

Orthoclase. See feldspars.

Orthohydrogen. See hydrogen.

Orthophosphoric Acid. See phosphoric(V) acid.

Orthophosphorous Acid. See phosphonic acid.

Orthoplumbate. See plumbate.

Orthosilicate. See silicate.

Orthostannate. See stannate.

Osmic Acid. The name given to osmium tetroxide and its solution in water. This solution contains free OsO_4. In alkaline solution $[OsO_4(OH)_2]^{2-}$ ions are present

Osmiridium. A hard white naturally occurring alloy consisting principally of osmium (17-48%) and iridium (49%). It also contains small quantities of platinum, rhodium, and ruthenium. It is used for making small items subject to wear, *e.g.* electrical contacts or the tips of pen nibs.

Osmium, Os. At.no. 76, at.wt. 190.2, m.p. 3045°C, b.p. 5027°C, d 22.6. Osmium is a platinum group metal. It occurs as a sulphide or as osmiridium (extraction see Ru). The metal is hcp and has very similar properties to Ru. Oxygen at high temperatures gives OsO_4. Os is used for hardening purposes in alloys with other plantinum metals.

Osmium(IV) Oxide (Osmium Tetroxide) A yellow solid, OsO_4, which is prepared by heating osmium in air. It is used as an oxidizing agent in organic chemistry, as a catalyst, and as a fixative in electron microscopy.

Oxidant. An oxidizing agent. In rocket fuels, the oxidant has been the compound that provides the oxygen for combustion (*e.g.* liquid oxygen or hydrogen peroxide).

Oxidation. Any process by which the proportion of the electronegative constituent in a compound is increased, *e.g.* $Cu_2O \rightarrow CuO$. In oxidation electrons are removed from the species oxidized, *e.g.*

$$Cu^+ \rightarrow Cu^{2+} + e$$

and the oxidation state is increased. In organic chemistry particularly the most obvious effect is to increase the proportion of oxygen, *e.g.* $C_2H_5OH \rightarrow CH_3CO_2H$. For specific selective oxidizing agents see *e.g.* Oppenauer oxidation, selenium dioxide, chromium trioxide.

Oxidation Number (Oxidation State). See oxidation—reduction.

Oxidation—Reduction (Redox). Originally, oxidation was simply considered as a chemical reaction with oxygen. The reverse process—loss of oxygen—was termed as reduction. Reaction with hydrogen also came to be regarded as reduction. Later, a more general idea of oxidation and reduction got developed in which oxidation was loss of electrons and reduction was gain of electrons. This wider definition covered the original one. For example, in the reaction

$$4Na(s) + O_2(g) \rightarrow 2Na_2O(s)$$

the sodium atoms lose electrons to give Na^+ ions and are oxidized. At the same time, the oxygen atoms gain electrons and get reduced. These definitions of oxidation and reduction are also applicable to reactions that do not involve oxygen. For instance in

$$2Na(s) + Cl_2(g) \rightarrow 2NaCl(s)$$

the sodium gets oxidized and the chlorine reduced. Oxidation and reduction also occurs at the electrodes in cells.

This definition of oxidation and reduction is applicable only to reactions in which electron transfer occurs—*i.e.* to reactions involving ions. It could be extended to reactions between covalent compounds by using the concept of oxidation number (or state). This is a measure of the electron control that an atom has in a compound compared to the atom in the pure element. An oxidation number is having two parts :

(1) Its sign, which indicates whether the control gets increased (negative) or decreased (positive).

(2) Its value, which gives the number of electrons over which control has changed.

The change of electron control may be complete (in ionic compounds) or partial (in covalent compounds). For example, in SO_2 the sulphur is having an oxidation number +4, having gained partial control over 4 electrons compared to sulphur atoms in pure sulphur. The oxygen is having an oxidation number —2, each oxygen having lost partial control over 2 electrons compared to oxygen atoms in gaseous oxygen. Oxidation is a reaction which involves an increase in oxidation number and reduction involves a decrease. Thus in

$$2H_3 + O_2 \rightarrow 2H_2O$$

the hydrogen in water is +1 and the oxygen −2. The hydrogen is oxidized and the oxygen is reduced.

The oxidation number finds use in naming inorganic compounds. Thus in H_2SO_4, sulphuric (VI) acid, fthe sulphur is

having and oxidation number of + 6. Compounds that tend to undergo reduction readily are oxidizing agents; those that under go oxidation are reducing agents.

Oxidation State. The difference (+ve or —ve) between the number of electrons associated with an atom in a compound as compared with an atom of the element. In ions corresponds to the ionic charge, *e.g.* NaC([Na + 1), Cl (—1); in covalent compounds to the formal change assuming an ionic structure taking account of electronegativity to determine charge, *e.g.* CCl_4 is C^{4+} $4Cl^-$ and thus C (+4). Co-ordinate bonds do not affect the oxidation state, *e.g.* $[Co(NH_3)_6]^{3+}$ contains Co (+3) and $Fe(CO)_5$ contains Fe (0). Elements are in the zero oxidation state. The oxidation state, Roman numerals, is used in systematic nomenclature, *e.g.* MnO_4^- is manganate(VII).

Oxides. Binary compounds which are formed between elements and oxygen. Oxides of nonmetals are covalent compounds which are having simple molecules (*e.g* CO, CO_2, SO_2) or giant molecular lattices (*e.g.* SiO_2). They have been typically acidic or neutral. Oxides of metals are ionic, containing the O^{2-} ion. They have been generally basic or amphoteric. Various other types of ionic oxide exist.

Oxidizing Acid. An acid which can act as a strong oxidizing agent as well as an acid. Nitric acid is a common example. It has been able to attack metals, such as copper, that have been below hydrogen in the electromotive series, by oxidizing the metal :

$$2HNO_3 + Cu \rightarrow CuO + H_2O + 2NO_2$$

This has been followed by reaction between the acid and the oxide :

$$2HNO_3 + CuO \rightarrow Cu(NO_3)_2 + H_2O$$

Oxidizing Agent (Oxidant). Refers to a substance that brings about oxidation in other substances. It achieves this by being itself reduced. Oxidizing agents are having atoms with high oxidation numbers; that is the atoms have suffered electron loss. In oxidizing other substances these atoms gain electrons.

Oxo. Prefix which indicates the presence of oxygen in a chemical compound.

Oxo Acid. Refers to an acid in which the acidic hydrogen atom(s) have been bound to oxygen atoms. Sulphuric acid is an example: the two acidic hydrogens have been on the —OH groups bound to the sulphur.

Oxygen Symbol O. A colourless odourless gaseous element which belongs to group VI of the periodic table; a.n. 8; r.a.m. 15.9994; d. 1.429 g dm^{-3}; m.p, −214.4°C; b.p. −183°C. It is the most abundant element in the earth's crust (49.2% by weight) and occurs in the atmosphere (28% by volume). Atmospheric oxygen has been of vital importance for all organisms that carry out aerobic respiration. For industrial purposes it may be prepared by fractional distillation of liquid air. It is used in metallurgical process, in high temperature flames (*e g.* for welding), and in breathing apparatus. The common form has been diatomic (dioxygen, O_2); there is also a reactive allotrope ozone (O_3). Chemically, oxygen reacts with most other element was discovered by Priestley in 1774.

Ozone (trioxygen, O_3). Refers to an allotrope of oxygen which is prepared by passing oxygen through a silent electric discharge. Ozone is unstable and decomposes to oxygen on warming. It is found in the upper layers of the atmosphere, where it screens the Earth from harmful short-wave ultraviolet radiation.

Ozone acts a strong oxidizing agent. Its concentration could be determined by reacting it with iodide ions and titrating the iodine formed with standard sodium thiosulphate (VI) It gives rise to the formation of ozonides with alkenes.

Ozone Layer (Ozonosphere). Refers to a layer of the earth's atmosphere in which most of the atmosphere's ozone is concentrated. It is found 15—50 km above the earth's surface and is virtually synonymous with the stratosphere. In this layer most of the sun's ultra-violet radiation gets absorbed by the ozone molecules, causing a rise in the temperature of the stratosphere and preventing vertical mixing so that the stratosphere forms stable layer. By absorbing most of the solar ultraviolet radiation the ozone layer protects living organisms on earth. The fact that the ozone layer has been thinnest at the equator has been believed to account for the high equatorial incidence of the skin cancer because of exposure to unabsorbed solar ultraviolet radiation.

P

Packing Fraction. The mass of an atom is not the simple sum of the masses of the elementary particles making up the atom, since some of the mass is converted to energy [$E=mc^2$) which is used to hold the nucleus together. The difference between the actual atomic mass and the mass number (the total mass of the protons and neutrons which constitute the nucleus) is termed the mass defect. The packing fraction is defined as the mass defect divided by the mass number. For the lightest and heaviest elements, the packing fraction is positive; intermediate elements have a negative packing fraction.

Palladium Symbol Pd. A soft white ductile transition element (see also platinum metals); a n. 46; r.a.m. 106.4; r.d. 12.26; m.p. 1552)$\pm$1°C, b.p. 3140$\pm$1°. It is found is in some copper and nickel ores and is used in jewellery and as a catalyst for hydrogenation reactions. Chemically, it does not react with oxygen at normal temperatures. It dissolves slowly in hydrochloric acid. Palladium is capable of occluding 900 times its own volume of hydrogen. It forms few simple salts, most compounds being complexes of palladium (II) with some palladium (IV). It was discovered by Woolaston in 1803.

Palladium Ammines. An extensive range of palladium (II) complexes; generally 4-co-ordinate, containing $Pd-NH_3$ (or amine) bonds.

Palladium Black. A finally divided precipitate obtained by reducing solutions of Pd salts. Used as a hydrogenation catalyst.

Para. Designating the form of diatomic molecule in which both nuclei are having opposite spin directions; *e.g.* parahydrogen, paradeuterium.

Parahydrogen. See hydrogen.

Paramagnetism. See magnetism.

Partial Ionic Character. The electrons of a covalent bond between atoms or groups with different electronegativities will be polarized towards the more electronegative constituent; the magnitude of this effect may be measured by the ionic character of the bond. If the effect is small the bond may be referred to simply as a polar bond and is adequately treated using dipole moments;

as the effect grows the theoretical treatment requires other contributions to ionic character. Examples of partial ionic character (from nuclear quadrupole resonance) have been H-I 21%; H—Cl, 40%; Li-Br, 90%, compare NaCl, 100%.

Parting. The process of separating Au from Ag. As the chemical properties of the elements are very different this is an easy matter and methods used include dissolving the Ag in conc. H_2SO_4 or HNO_3 or the preferential reaction of Ag with Cl_2 to give AgCl.

Paschen Series. The term used for a series of lines in the infrared spectrum emitted by excited hydrogen atoms. The lines correspond to the atomic electrons falling into the third lowest energy level and emitting energy as radiation. The wavelength (λ) of the radiation in the Paschen series has been given by $1/\lambda = R(1/3^2 - 1/n^2)$ where n is an integer and R is the Rydberg constant.

Passive. Describing a solid that has reacted with another substance to form a protective layer, so that further reaction stops. The solid has been considered to have been 'rendered passive'. For example, aluminium spontaneously with oxygen in air to form a thin layer of aluminium oxide, which prevents further oxidation. Similarly, pure iron is known to form a protective oxide layer with concentrated nitric acid and is not dissolved further.

Passivity. Many metals (*e.g.*, Fe, Co, Ni, Cr) form a surface layer of oxide in contact with strong oxidizing agents (*e.g*, conc. HNO_3, chromic acid, H_2O_2). Further solution does not occur and many properties are altered. Other metals, *e.g.* Al and Mg normally have a tough surface film of oxide which prevents further attack so that although electropositive they do not dissolve in water. Passivity can also be conferred by fluorine which forms a layer of fluoride on, *e.g.*, Cu or Ni.

Patina. The decorative and corrosion-resistant surface given to bronze, iron, etc., as an oxide film formed on warming the metal.

Pauli Exclusion Principle. In any atom no two electrons can have all four quantum numbers the same.

Pauling Slater Theory. Pauling and Slater developed Heitler-London theory to account for the directional characteristics in covalent bonds. To take the geometrical shapes of orbitals into account and to relate structural chemistry to the covalent bond, the Pauling-Slater's theory suggests that :

(a) The atoms which unite to form a molecule completely retain their identities in the resulting molecule.

(b) A covalent molecule is formed due to the overlapping of atomic orbitals. If there are two atoms, each having unpaired electron, the atomic orbitals having unpaired electrons overlap resulting in the formation of a covalent bond between these two atoms in which the spins of two electrons get mutually neutralised. If these atomic orbitals have parallel spins, repulsion will take place and no molecule will, therefore, be formed.

(c) If the atomic orbitals are having more than one unpaired electron, multiple bond formation is possible. In nitrogen molecule, for example, three bonds are possible because N atom has three unpaired electrons.

(d) The number of covalent bonds formed by an atom is equal to the number of unpaired electrons it has.

(e) The strength of the covalent bond is proportional to the extent of the overlapping of the two atomic orbitals. The greater the overlap between the atomic orbitals, the greater is the strength of the resulting covalent bond. The extent of overlap of two atomic orbitals of wave function ψ_A and ψ_B is represented by overlap integral, S, which is as follows :

$$S = \int \psi_A \psi_B \, dr$$

From the above relation, it is event that :

(i) If S is zero, there is no overlap, *i.e.*, there is neither attraction nor repulsion between the combining atoms. Thus, no bond formation is possible.

(ii) If S is positive, there is an increase in the electron density between two nuclei which results in an increase in attraction between the two combining atoms, forming a covalent bond.

(iii) If S is negative, there occurs a decrease in electron density between the nuclei of two combining atoms. Thus, the repulsion between them is increased and hence no bond formation is possible.

(f) Overlapping takes place between orbitals of only those electrons, which take part in bond formation and not between other orbitals of the atoms.

(g) Between two orbitals of the same energy content, the overlapping axially will form a stronger bond.

(h) A spherically symmetrical orbital, *i.e.*, s-orbital does not show any directional preference whereas non-spherical orbitals, *viz.*, p.d f etc., tend to form a bond in the direction of maximum electron density within the orbital, *i.e*, along their axes.

For example :

(i) As the charge density of s-orbital is spherically symmetrical, the two s-orbital will overlap to the same extent in all directions.

(ii) The three p-orbitals lie on mutually perpendicular axes having nucleus at the centre. Whenever p-orbitals overlap, they tend to form a stronger bond in the direction of their maximum electron density.

(iii) When a s-orbital overlaps p-orbital, the bond is formed but it is weaker than p-p bond but stronger than s-s bond.

P-block Elements. The block of elements in the periodic table which are including the main groups III (B to Tl), IV (C to Pb), V (N to Bi), VI (O to Po), VII (F to At) and 0 (He to Rn). The outer electronic configurations of these elements all have the form $ns^2\ np^x$ where $x=1$ to 6. Members at the top and on the right of the p-block are nonmetals (C,N,P,O,F,S,Cl,Br,I, At). Those on the left and at the bottom have been metals (Al, Ga, In, Tl, Sn, Pb, Sb, Bi, Po). Between the two, from the top left to bottom right, lie an ill-defined group of metalloid elements (B, Si, Ge, As, Te).

Peracids. A peroxy acid or peracid is one which contains the peroxide linkage (—O—O) and either it is formed by the action of H_2O_2 on the normal acid or it gives H_2O_2 on treatment with dil. H_2SO_4.

Unfortunately, "per" names for a few acids and salts which are not derivatives of H_2O_2, have become so familiar that their names could not be changed. For this reason, the acids and salts derived from H_2O_2 have been described by the prefix 'peroxy' and not by per-acids.

Peacock Ore. See bornite.

Pearl Ash. See potassium carbonate.

Pearlite. The ferrite (c. 0.06% C in α-Fe)-cementite (FeC_3) eutectic formed on cooling austenite.

Peat. A naturally-occurring, brown, fibrous mass of partially decayed plant material. The composition of peat depends on age and type; bog peat being largely derived from mosses and forest peat from trees. The moisture content is initially about 95% decreasing to about 25% on air-drying. On a dry ash-free basis a typical peat contains about 60% C, 6% H and 32% O, the remainder being N and S. The calorific values of peat on a dry basis range from about 16-20 MJ/kg. Air-dried peat may be used for firing power stations or it can be converted to peat charcoal by low temperature carbonization. Producer gas has been made from air-dried peat using conventional producers.

Pentahydrate. A crystalline hydrate having five molecules of water per molecule of compound.

Pentavalent (quinquevalent). Having a valence of five.

Pentlandite. A mineral consisting of a mixed iron-nickel sulphide, $(Fe, Ni)_9S_8$, crystallizing in the cubic system; the chief ore of nickel. It is yellowish-bronze in colour with a metallic lustre. The chief occurreance of the mineral has been at Sudbury in Ontario, Canada.

Per. A prefix which indicates that a chemical compound is having an excess of an element, *e.g.*, a peroxide.

Per-acid. Oxy-acids in which $[O{-}O]^{2-}$ or $[O{-}OH]^-$ groups have replaced oxide in oxy-anions (compounds containg the superoxide ion $[O_2]^-$ are formally not per-acids but the distinction is often difficult to make). The O—O group may be bonded end on [M—O—O] or symmetrically $\left[M \leftarrow \begin{matrix} O \\ \vert \\ O \end{matrix}\right]$. Prepared by use of H_2O_2 or by electrolytic oxidation. Persulphuric acids, perborates, are of importance. Permanganates, per-chlorates and periodates are not salts of per-acids. Organic per-acids are prepared similarly. The more stable per-acid such as perbenzoic acid, monoperphtalic acid convert olefines to epoxides and by hydrolysis to trans-glycols. Perethanoic and permethanoic acids give glycol monoesters with olefines.

Perborates, Perborax. Sodium perborate, $NaBO_2.3H_2O.H_2O_2$ (borax plus Na_2O_2 or H_2O_2 plus NaOH) contains H_2O_2 of crystallization. It is quite stable and is used as a bleaching agent and antiseptic.

Perbromates, M $[BrO_4]$. See bromates (VII).

Perbromic Acid, $HBrO_4$. See bromates.

Percarbonates. It is doubtful whether true percarbonates exist but sodium carbonate with varying amounts of H_2O_2 of crystallization are stable as solids and $2Na_2CO_3, 3H_2O_2$ is used in laundering and for cleaning dentures.

Perchlorate. See chlorates.

Perchloric Acid. See chloric (VII) acid.

Perdisulphuric acid. See peroxosulphuric (VI) acid.

Perfect Complexes. Those coordination compounds which retain their complex character in solid as well as in solution state. A numberless complex compounds such as $K_4[Fe(CN)_6]$,$[Co(NH_3)_6]Cl_2$,$[Cu(NH_3)_4]$,$SO_4]$,$K_3[Fe(CN)_5]$, etc, fall under this class of coordination compounds.

Period. A series of elements in the periodic table. a period being elements from an alkali metal to a noble gas. The periods are (1) the very short period H, He, (2) two short periods Li—Ne and Na—Ar, (3) two long periods containing transition elements K—Kr and Rb—Xe, (4) a very long period containing the rare-earth elements Cs—Rn, (5) an unfinished period Fr—.

Periodates. See iodates(VII).

Periodic Acids. See iodates(VII).

Periodic Law. Mendeleeff (1869), independently of Newlands (octaves law), but forward the law that 'the properties of the elements are periodic functions of the atomic weights', *i.e.* by arranging the elements in order of increasing atomic weights, elements having similar properties occur at fixed intervals. This principle, with the modification that atomic numbers are used instead of atomic weights, forms the basis of the periodic table.

Periodic Table. A table of the elements arranged in order of atomic numbers, the arrangement being such as to emphasize the chemical relationships between the elements and to emphasize the relationship between elements with similar configurations.

Permalloys. A group of alloys of high magnetic permeability consisting of iron and nickel (usually 40—80%) often with small amounts of other elements (*e.g.* 3—5% molybdenum. copper, chromium, or tungsten). They have been used in thin foils in electronic transformers, for magnetic shielding, and in computer memories.

Permanganate. See manganate (VII).

Permonosulphuric Acid, Caro's Acid, $HOS_6(O)_2OOH$. Powerful oxidizing agent (H_2O_2 on conc. H_2SO_4 or HSO_3Cl; hydrolysis of $H_2S_2O_6$ with H_2SO_4). The pure acid can be obtained crystalline but the salts are unstable.

Permutit (Tradename). A substance which is used to remove unwanted chemicals that have dissolved in water.

It is a zeolite consisting of a complex chemical compound, sodium aluminium silicate (Na^+Alsilicate). When hard water is allowed to pass over this material, calcium and magnesium ions exchange with sodium ions in the Permutit. This has been a good example of ion exchange. Once all the available sodium ions have been used up, the Permutit can be regenerated by washing it with a saturated solution of sodium chloride. The excess of sodium ions now exchange with the calcium and magnesium ions in the Permutit.

Permutite Process. The process of softening hard water by replacing Ca^{2+} and Mg^{2+} by Na^+ by passage through a zeolite. The zeolite is regenerated with NaCl solution.

Perovskite. Mineral $CaTiO_3$.

Perovskite Structure. A structure adopted by many ABX_3 compounds in which A and B are positive ions and X is F^- or O_2^-. The structure has A at the corners of a cube, B at the body centre and X (A and X are ccp) at the face centres. A has 12X neighbours, B has 6X (octahedral) neighbours). Examples are $KNiF_3$, $BaTiO_3$, $KNbO_3$. Many perovskites show distortions from the idealized cubic structure.

Peroxide. A peroxide may be regarded as a salt of hydrogen peroxide which contains at least one pair of oxygen atoms bonded by a single covalent bond $(O-O)^{2-}$, and which in addition to its oxidising and reducing properties, behaves as an extremely weak acid.

They yield H_2O_2 with dilute acids. BaO_2 and Na_2O_2 are typical peroxides. PbO_2, NO_2, MnO_2 are sometimes erroneously termed peroxides.

Peroxodisulphuric Acid. See peroxosulphuric (VI) acid.

Peroxosulphuric (VI) Acid (Caro's acid H_2SO_5) A white crystalline solid which is obtained by reacting hydrogen peroxide with sulphuric acid. It is a powerful oxidizing agent.

Peroxychromium Compounds. H_2O_2 reduces $[CrO_4]^{2-}$ to Cr^{3+} but gives an intermediate green $[Cr_2(O_2)]^{4+}$ and $[Cr_3(O_2)_2]^{5+}$ and blue $CrO(O_2)_2$ which can be extracted into ether (test for Cr or after distillation of CrO_2Cl_2 for C_2^-) and stabilized as pyridine. CrO_3. Blue-violet salts $[CrO(O_2)_2OH]^-$ which are very explosive from neutral solution, red-brown $[CrO_8]^3$, brown $[NH_3)_3Cr(O_2)_2]$ and $[CrO_4(CN)_3]^{3-}$ (from alkaline solution) are also known and are relatively stable.

Persulphuric Acid, Perdisulphuric Acid, $HOS(O)_2OOS(O)_2OH$. Dibasic acid formed as salts by electrolysis of sulphates at low temperatures and high current density. The acid and persulphates are strong oxidizing agents ($E[S_2O_8]^{2-}$ to SO_4^{2-} +2.01 volts in acid) but the reactions are often slow.

Perturbation Method. This perturbation principle is another way for approximate solution of the problem. This method is based on the principle that the actual problem can be treated as a slight modification or perturbation of another problem capable of exact solution. If $H\psi=E\psi$ represents the actual Schrodinger equation to be solved then it is assumed that an exact solution is available for $H_0\psi_0=E_0\psi_0$, where H_0 differs only slightly from the Hamiltonian H for the actual problem.

Peta-Symbol : P. A prefix denoting 10^{15}.

Pewter. Refers to an alloy of lead and tin. It usually contains 63% tin; pewter tankards and food containers should have less than 35% of lead so that the lead remains in solid solution with the tin in the presence of weak acids in the food and drink. Copper has been sometimes added to increase ductility and antimony is added if a hard alloy is required.

Pharaoh's Serpents' Eggs. Pillules of mercury(II) thiocyanate, $Hg(CNS)_2$, with gum as a binding agent. When ignited they burn freely and evolve a very voluminous, snake-like ash.

Phlogiston Theory. A former theory of combustion (disproved in the eighteenth century by Lavoisier). Flammable compounds were supposed to contain a substance (phlogiston), which was released when they burned, leaving ash.

Phosgene. Carbonyl chloride.

Phosphates. Salts which are based formally on phosphorus (V) oxoacids and in particular salts of phosphoric (V) acid, H_3PO_4. A large number of polymeric phosphates also exist, containing P—O—P bridges. These are obtained by heating the free acid and its salts under a variety of conditions; as well as linear polyphosphates, cyclic polyphosphates and or cross-linked polyphosphates ultra-phosphates are known.

Phosphides. Compounds of the elements with phosphorus. The non-metals form covalent materials (*e.g.* S). The most electropositive elements form ionic derivatives, *e.g.* Na_3P rapidly hydrolysed to PH_3 by water. The transition elements form derivatives containing P_n groups (*e.g.* CdP_2) or hard materials with complex structures.

Phosphine. A colourless highly toxic gas, PH_3; m.p.—133°C; b.p.—87.7C; slightly soluble in water. Phosphine may be obtained by reacting water or dilute acids with calcium phosphide or by reaction between yellow phosphorus and concentrated alkali. Solutions of phosphine are neutral but phosphine does react with some acids to give phosphonium salts having PH_4 ions, analogous to the ammonium ions. Phosphine prepared in the laboratory is usually contaminated with diphosphine and is spontaneously flammable but the pure compound is not so. Phosphine can function as a ligand in binding to transition-metal ions. Dilute gas mixtures of very pure phosphine and the rare gases are used for doping semi-conductors.

Phosphinic Acid (hypophosphorous acid, H_3PO_2). A white crystalline solid. It is a monobasic acid forming the anion $H_2PO_2^-$ in water. The sodium salt, and hence the acid, can be obtained by heating yellow phosphorus with sodium hydroxide solution. The free acid and its salts are powerful reducing agents.

Phosphite. See phosphonic acid.

Phosphomolybdates. Salts of the hetero-phosphomolybdic acids (see molybdates) but particularly of $H_3PMo_{12}O_{40}$. The yellow ammonium salt is used for the estimation of Mo or P.

Phosphonate. See phosphonic acid.

Phosphonic Acid. (phosphorous acid, orthophosphorus acid, H_3PO_3). A colourless deliquescent solid that can be obtained by the action of water on phosphorus(III) oxide or phosphorus(III) chloride. It is a dibasic acid producing the anions H_2PO_3 and HPO_3^{2-} in water. The acid and its salts have been low reducing agents. On warming, phosphonic acid decomposes to phosphine and phosphoric(V) acid.

Phosphonitrilic Derivatives, Phosphazenes, $(PNX_2)_n$. A group of polymers. The simplest compounds, the chlorides, are formed by heating PCl_5 with NH_4Cl. Various polymers are known;

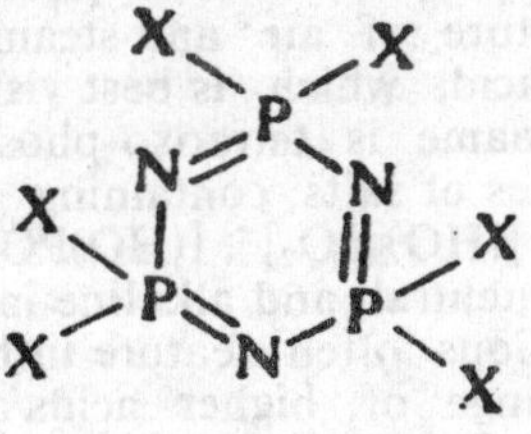

most compounds have cyclic structures as shown but some linear polymers are known. On heating the chlorides to 300°C they are converted to materials possessing some rubber-like properties. The chlorine atoms may be substituted by many other groups. The rings have low hydrolytic stability.

Phosphonium Ion. The ion PH_4^+, or the corresponding organic derivatives of the type R_3PH^+, RPH_3^+. The phosphonium ion PH_4^+ is formally analogous to the ammonium ion NH_4^+ but PH_3 has a much lower proton affinity than NH_3 and reaction of PH_3 with acids is necessary for the production of phosphonium salts.

Phosphorus Sulphides. Yellow solids P_4S_3, P_4S_4, P_4S_5, P_4S_7 and P_4S_{10} formed from red phosphorus and sulphur. The structures are based on a P_4 tetrahedron with some bridging S atoms and, in P_4S_7 and P_4S_{10}, terminal S atoms. P_4S_3 is used in matches; P_4S_{10} in the preparation of organo-sulphur compounds (particularly pesticides and oil additives).

Phosphor. A substance that shows luminescence or phosphorescence.

Phosphor Bronze. An alloy of copper which are having 4% to 10% of tin and 0.05% to 1% of phosphorus as a deoxidizing agent. It has been used particularly for marine purposes and where it gets exposed to heavy wear, as in gear wheels.

Phosphoric (V) Acid (orthophosphoric acid). A white rhombic solid, H_3pO_4; r.d. 1.834; m.p. 42.35°C: loses water at 213°C; very soluble in water and soluble in ethanol. Phosphoric(V) acid has been very deliquescent and has been generally supplied as a concentrated aqueous solution. It is the most commercially important derivative of phosphorus, accounting for over 90% of the phosphate rock mined. It may be manufactured by two methods; the wet process, in which the product contains some of the impurities originally present in the rock and applications are largely in the fertilizer industry, and the thermal process, which produces a much purer product suitable for the foodstuffs and detergent industries. In the wet process the phosphate rock, $Ca_3(PO_4)_2$, has been treated with sulphuric acid and the calcium sulphate removed either as gypsum or the hermihydrate. In the thermal process, molten phosphorus is sprayed and burned in a mixture of air and steam. Phosphoric(V) acid is a weak tribasic acid, which is best visualized as $(HO)_3PO$. Its full systematic name is tetraoxo-phosphoric(V) acid. It gives rise to three series of salts containing phosphate(V) ions based on the anions $[(HO)_2PO_2]^-$, $[(HO)PO_3]^{2-}$ and PO_4^{3-}. These salts are acidic, neutral, and alkaline in character respectively and phosphate ions often feature in buffer systems. There is also a wide range of higher acids and acid anions in which there is some P—O—P chain formation. The simplest of these is pyrophosphoric acid (technically heptaoxodiphosphoric(V) acid), $H_4P_2O_7$, produced by heating phosphoric (V) acid (solid) and phosphorus(III) chloride oxide. Metaphosphoric acid is a glassy polymeric solid $(HPO)_3$).

Phosphorus. Symbol P. A nonmetallic element which belongs to group V of the periodic table; a.m. 15; r.a.m. 30.9738; r.d. 1.82 (white), 2.20 (red); m.p. 44.1°C (α-white) b.p. 280°C (α-white). It is found in various phosphate rocks, from which it is extracted by heating with carbon (coke) and silicon (IV) oxide in an electric furnace (1500°C). Calcium silicate and carbon monoxide are also produced. Phosphorus is having a number of allotropic forms. The α-white form consists of P_4 tetrahedra (there is also β-white form stable below-77°C). If α-white phosphorus gets dissolved in lead and heated at 500°C a violet form is obtained. Red phosphorus, which is a combination of violet and white phosphorus, is prepared by heating α-white phosphorus at 250°C with air excluded. There is also a black allotrope, which has a graphite-like structure, made by heating white phosphorus at 300°C with a mercury catalyst. The element is highly reactive. It is known to form metal phosphides and covalently bonded phosphorus(III) and phosphorus(V) compounds. Phosphorus is an essential element for living organisms. It was discovered by Brandt in 1669.

Phosphorous Acid. See phosphonic acid.

Phosphorus(III) Bromide (phosphorus tribromide, PBr_3). A colourless liquid which is prepared by reacting phosphorus with bromine. It is readily hydrolysed by water to phosphonic acid any hydrogen bromide. Phosphorus(III) bromide has been important in organic chemistry, being used to replace a hydroxyl group with a bromine atom.

Phosphorus(V) Bromide (phosphorus pentabromide, PBr_5). A yellow crystalline solid which sublimes easily. It can be prepared by the reaction of bromine and phosphorus(III) bromide. Phosphorus(V) bromide readily gets hydrolysed by water to phosphoric(V) acid and hydrogen bromide. Its main use has been in organic chemistry to replace a hydroxyl group with a bromine atom.

Phosphorus(III) Chloride (Phosphorus Trichloride). Refers to a colourless ruming liquid, PCl_3; r.d. 1.57; m.p. -112°C; b.p. 75.5°C. It is soluble in ether and in carbon tetrachloride but reacts with water and with ethanol. It may be obtained by passing chlorine over excess phosphorus (excess chlorine contaminates the product with phosphorus(V) chloride). The molecule has been pyramidal in the gas phase and possesses weak electronpair donor properties. It gets hydrolysed violently by water to phosphonic acid and hydrogen chloride. Phosphorus(III) chloride has been an important starting point for the synthesis of a variety of inorganic and organic derivatives of phosphorus.

Phosphorus(V) Chloride (phosphorus pentachloride. PCl_5) A white easily sublimed solid which is obtained by the action of chlorine on phosphorus(III) chloride. It gets hydrolysed by water to phosphoric(V) acid and hydrogen chloride. Its main use has been as a chlorinating agent in organic chemistry to replace a hydroxyl group with a chlorine atom.

Phosphorus(III) Chloride Oxide (Phosphorus Oxychloride; Phosphoryl Chloride) A colourless fuming liquid, $POCl_3$; r.d. 1.67; m.p. 2°C; b p. 105.3°C. It may be obtained by the reaction of phosphorus(III) chloride with oxygen or by the reaction of phosphorus(V) oxide with phosphorus(V) chloride. Its reactions have been very similar to those of phosphorus(III) chloride, Hydrolysis with water gives phosphoric(V) acid. Phosphorus(III) chloride oxide is having a distorted tetrahedral shape and can act as a donor towards metal ions, thereby giving rise to a series of complexes.

Phosphorus(III) Oxide (phosphorus trioxide. P_2O_3). A white waxy solid having a characteristic smell of garlic; it usually exists as P_4O_6 molecules. It is soluble in organic solvents, like benzene and carbon disulphide, and is obtained by burning phosphorus in a limited supply of air. Phosphorus(III) oxide oxidizes in air to phosphorus(V) oxide at room temperature and inflames above 70°C It dissolves slowly in cold water or dilute alkalis to yield phosphonic acid or one of its salts. Phosphorus(III) oxide reacts violently with hot water to form phosphine and phosphoric(V) acid.

Phosphorus(V) Oxide (phosphorus pentoxide, P_2O_5). A white powder that is soluble in organic solvents. It usually exists as P_4O_{10} molecules. Phosphorus(V) oxide can be obtained by burning phosphorus in a plentiful supply of oxygen It readily reacts with water to form phosphoric(V) acid and is therefore used as a drying agent for gases. It acts as a useful dehydrating agent because it has been able to remove the elements of water from certain oxyacids and other compounds containing oxygen and hydrogen, for example, it reacts with nitric acid (HNO_3) to yield nitrogen pentoxide (N_2O_5) on heating.

Phosphorus Oxychloride. See phosphorus(III) chloride oxide.

Phosphorus Pentabromide. See phosphorus(V) bromide.

Phosphorus Pentachloride. See phosphorus(V) chloride.

Phosphorus Pentoxide. See phosphorus(V) oxide.

Phosphorus Tribromide. See phosphorus(III) bromide.

Phosphorus Trichloride. See phosphorus(III) chloride.

Phosphorus Trichloride Oxide. See phosphorus(III) chloride oxide.

Phosphorus Trioxide. See phosphorus(III) oxide.

Phosphoryl Chloride. See phosphorus(III) chloride oxide.

Photographic Developers. Chemical reducing agents which specifically reduce silver halide grains in the pnotographic plate which have been exposed to light, Examples include aromatic derivatives of H_2O_2, H_2 NOH, N_2H_4.

Photoionization. Refers to the ionization of an atom or molecule because of irradiation by electromagnetic radiation. For a photoionization to take place the incident photon of the radiation must be having an energy in excess of the ionization potential of the species being irradiated. The ejected photoelectron will having an energy, E, given by $E=hf-I$, where h is the Planck constant, f is the frequency of the incident radiation, and I is the ionization potential of the irradiated species.

Photolysis. A chemical reaction which is produced by exposure to light or ultraviolet radiation Photolytic reactions generally involve free radicals, the first step being homolytic fission of a chemical bond.

Photon. A particle having zero rest mass consisting of a quantum of electromagnetic radiation. The photon may also be considered as a unit of energy equal to hf, where h is the Planck constant and f is the frequency of the radiation in hertz. Photons travel at the speed of light. They are needed to explain the photoelectric effect and other phenomena that require light to have particle character.

Physical Development. A photographic process in which a soluble silver complex present in the developer solution is reduced to Ag which is deposited specifically on the exposed halide grains or on heavy metal or other foreign particles.

Pi Bonding, π-Bonding. Bonding by interaction of orbitals of such shape that there are two regions of overlap. Thus ethene, C_2H_4, is considered to have a σ-bond formed by overlap of

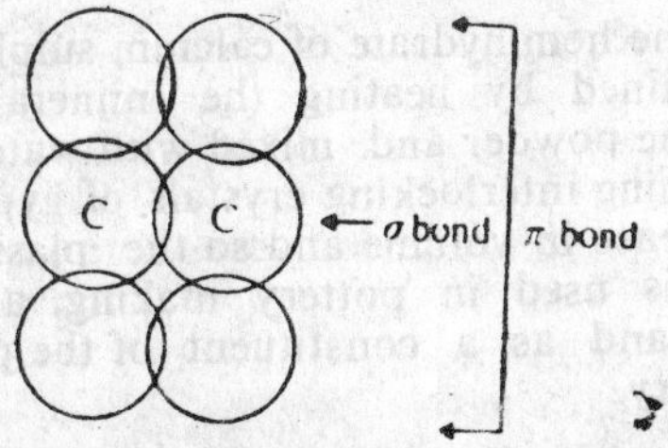

s-orbitals and a π-bond formed by overlap of p-orbitals. π-Bonding is of importance in co-ordination compounds where co ordinate bonds may have π-character. The electrons can be donated from the donor or there may be electron transfer from the acceptor to the donor (back-bonding).

Pico. Symbol : p A prefix denoting 10^{-12}. For example, 1 picofarad (pF)$=10^{-12}$ farad (F).

Pi Electron. An electron in a pi orbital.

Pig Iron. The impure form of iron which is produced by a blast furnace. which is cast into pigs (blocks) for converting at a later date into cast iron, steel, etc. The composition depends on the ores used, the smelting procedure, and the use to which the pigs will later be put.

Pi Orbital. See orbital.

Pirsonnite. A mineral having a hydrated mixed carbonate of sodium and calcium, $Na_2CO_3.CaCO_3.2H_2O$.

Pitchbleade. $UO_{2-2.25}$. An important uranium ore also containing useful quantities of Ra. Th, Ce, Pb are also present.

Planck Constant Symbol h. The fundamental constant which is equal to the ratio of the energy of a quantum of energy to its frequency. It has the value 6.626196×10^{-34} Js. It has been named after Max Planck (1858—1947).

Plagioclase Feldspars. A group of feldspars containing mainly Na^+ and Ca^{2+} as cations, E.g. albite, $NaAlsi_3O_8$, and anorthite, $CaAl_2Si_2O_8$, which also form minerals of mixed composition (*e.g.* labradorite).

Planar Complexes. Complexes in which the acceptor atom is surrounded by four ligands in a plane. If predominantly covalent the hybridization is normally considered as dsp^2. E.g. $[Pt(NH_3)_4]^{2+}$, $[Ir(CO)Cl(PPh_3)_2]$ and $[PtCl_4)^{2-}$.

Plasma. Refers to a mixture of ions and electrons, as in an electrical discharge. Sometimes, a plasma is termed as a fourth state of matter.

Plaster of Paris. The hemihydrate of calcium sulphate, $2CaSO_4.H_2O$, which is obtained by heating the mineral gypsum. When ground to a fine powder and mixed with water, plaster of Paris sets hard, forming interlocking crystals of gypsum. The setting causes an increase in volume and so the plaster fits tightly into a mould. It is used in pottery making, as a cast for setting broken bones, and as a constituent of the plaster used in the building industry.

Platinic Chloride. See chloroplatinic acid.

Platinum Symbol Pt. A silvery white metallic transition element (see also platinum metals); a.n. 78; r.a.m. 195.09; r.d. 2137; m.p. 1772°C; b p. —3800°C. It is found in some nickel and copper ores and is also found native in some deposits. The main source has been the anode sludge obtained in copper—nickel refining. The element finds use in jewellery, laboratory apparatus (*e.g* thermo-couples, electrodes, etc,), electrical contacts, and in certain alloys (*e.g.* with iridium or rhodium). It is also a hydrogenation catalyst. The element does not oxidize nor dissolve in hydrochloric acid. Most of its compounds have been platinium(II) or platinium(IV). complexes.

Platinum Black. Black finely divided platinum metal which is produced by vacuum evaporation and used as an absorbent and a catalyst.

Platinum (II) Chloride ($PtCl_2$). A grey-brown power which is obtained by the partial decomposition of platinum(IV) chloride. It may also be prepared by passing chlorine over heated platinum. †Platinium(II) chloride is insoluble in water but dissolves in concentrated hydrochloric acid to form a complex acid.

Platinum(IV) Chloride ($PtCl_4$). A reddish hygroscopic solid which is obtained by the action of heat on chloroplanic acid. Crystals of the pentahydrate, $PtCl_4 \cdot 5H_2O$, are formed; anhydrous platinum(IV) chloride is obtained by treating the hydrated crystals with concentrated sulphuric acid. The chloride dissolves in water to produce a strongly acidic solution. †Platinum (IV) chloride forms a series of hydrates having 1, 2, 4, and 5 molecules of water; the tetrahydrate is the most stable. The strongly acidic solution produced on dissolving in water probably is having $H_2[PtCl_4(OH_2]$.

Platinum-iridium. An alloy of platinum having up to 30% iridium. Hardness and resistance to chemical attack increase as the iridium content gets increased. It is used in jewellery, electrical contacts, the hypodermic needles.

Platinum Metals. The three members of the second and third transition series which are immediately proceeding silver and gold; ruthenium (Ru). rhodium (Rh), and palladium (Pd); and osmium (Os). iridium (Ir), and platinum (Pt). These elements, together with iron, cobalt, and nickel, were formerly classed as group VIII of the periodic table. The platinum-group metals have been relatively hard and resistant to corrosion and are used in jewellery and in some industrial applications (*e.g.* electrical contacts). They are having certain chemical similarities that justify classifying them together. All are resistant to chemical attack. In solution they form a vast range of complex ions. They also form coordination compounds with carbon monoxide and other pi-bonding ligands. A number of complexes can be prepared in which a hydrogen atom is linked directly to the metal. The metals and their organic compounds are having considerable catalytic activity.

Platinum Oxides. Brown PtO_2 is formed by dehydration of the hydrated oxide precipitated on hydrolysis of $PtCl_5^{2-}$. PtO ($PtCl_2$ plur KNO_3) and Pt_3O_4 (Pt electrode or Pt plus pressure of O_2) are more doubtful. Mixed oxides, *e.g.* $Tl_2{}^3Pt_2{}^4O_7$, are known.

Platinum Sulphides. The known sulphides are PtS_2 and PtS obtained from solution by precipitation with sulphides or H_2S.

Pleochromism. The property of a crystal of having a different colour depending upon the direction of transmitted light through the crystal. Cyanoplatinates, *e.g.* $K_2Pt(CN)_4$, are strongly pleochroic.

Plumbane (Lead(IV) Hydride). An extremely unstable gas, PbH_4 which is formed but the action of acids on magnesium lead alloys. It was first reported in 1924, although doubts have since been expressed about the existence of the compound. It demonstrates the declining stability of the hydrides in group IV. More stable organic derivatives have been known; *e.g.* trimethyl plumbane, $(CH_3)_3$ PbH.

Plumbate. A compound which is obtained by reaction of lead oxides (or hydroxides) with alkali. The oxides of lead are amphoteric (weakly acidic) and react to yield plumbate ions. With the lead (IV) oxide, reaction with molten alkali gives the plumbate (V) ion.

$$PbO_2 + 2OH^- \rightarrow PbO_3^{2-} + H_2O$$

In fact, various ions have been found in which the lead is bound to hydroxide groups, the principal one being the hexahydroxoplumbate(IV) ion $Pb(OH)_6^{2-}$ This is the negative ion present incrystalline 'trihydrates' of the type $K_2PbO_3.3H_2O$ Lead(II) oxide yields the trihydroxoplumbates(II) ion in alkaline solutions

$$PbO(s) + OH^-(aq) + H_2O(l) \rightarrow Pb(OH)_3^{2-}(aq)$$

Plumbate(IV) compounds were formerly referred to as orthoplumbates ($Pb_3{}^{4-}$) or metaplumbates (PbO_3^{2-}). Plumbate(II) compounds were known as plumbites.

Plumbic. Designating a lead (IV) compound.

Plumbic Compounds. Compounds of lead in its higher (+4) oxidation state; *e.g.* plumbic oxide has been lead(IV) oxide, PbO_2.

Plumbous. Designating a lead(II) compound.

Plumbous Compounds. Compounds of lead in its lower (+2) oxidation state; *e.g.* plumbous oxide has been lead (II) oxide, PbO.

Plutonium. A highly toxic radioactive silvery element of the actinoid series of metals. It is a transuranic element which occurs in Earth only in minute quantities in uranium ores but readily obtained, as plutonium−239, by neutron bombardment of natural uranium. The readily fissionable ^{239}Pu has been a major nuclear fuel and nuclear explosive.

Symbol : Pu; m.p. 641°C; b.p. 3232°C; r.d. 19.84; p.n. 94; stablest isotope ^{244}Pu.

Pm. Promethium.

Pnictogens. The elements nitrogen, phosphorus, arsenic, antimony and bismuth of group V. Pnictides contain X^{3-} species.

Po. Polonium.

Polar. Describing a compound having molecules that have a permanent dipole moment. Hydrogen chloride and water have been examples of polar compounds.

Polar Bond. A covalent bond in which the bonding electrons are not shared equally between the two atoms. A bond between two atoms of different electronegativity has been said to be polarized in the direction of the more electronegative atom, *i.e.* the electrons are drawn preferentially towards the atom. This gives rise to a small separation of charge and the development of a bond dipole moment as in, for example, hydrogen fluoride, represented as H ← F or as $H\delta^{+}—F\partial^{-}$ (F is more electronegative).

The charge separation has been much smaller than in ionic compounds; molecules in which bonds are strongly polar are said to display partial ionic character. The effect of the electronegative element can get transmitted beyond adjacent atoms, thus the C−C bonds in, for example, $CCl_3.CH_3$ and $CH_3.CHO$ are slightly polar.

Polar Compound. A compound that has been either ionic (*e.g.* sodium chloride) or that has molecules with a large permanent dipole moment (*e.g.* water).

Polarizability. Refers to the ease with which an electron cloud gets deformed (polarized). In ions, an increase in size or negative charge leads to an increase in polarizability. The concept has been of particular use in the treatment of covalent contributions to predominantly ionic bonds embodied in Fajans' Rules. Thus ions, like I^- Se^{2-}, Te^{2-} have been especially prone to covalent character, particularly in combination with ions of small size and high positive charge (*i.e.* high ionic potential). Molecular polarizability has been a measure of the ease with which electrons in the molecule are deformed, particularly by electromagnetic radiation.

Polar Molecule. A molecule that is having a dipole moment; *i.e.* one in which there has been some separation of charge in the chemical bonds, so that one part of the molecule is having a positive charge and the other a negative charge.

Polar Solvent. A solvent in which the molecules are having a moderate to high dipole moment and in which polar and ionic compounds are easily soluble. Polar solvents are usually poor solvents for non-polar compounds. For example, water has been a good solvent for many ionic species, such as $NaCl_3$, KNO_3 or polar molecules, like the sugars, but does not dissolve paraffin wax.

The dipole moment of the polar solvent molecule allows moderately strong intermolecular attraction between it and the charged regions of the solute molecules (solvation): if the energy of solvation exceeds the lattice energy of the solute, dissolution will occur.

Pollucite, $Cs(AlSi_2O_6),xH_2O$. The major mineral source of Cs. Contains (Al, Si)O_4 tetrahedra linked in a three-dimensional network.

Polonium. A radioactive metallic element belonging to group VI of the periodic table. It is found in very minute quanities in uranium ores. Over 30 radioisotopes are known, nearly all alpha-particle emitters. A suggested use for polonium-210 has been as a thermoelectric power source in satellites.

Symbol : Po; m.p. 254°C; b.p. 962°C; r.d. 9; p.n. 84; stablest isotope ^{209}Po.

Poly Acids. The weak acids formed from the amphoteric metals of groups VB and VIB of the periodic classification are characterised by the ease with which they condense to form anions containing several molecules of the acid anhydride. These are known as polyacids. For example, the formulae of molybdates in alkaline solutions are represented by R_2MoO_4 whereas in acid solution the formation of more complex anions takes place which are represented by $R_2O.nMoO_3$ aq. where n=1, 2, 3, 4, 8, 10 and 16 depending upon the pH of solutions.

Types of Poly Acids. There are two types of poly acids :

(a) Isopoly acids. These are the poly acids containing only one type of acid anhydrides The polychromic acids, $H_2O.yCrO_3$ (y>1) may be cited examples of this class.

(b) Heteropoly acids. These are the poly acids containing more than a single type of anhydrides. It means that they contain different series of anhydrides. For example, polymolybdophosphoric acids, $mH_2O.P_2O_5\,yMoO_3$. y=12 or 24, are most common heteropolyacids.

Polyatomic. Describing a molecule (or ion or radical) that consists of several atoms (three or more). Examples are benzene (C_6H_6) and methane (CH_4).

Polycrystalline. Describing a substance composed of very many minute interlocking crystals that have solidified together.

Polydentate Ligands. When a ligand has two or more donor atoms which may simultaneously coordinate to a metal atom, it is termed as a multi or polydentate ligand. Depending upon the number of donor sites, these ligands may be referred to as bidentate (two donor atoms), tridentate (three donor atoms) tetradentate (four donor atoms), and so on.

Polydioxoboric(III) acid. See boric acid.

Polyhalides. When halide ions associate with molecules of halogens or inter-halogens, univalent ions are obtained. These are known as polyhalide ions and the compounds of these are known as polyhalides. Some examples of polyhalides are :

(i) It has been suggested that the solubility of iodine in water is increased by the addition of iodide ions which is due to the formation of the tri-iodide ion I_3^- (Polyhalide ion).

$$I_2+I^- \rightleftharpoons I_3^-$$

The formation of more complex polyhalide ions like I_5^-, I_7^-, I_9^- have also been confirmed by spectral studies.

(ii) When bromine is added to a metal halide, mixed polyhalide is obtained. This is carried out in the absence of a solvent.

Polyhalite. $K_2Ca_2Mg(SO_4)_4$. A mixed mineral sulphate found in beds of other potassium salts. Used for production of KOH.

Polymorphism. Refers to the ability of certain chemical substances to exist in more than one physical form. For elements this is called allotropy. The existence of two forms is termed as dimorphism.

Polysulphides. See sulphides.

Polythionate. A salt of a polythionic acid.

Polythionic Acids, Polythionates, $H_2S_nO_6$. A series of more or less unstable dibasic acids, obtained by the action of iodine solution on sodium sulphite, thiosulphate and mixtures of these, by the action of hydrogen sulphide on other members of the same series and of H_2S on H_2SO_3 at 0°C. The best known as alkali salts are: dithionic acid $H_2S_2O_6$, trithionic acid $H_2S_3O_6$, tetrathionic acid $H_2S_4O_6$, pentathionic acid $H_2S_5O_6$, and hexathionic acid $H_2S_6O_6$.

Aqueous solutions of the free acids decompose on concentration, yielding sulphuric acid, sulphur dioxide and sulphur. Aqueous solutions of their alkali salts decompose slowly. Solid barium and alkali salts of various acids of the series have been prepared. The anions contain S_n chains and formally $H_2S_2O_6$ is not a member of the series. Species containing Se and Te *e.g.* $[O_3SSSeSSO_3]^{2-}$ are also known.

Positron. β^+. The positive counterpart of the electron. The ultimate indivisible unit of positive electricity.

Positronium. One electron and one position formally analogous to hydrogen.

Post-actinide Elements. The elements following $_{103}$Lw which should be part of a 6d transition series. Element 104 has been formed from ^{242}Pa with accelerated ^{22}Ne (^{260}104, $t_{\frac{1}{2}}$~0.3 sec and ^{259}104, $t_{\frac{1}{2}}$ 4.5 sec), from ^{249}Cf with ^{13}C and ^{12}C) (^{257}104 and ^{259}104), or from ^{248}Cm with ^{18}O (^{261}104, $t_{\frac{1}{2}}$=70 sec). The element has been variously named Kurchatovium (Ku) and Rutherfordium (Rf). Element 105 has been formed similarly and named Hahnium (Ha). Element 106 has been claimed. Each of these elements have very short half-lives.

Potash. Any of a number of potassium compounds, such as the carbonate or the hydroxide.

Potash Alum. See aluminium potassium sulphate; alums.

Potash Mica. See muscovite.

Potassium. Symbol K. A soft silvery metallic element which belongs to group I of the periodic table (see alkali metals); a.n. 19;

r.a.m. 39.098; r.d. 0.87; m.p. 63.7°C b.b. 774°C. The element is found in seawater and in a number of minerals, such as sylvite (KCl), carnallite ($KCl.MgCl_2,6H_2O$), and kainite ($MgSO_4.KCl.3H_2O$). It is obtained by electrolysis. The metal has few uses but potassium salts are used for a wide range of applications. Potassium is an essential element for living organisms. Chemically, it has been highly reactive, resembling sodium in its behaviour and compounds. It also forms an orange coloured superoxide, KO_2, which is having the O_2 ion. Potassium was discovered by Sir Humphry Davy in 1807.

Potassium-argon Dating. Refers to a dating technique for certain rocks that depends on the decay of the radioisotope potassium-40 to argon-40, a process with a half-life of about 1.27×10^{10} years. It regards that all the argon-40 formed in the potassium bearing mineral accumulates within it and that all the argon present is formed by the decay of potassium-40. The mass of argon-40 and potassium-40 in the sample is estimated and the sample is then dated from the equation :

$$^{40}Ar=0.1102\ ^{40}K(e^{\lambda t}-1),$$

where λ is the decay constant and t is the time in years since the mineral cooled to about 300°C, when the ^{40}Ar became trapped in the crystal lattice. The method has been found to be effective for micas, feldspar and some other minerals.

Potassium Carbonate. See potassium hydrogencarbonate.

Potassium Bichromate. See potassium dichromate.

Potassium Bromide (KBr). A white solid which is extremely soluble in water. It is obtained by the action of bromine on hot potassium hydroxide solution or by the neutralization of the carbonate with hydrobromic acid. Potassium bromide forms colourless cubic crystals. It finds use extensively in the manufacture of photographic plates, films, and papers and (formerly) as a sedative in medicine.

Potassium Carbonate (Pearl Ash; Potash). A translucent (grangular) or white (powder) deliquescent solid which is known in the anhydrous and hydrated forms. K_2CO_3 (monoclinic; r.d. 2.4; m.p. 891°B) decomposes without boiling. $2K_2CO_3.3H_2O$ monoclinic; r.d 2.04) dehydrates to $K_2CO_3H_2O$ above 100°C. and to K_2CO_3 above 130°C It is prepared by the Engel-precht process in which potas siumchloride and magnesium oxide react with carbon dioxide to give the compound Engel's salt. $MgCO_3.KHCO_3.4H_2O$. This gets decomposed in solution to give the hydrogencarbonate, which can then be calcined to K_2CO_3. Potassium carbonate is soluble in water (insoluble in alcohol) with significant hydrolysis to produce basic solutions. Industrial uses include glasses and glazes, the manufacture of soft soaps, and in dyeing and wool finishing. It finds use in the laboratory as a drying agent.

Potassium Chlorate ($KClO_3$). A white soluble solid which is obtained by the electrolysis of a concentrated solution of potassium chloride. Industrially, it is obtained by the fractional crystallization of a solution containing sodium chlorate and potassium chloride. When heated, it decomposes to yield oxygen and potassium chloride. Potassium chlorate is a powerful oxidizing agent and is used in explosives, matches, weedkillers, and fireworks and as a disinfectant. It is able to oxidize iodide ions to iodine when in an acidic medium. When heated just above its melting point, potassium perchlorate is formed.

Potassium Chloride It is a white crystalline solid, KCl, which is soluble in water and very slightly soluble in ethanol: cubic; r.d. 1.98; m.p. 772°C; sublimes at 1500°C. Potassium chloride occurs naturally as the mineral sylvite (KCl) and as carnallite ($KCl.MgCl_2.6H_2O$); it is formed industrially by fractional crystallization of these deposits or of solutions from lake brines. It is having the interesting property of being more soluble than sodium chloride in hot water but less soluble in cold. It finds use as a fertilizer, in photography, and as a source of other potassium salts, such as the chlorate and the hydroxide. It is having low toxicity.

Potassium Chromate. It is a bright yellow crystalline solid, K_2CrO_4, soluble in water and insoluble in alcohol; rhombic; r d. 2.73; m.p. 971°C; decomposes without boiling. It is obtained industrially by roasting powdered chromite ore with potash and limestone and leaching the resulting cinder with hot potassium sulphate solution. Potassium chromate finds use in leather finishing, as a textile mordant, and in enamels and pigments. In the laboratory it finds use as an analytical reagent and as an indicator. Like other chromium(III) compounds it has been toxic when ingested or inhaled.

Potassium Chromium Sulphate (chrome alum). A violet or ruby-red crystalline solid, $K_2SO_4.Cr_2(SO_4)_3.24H_2O$ which is soluble in water and insoluble in ethanol: cubic or octahedral; r.d. 1.826; m.p. 89°Cs loses $10H_2O$ at 100°C, $12H_2O$ at 400°C. Six water molecules surround each of the chromium(III) ions and the remaining ones have been hydrogen bonded to the sulphate ions. Like all alums, the compound may be have been by mixing equimolar quantities of the constituent sulphates.

Potassium Cyanide (KCN). A white ionic solid that is very soluble in water and extremely poisonous. It is obtained industrially by the Castner process and is used as a source of cyanide and hydrocyanic acid. Aqueous solutions of potassium cyanide strongly get hydrolysed the solution being alkaline. On standing these solutions slowly evolve hydrocyanic acid.

Potassium Dichromate, $K_2Cr_2O_7$. M.p. 396°C.' Orange red crystals prepared from chromite, K_2CO_3 and CaO followed by acidification. Used in the preparation of chrome pigments, bleaching, as an oxidizing agent, in organic chemistry and in volumetric analysis.

Potassium Dioxide. See potassium superoxide.

Potassium Hydride. It is a white or greyish white crystalline solid, KH; r.d. 1.43–1.47. It is obtained by passing hydrogen over heated potassium and marketed as a light grey powder dispersed in oil. The solid decomposes on heating and in contact with moisture and is an excellent reducing agent. Potassium hydride has been a fire hazard because it produces hydrogen on reaction with water.

Potassium Hydrogencarbonate (potassium bicarbonate, $KHCO_3$). A white solid which is obtained by passing carbon dioxide through a saturated solution of potassium carbonate. It is found naturally in the minera calcinite. Potassium hydrogencarbonate is more soluble in water than sodium hydrogencarbonate. When heated, it undergoes thermal decomposition to yield the carbonate, water, and carbon dioxide. The salt forms monoclinic crystals. Its solutions have been strongly alkaline due to salt hydrolysis.

Potassium Hydrogentartrate (cream of tartar). A white crystalline acid salt, HOOC $(CHOH)_2$COOK which is obtained from deposits on wine vats (argol) and used in backing powders.

Potassium Hydroxide (caustic potash, KOH). A white solid which is obtained by the electrolysis of potassium chloride solution in a mercury-cathode cell. It can also be obtained by heating either potassium carbonate or potassium sulphate with slaked lime. The process closely resembles the Gossage process for making sodium hydroxide. In the laboratory, it can be obtained by reacting potassium, potassium monoxide or potassium superoxide with water. Potassium hydroxide resembles sodium hydroxide but is more soluble in water and alcohol. It is preferred to sodium hydroxide for the absorption of carbon dioxide and sulphur dioxide due to its greater solubility. Potassium hydroxide finds use as an electrolyte in the Ni-Fe electric storage battery and in the production of salt soaps.

Potassium Iodate. A white crystalline solid, KIO_3, soluble in water and insoluble in ethanol; monoclinic; r.d. 3.9; m.p. 561°C. It may be obtained by the reaction of iodine with hot concentrated potassium hydroxide or by careful electrolysis of potassium iodide solution. It is an oxidizing agent and finds use as an analytical reagent. Some potassium iodate finds use as a food additive.

Potassium Iodide (KI). It is a white ionic solid readily soluble in water. It is obtained by dissolving iodine in hot concentrated potassium hydroxide solution. Both the iodide and iodate have been formed but the latter may be removed by fractional crystallization. Potassium iodide is having a sodium chloride cubic lattice. In solution with iodine it forms potassium tri-iodide. It finds use in medicine, particularly in the treatment of goitre resulting from iodine deficiency. Dilute acidified solutions of potassium iodide can act as reducing agents, manganate(VII) ions being reduced to manganese(II) ions, copper(II) ions to copper(I) ions and iodate ions to iodine.

Potassium Manganate(VII) (potassium permanganate, $KMnO_4$). A purple solid soluble in water. It is obtained by oxidizing potassium manganate(VI) with chlorine. Potassium permanganate finds use in volumetric analysis as an oxidizine agent, as a bactericide, and as a disinfectant. In aqueous solution its behaviour as an oxidizing agent has been found depend on the pH of the solution.

Potassium Monoxide- A grey crystalline solid, K_2O; cubic; r.d. 2.32; decomposition occurs at 350°C. It may be obtained by the oxidation of potassium metal with potassium nitrate. It reacts with ethanol to yield potassium ethoxide (KOC_2H_5), and with liquid ammonia to form potassium hydroxide and potassamide (KNH_2).

Potassium Nitrate (saltpetre, nitre, KNO_3). A white solid, soluble in water, which is obtained by fractional crystallization of sodium nitrate and potassium chloride solutions. It is found naturally as nitre (saltpetre) in rocks in India, South Africa, and Brazil. When heated it decomposes to yield the nitrite and oxygen. Unlike sodium nitrate it is non-deliquescent, Potassium nitrate finds use in gunpowder, fertilizers, and in the laboratory preparation of nitric acid.

Potassium Nitrite. A white or slightly yellow deliquescent solid, KNO_2 which is soluble in water and insoluble in ethanol; r.d. 1.91; m.p. 300°C; may explode at 600°C. Potassium nitrite reacts with cold dilute mineral acids to yield nitrous acid and is also able to behave as a reducing agent (if oxidized to the nitrate) or as an oxidizing agent (if reduced to nitrogen). It finds use in organic synthesis because of its part in diazotization and in detecting the presence of the amino groups in organic compounds.

Potassium Permanganate. See potassium manganate(VII).

Potassium Sulphate (K_2SO_4). A white solid which is obtained by the neutralization of either potassium hydroxide or potassium carbonate with dilute sulphuric acid. It is found naturally in Stassfurt deposits as schonite. Potassium sulphate is soluble in water, forming a neutral solution. It finds use as a fertilizer, particularly for tobacco plants, and in the chemical industry in the preparation of alums. The anhydrous form is known to crystallize in the rhombic form.

Potassium Sulphide (K_2S). A yellowishbrown solid which is obtained by saturating an aqueous solution of potassium hydroxide with hydrogen sulphide and then adding an equal volume of potassium hydroxide. Industrially it is obtained by heating potassium sulphate and carbon at high temperature. The sulphide crystallizes out from aqueous solution as $K_2S.5H_2O$. Its aqueous solutions undergo hydrolysis, the solution being strongly alkaline.

Potassium Sulphite. A white crystalline solid, K_2SO_3 which is soluble in water and very sparingly soluble in ethanol; r.d. 1.51; decomposes on heating. It is a reducing agent and finds use as such in photography and in the food and brewing industries, where it prevents oxidation.

Potassium Superoxide (potassium dioxide). A yellow paramagnetic solid, KO_2, which is obtained by burning potassium in an excess of oxygen; it is very soluble (by reaction) in water, soluble in ethanol, and slightly soluble in diethyl ether m p. 490°C. When treated with cold water or dilute mineral acids, hydrogen peroxide is obtained. The compound is a powerful oxidizing agent and on strong heating releases oxygen with the formation of the monoxide, K_2O.

Potter's Clay. A term used for any clay or earth which can be used for making pottery, either alone or after admixture with non-plastic materials, though the term is usually restricted to the clays used for fine pottery. The composition varies greatly, and is usually of minor importance; the essential properties are sufficient plasticity to enable the desired articles to be shaped readily and sufficient refractoriness to enable the pottery to be burned in a potter's oven or kiln. Other properties, such as smoothness of working during the shaping, the colour of the finished ware, brittleness, porosity of the ware, etc., are also important in some instances. The red colour of terra-cotta flowerpots and some other pottery is due to iron compounds (probably nontronite) in the clay; these decompose on heating and liberate Fe_2O_3 or Fe(III) complexes which are red in colour. Other colours are produced by firing the kiln or oven under reducing conditions, or by adding coloured oxides to the clay.

Powder Metallurgy. The techniques of sintering powders under pressure but below the melting point to produce solid objects. Used for metals, alloys and some plastics.

Pr. Praseodymium.

Praseodymium. A soft ductile melleable silvery element of the lanthanoid series of metals. It is found in association with other lanthanoids. Praseodymium finds use in several alloys, as a catalyst, and in compound form in carbon-arc searchlights, etc., and in the glass and enamel industries.

Symbol : Pr; m.p. 931°C; b.p. 3212°C; r.d. 6.77; p n. 59; r.a.m. 140.91.

Precursor. A compound that leads to another compound in a series of chemical reactions.

Principal Quantum Number. This number is designated as 'n' and is identical with the Bohr's integers to describe stationary states in atom.

This gives the number of the principal shell in which the electron is revolving around the nucleus. It cesignates the average distance of the electron from the nucleus. Hence this quantum number represents the size of the electron orbit.

The principal quantum number not only determines the size of the orbit but also governs the allowed energy levels in an atom.

The principal quantum, number given by Bohr's Theory, can be described mathenatically in the equation.

$$mur=\frac{nh}{2\pi}$$ where h is the plancks, constant.

n always assumes the integral values 1, 2, 4,............ $\propto$ (not zero).

Thus n-1 for the K-level, 2 for L-level, 3 for the M-level, and so on. The first energy level, n=1, is the one which is nearest to the nucleus, n=2 the next energy level is away from the nucleus, and so on. As the first energy level is nearest to the nucleus, it binds strongly to then ucleus. The next energy level is farther from the nucleus and hence binds to the nucleus with a lesser binding force and so on. Thus, the principal quantum number describes roughy :

(i) The size of the electron orbit, *i.e.*, distance between the nucleus and the electron.

(ii) The binding force between the nucleus and the electron. As n increases, the energy of orbit increase until it is zero at $n = \infty$ Beyond this energy is no longer quantised; the electron is not bound to the nucleus and can have any arbitray amount of kinetic energy.

Accordingly, the spectrum in this region does not contain lines but is continuous and is called continum. The difference in energy between E_{n-1} and $E_n{}^{-}\infty$ is the energy needed to ionise the atom and is called the ionisation energy of the hydrogen atom.

The values of n also fix the number of nodal surfaces. The total number of nodal surfaces in an atom is n including the one∞. At a nodal surface the wave function ψ changes phase and ψ^2 is zero, *i.e.*, an electron is not found at this surface.

Producer Gas. A mixture of CO and N_2 produced by passing air through a bed of incandescent coke or coal in a producer. Carbon-air reactions are exothermic and in order to control the temperature in the producer the air is saturated with steam at about 50°C. This results in endothermic carbon-steam reactions—the water gas reaction—which yields CO and H_2. The gas also contains some carbon dioxide the methane. The calorific value of producer gas from coke is of the order of 4600kJ/m³ and from coal, about 6400kJ/m³.

Promethium Symbol Pm. A soft silvery metallic element belonging to the lanthanoids; a.n. 61; r.a.m. 145; r.d. 7.26 (20°C); m.p. 1042°C; b.p. 3000°C. The only naturally occurring isotope, promethium—147, has a half-life of only 2.52 years. Eighteen other radioisotopes are but they are having very short half-lives. The only known source of the element has been nuclear waste material Promethium—147 is of interest as a beta-decay powersource but the —promethium—146 and —148, which emit penetrating gamma radiation, must first be removed. It was discovered by J. A Marinsky, L. E. Glendenin, and C. D. Coryell in 1947.

Protectinium Symbol Pa. A radioactive metallic element which belongs to the actinoids, a.n. 91; r.a.m. 231.036; r.d. 15.37 calculated); m.p. < 1600°C (estimated). The most stable isotope, protactinium—231, is having a half-life of 3.43×10^4 years; ten other radioisotopes are known. Protactinum—231 is found in all uranium ores as it is a member of the uranium series. Protactinium has no practical applications; it was discovered by Lise Meitner and Otto Hahn in 1917.

Proton. An elementary particle having a positive charge (+1.602192 C) and rest mass 1.672614 × 10^{27} kg. Protons have been nucleons, found in all nuclides.

Protonic Acid. An acid that yields positive hydrogen ions (or, strictly, oxonium ions) in aqueous solution. The term is used to distinguish 'traditional' acids from Lewis acids or from Lowry—Br o/nsted acids in non-aqueous solvents.

Proton Number (atomic number) Symbol: Z. It is equal to the number of protons in the nucleus of an atom. The proton number determines the chemical properties of the element because the electron structure, which determines chemical vonding, depends on the electrostatic attraction to the positively charged nucleus.

Prussic Acid. See hydrogen cyanide.

Pseudohalogens. A group of compounds, including cyanogen $(CN)_2$ and thiocyanogen $(SCN)_2$ that are having some resemblance to the halogens. Thus, they form hydrogen acids (HCN and HSCN) and ionic salts containing such ions as CN^- and SCN^-.

Pumice. Refers to a porous volcanic rock that is light and full of cavities due to expanding gases that were liberated from solution in the lava while it solidified. Pumice has been often light enough to float on water. It is usually acid (siliceous) in composition, and finds use as an abrasive and for polishing.

Pyrites FeS_2. A common mineral, brass-yellow with a metallic lustre. The structure contains S_2^{2-} species, see marcasite. Used for production of H_2SO_4.

Pyro Acids. Acids or anions derived from hypothetical acids containing two oxygroups with a bridging (μ) oxygen, *e.g.* $[O_3SOSO_3]^{2-}$, pyrosulphate.

Pyro-Prefix which denotes an axo acid that could be obtained from a lower acid by dehydration of two molecules. For example, pyrosulphuric acid has been $H_2S_2O_7$ (*i.e.* $2H_2SO_4$ minus H_2O).

Pyroboric Acid. See boric acid.

Pyrophoric. Igniting spontaneously in air. Pyrophoric alloys are alloys that give sparks when struck.

Pyrophosphoric Acid. See phosphoric(V) acid.

Pyrosilicate. See silicate.

Pyrosulphuric Acid, $H_2S_2O_7$. Present in fuming sulphuric acid. Pyrosulphates contain $[S_2O_7]^{2-}$ ions.

Pyrosulphurous Acid, $H_2S_2O_5$. The free acid is unknown but pyrosulphites contain $[S_2S_5]^{2-}$, $[O_3SS(O)_2]^{2-}$ ions. See sulphur oxyacids. Formed from sulphites, $[SO_3]^{2-}$, and SO_2.

Pyroxens. Refers to a group of ferromagnesian rock-forming silicate minerals. They are common in basic igneous rocks but may also be developed by metamorphic processes in gneisses, schists, and marbles. Pyroxens have a complex crystal chemistry; they are composed of continuous chains of silicon and oxygen atoms linked by a variety of other elements. They are related to the amphiboles, from which they differ in cleavage angles. The general formula has been $X_{1-p}Y_{1+p}Z_2O_6$, where X=Ca,Na; Y=Mg, Fe^{2+}, Mn,Li,Al,Fe^{3+},Ti; and Z=Si,Al. Orthorhombic pyroxenes (orthopyroxenes), $(Mg,Fe)_2Si_2O_6$, have been found to vary in composition between the end-members enstatite ($Mg_2Si_2O_6$) and orthoferrosilite ($Fe_2Si_2O_6$). Monoclinic pyroxenes (clinopyroxenes), the larger group, include diopside, $CaMgSi_2O_6$; hedenbergite, $CaFe^{2+}Si_2O_6$; johannsenite, $CaMnSi_2O_6$; augite, $(Ca,Mg,Fe,Ti,Al)_2$ $(Si,Al)_2O_6$; aegirine, $NaFe^{3+}Si_2O_6$; jadeite (see jade), and pigeonite (Mg,Fe^{2+},Ca) (Mg,Fe^{2+}) Si_2O_6.

Pyrrhotite, Magnetic Pyrites. An iron sulphide mineral with composition ranging from $Fe_{0\cdot877}S$ to FeS. There is a series of phases.

Q

Quadrivalent. Having a valence of four.

Quantized. Describing a physical quantity that can only take certain discreet values, and not a continuous range of values. Thus, in an atom or molecule the electrons around the nucleus can have certain energies, E_1, E_2, etc., and cannot have intermediate values. Similarly, in atoms and molecules, the electrons have quantized values of spin angular momentium and orbital angular momentum.

Quantum. The minimum amount by which certain properties, such as energy or angular momentum, of a system can change. Such properties do not, therefore, vary continuously, but in integral multiples of the relevant quantum. This concept from the basis of the quantum theory. In waves and fields the quantum can be regarded as an excitation, giving a particle-like interpretation to the wave or field. Thus, the quantum of the electromagnetic field is the photon and the graviton is the quantum of the gravitational field.

Quantum Mechanics. May be defined as a system of mechanics that was developed from quantum theory and is used to explain the properties of atoms and molecules. Using the energy quantum as a starting point it incorporates Heisenberg's uncertainty principle and the de Broglie wavelength to establish the wave-particle duality on which Schrodinger's equation has been based. This form of quantum mechanics is known as wave mechanics. An alternative but equivalent formalism, matrix mechanics, has been based on mathematical operators.

Quantum Number. The planetary electrons in an atom or the atoms in a molecule, possess energy due to type, retation, vibration or spin. According to the quantum theory the energy possessed by such a particle is quantized, *i e.* can only have certain definite values. Each such value for a particular form of energy is a multiple-which is in variably a small whole number, or sometimes one half-of a quantum, which is characteristic of the form of energy under consideration. This multiple which defines the energy of the particle, is called the quantum number.

Quantum State. Refers to the state of a quantized system as described by its quantum numbers. For instance, the state of a hydrogen atom is described by the four quantum numbers n, l, m, m_5. In the ground state they have values 1, 0, 1, and $\frac{1}{2}$ respectively.

Quantum Theory. The theory devised by Max Planck (1858-1947) in 1900 to account for the emission of the black-body radiation from hot bodies. According to this theory energy is emitted in quanta (see quantum), each of which has an energy equal to hv, where h is the Planck constant and v is the frequency of the radiation. This theory led to the modern theory of the interaction between matter and radiation known as quantum mechanics, which generalizes and replaces classical mechanics and Maxwell's electromagnetic theory. In nonrelativistic quantum theory particles are assumed to be neither created nor destroyed, to move slowly relative to the speed of light, and to have a mass that does not change with velocity. These assumptions apply to atomic and molecular phenomena and to some aspects of nuclear physics. Relativistic quantum theory applies to particles that have zero rest mass or travel at or near the speed of light.

Quartz. SiO_2. The low temperature form of silica, low temperature quartz crystallizes in enantiomorphous pairs. Quartz is used extensively for optical components, it is more transparent to u.v. radiation than is glass. Rock crystal is colourless and glassy, amethyst is violet (yellow on heating), brown and yellow quartz are semi-precious stones-cairngorms.

Quartz Glass. See silica vitreous.

Quaternary Ammonium Compound. A compound which is obtained formed from an amine by addition of a proton to produce a positive ion. Quarternary compounds have been salts, the simplest example being ammonium compounds formed from ammonia and an acid, for example :

$$NH_3 + HCl \rightarrow NH_4^+ \, Cl^-$$

Other amines can also add protons to give analogous compounds. For instance, methylamine (CH_3NH_2) forms the compound.

$$[CH_3NH_3]^+X^-$$

where X is an acid radical.

†The formation of quarternary compounds occurs because the lone pair on the nitrogen atom can form a coordinate bond with a proton. This are also found with heterogeneous nitrogen compounds, like adenine, cytosine, thymine, and guanine. Such compounds are could nitrogenous bases.

Quenching. A method which is used to alter the mechanical properties of metals. Hot metal is lowered quickly into a bath of oil, water, or brine and the rapid cooling results in a fine grain structure. This treatment tends to increase the hardness of a metal but often makes it brittle.

Quicklime. See calcium oxide.

R

Racemization. A chemical reaction in which an optically active compound is converted into a racemic mixture.

Rad. See radiation units.

Radial Probability Distribution. By the radial probability distribution, it means the probability of finding and electron within a small radial space around the nucleus.

Suppose the space around the nucleus is divided into a large number of uniform concentric shells. The volume of a spherical shell between radii r and r+dr would be given by $4\pi r^2 dr$. Thus, the probability of finding electron within this spherical shell would be $4\pi^2\psi r^2$. This value also represents the radial probability at a distance r from the nucleus. In the function $4\pi r^2 dr\psi^2$, the probability factor ψ^2 decreases whereas volume factor $4\pi r^2 dr$ increases with increase in the value of r. In order to get the radial probability distribution of the electron the function $4\pi r^2 dr\psi^2$ is plotted against the distance (r) from the nucleus. The graphs obtained in the case of 1s, 2s and 3s electrons of the hydrogen atom are shown in Fig. The most important feature of the radial distribution plots is fact that it determines the spatial extent of the orbital.

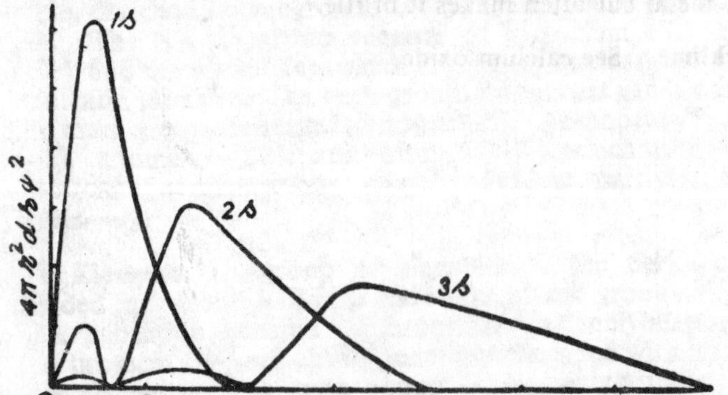

Fig. Radial distribution functions for the 1s, 2s, and 3s states of the hydrogen atom.

Radiation

1. Energy travelling in the form of electromagnetic waves or photons.
2. A stream of particles, especially alpha-or beta-particles from a radioactive source or neutrons from a nuclear reactor.

Radiation Units. Units of measurement which are used to express the activity of a radionuclide and the dose of ionizing radiation. The units curie, roentgen, rad, and rem have been not coherent with SI units but their temporary use with Si units has been approved while the derived SI units becquerel, gray, and sievert become familiar.

The becquerel (Bq), the SI unit of activity, refers to the activity of a radionuclide decaying at a rate, on average, of one spontaneous nuclear transition per second. Thus 1 $Bq=1^{-1}$. The former unit, the curie (Ci) is equal to 3.7×10^{10} Bq. The curie was originally chosen to approximate the activity of 1 gram of radium-226.

The gray (Gy), the SI unit of absorbed dose, is the absorbed dose when the energy per unit mass imparted to matter by ionizing radiation is 1 joule per kilogram. The former unit, the rad (rd), is equal to 10^{-2} Gy.

The sievert (Sv), the SI unit of dose equivalent; is the dose equivalent when the absorbed dose of ionizing radiation multiplied by the stipulated dimensionless factors is 1 J kg^{-1}. As different types of radiation cause different effects in biological tissue a weighted absorbed dose, called the dose equivalent, is used in which the absorbed dose is modified by multiplying it by dimensionless factors stipulated by the international Commission on Radiological Protection. The former unit of dose equivalent, the rem (originally an acronym for roentgen equivalent man), is equal to 10^{-2} Sv.

In SI units, exposure to ionizing radiation is expressed in coulombs per kilogram, the quantity of X-or gamma-radiation that produces ion pairs carrying 1 coulomb of charge of either sign in 1 kilogram of pure dry air. The former unit, the roentgen (R), is equal to 2.58×10^{-4} C kg^{-1}.

Radioactive. Describing an element or nuclide that exhibits natural radioactivity.

Radioactive Age. The age of an archaeological or geological specimen as determined by a process that depends on a radioactive decay. See carbon dating; fission-track dating; potassium-argon dating; rubidium-strontium dating; uranium-lead dating.

Radioactive Dating. (radiometric dating). Any method of measuring the age of materials that depends on radioactivity.

Radioactive Disintegration Series. Whenever a radioactive disintegration occurs with the emission of an alpha or a beta particle, the original atoms, called parent, change into something else, called the daughter. In 1950 Rutherford and Soddy proposed that the nature of the daughter could be inferred from the nature of the parent and the particle emitted. We can state Rutherford Soddy rules as :

1. The total electric charge (atomic number) or algedraic sum of the charges before the disintegration must be equal to the total electric charge after the disintegration.
2. The sum of the mass numbers of the initial particles must be equal to sum of the mass numbers of the final products.

There are four main series of radioactive elements : the thorium uranium, actinium and neptunium series. The first three natural disintegration series are sometimes referred to as the 4n (thorium), 4n+2 (uranium) and 4n+3 (actinium).

In all the series, a parent radioactive element of large atomic mass and very long life gives rise to a series of radioactive elements as a result of the successive emission of α or β particles. The transformation may be represented by the expression in the form.

$$_{z}X^{A} \rightarrow {}_{z-2}Y^{A-4} + \alpha$$

$$_{z}P^{A} \rightarrow {}_{z+1}Q^{A} + \beta$$

The daughter products Y and Q are themselves radioactive if they have atomic weights greater than 82, the atomic number of lead.

Thorium Seri s (or 4n series) : This series is named after the longest lived nuclide of thorium, Th—232 (half life=14,000 million years). The series ends with the stable $_{82}Pb^{208}$ isotope. The atomic masses of all members of this series are exactly divisible by 4; hence the name "4n series". The mode of decay is presented in Table 1.

Uranium Series (or 4n+2 series) : Beginning with longest lived $_{92}U^{238}$ (half-life=4,500 million years) and ending up with the stable $_{82}Pb^{206}$. These atomic masses of the numbers of this series are given by the general formula 4n+2, where n is an integer. For this reason it is, sometimes, also known as the (4n+2) series. The mode of decay is shown in Table 2.

Actinium Series (or 4n+3 series) : Derives its name not from the longest lived nuclide (U—235 having a half-life of 713 million years), which occurs in natural uranium to an exceedingly small extent, but form the nuclide, Ac—221, which is the sole constituent of natural occurring actinium. The atomic masses of all the members of this series correspond to the expression (4n+3).

TABLE 1

(Radioactive Series of Natural Elements)

Radioactive species	Nuclide	Radiation	Half-life
Thorium Series (4n)			
Thorium (Th)	$_{90}U^{232}$	α	1.39×10^{10}y
Mesothorium-I (Ms Th_1)	$_{88}Ra^{228}$	β	6.7 y
Meso thorium-II (Ms Th_2)	$_{89}Ac^{228}$	α, β, γ	6 13 h
Radiothorium (Rd Th)	$_{80}Th^{228}$	α	1.90 y
Thorium-X (Th X)	$_{88}Ra^{224}$	α	3.64 d
Thoron (Th)	$_{86}Rn^{220}$	α	54.5 s
Thorium-A (Th A)	$_{84}Po^{216}$	α	0.158 s
Thorium-B (Th B)	$_{82}Pb^{212}$	β	1.06 h
{Thorium-C (Th C)	$_{83}Bi^{212}$	α, β	60.5 m
Thorium-C′ (Th C′) (66.3)	$_{84}Po^{212}$	α	3×10^{-7}
Thorium C″ (Th C″) (33.7)	$_{81}Tl^{208}$	β	3.1 m
Thorium-D (Th D)	$_{82}Pb^{208}$	stable	stable
Uranium Series (4n+2)			
Uranium-I (UI)	$_{92}U^{238}$	α	4.5×10^{9} y
Uranium-X_1 (UX_1)	$_{92}Th^{234}$	β	24.1 d
Uranium-X_2 (UX_2)	$_{91}Pa^{234}$	β	1.18 m
Uranium-Z (UZ)	$_{91}Pa^{234}$	β	6.7 h
Uranium-II (UII)	$_{92}U^{234}$	α	2.5×10^{5} y
Ionium (Io)	$_{90}Th^{230}$	α	8×10^{4} y
Radium (Ra)	$_{88}Ra^{226}$	α	1.620 y
Radon (Rn)	$_{86}Rn^{222}$	α	3.82 d
Radium-A (RaA)	$_{84}Po^{218}$	α, β	3.05 m
{Radium-B (RaB) (-9.96)	$_{82}Pb^{214}$	β	26.8 m
Astatine (2.8)	$_{85}At^{218}$	α	2 s
Radium C (RaC)	$_{83}Bi^{214}$	α, β	19.7 m

Actinium Series (4n+3)			
Actinouranium (AcU)	$_{92}U^{235}$	α	7.1×10^{8} y
Uranium Y (UY)	$_{90}Th^{231}$	β	24.6 h
Protactinium (Pa)	$_{91}Pa^{231}$	α	3.43×10 y
Actinium (Ac)	$_{89}Ac^{227}$	α, β	22 y
{Radioactinium (Rd Ac) (98.8)	$_{90}Th^{227}$	α	18.6 d
{Actinium K (AcK) (1.2)	$_{87}Fr^{223}$	β	21 m
Actinium-X (AcX)	$_{88}Ra^{223}$	α	11.2 d
Actinon (An)	$_{86}Rn^{219}$	α	3.92 s
Actinium-A (AcA)	$_{84}Po^{215}$	α, β	1.83×10^{-3} s
{Actinium B (AcB) (99.9995)	$_{82}Pb^{215}$	β	36.1 m
{Astatine-215 (0.0005)	$_{85}At^{215}$	α	10^{-4} s
Actinium C (Ac C)	$_{83}Bi^{211}$	α, β	2.16 m
{Actinium C′ (AcC′)	$_{84}Po^{211}$	α	0.52 s
{Actinium-C″ (AcC″) (99.5)	$_{81}Tl^{207}$	β	4.79 m
Actinium-D (AcD)	$_{82}Pb^{207}$	Stable	—

Radioactive Tracing. See labelling.

Radioactivity. May be defined as the spontaneous disintegration of certain atomic nuclei accompanied by the emission of alpha particles (helium nuclei), beta particles (electrons), or gamma radiation (short-wave electromagnetic waves).

Natural radioactivity has been the result of the spontaneous disintegration of naturally occuring radioisotopes; these radioisotopes can be arranged in three radioactive series. The rate of disintegration is uninfluenced by chemical changes or any normal changes in their environment. However, radioactivity can be induced in many nuclides by bombarding them with neutrons or other particles.

Radiocarbon Dating. See carbon dating.

Radiochemistry. The chemistry of radioactive isotopes of elements. Radiochemistry involves such topics like the preparation of radioactive compounds, the separation of isotopes by chemical reactions the use of radioactive labels in studies of mechanisms, and experiments on the chemical reactions and compounds of transuranic elements.

Radioisotope (radioactive isotope). An isotope of an element which is radioactive.

Radiolysis. A chemical reaction which is produced by high-energy radiation (X rays, gamma rays, or particles.

Radiometric Dating. See radioactive dating.

Radionuclide (radioactive nuclide). A nuclide that is radioactive.

Radiolysis. Refers to the use of ionizing radiation to produce chemical reactions. The radiation used includes alpha particles, electrons, neutrons, X-rays, and gamma rays from radioactive materials or from accelerators. Energy transfer produces ions and excited species, which undergo further reaction. A particular feature of radiolysis has been the formation of short-lived solvated electrons in water and other polar solvents.

Radium. It is a white radioactive luminescent metallic element of alkaline-earth group. It has several short-lived radioisotopes and one long-lived isotope, radium-226 (half-life 1602 years). Radium occurs in uranium ores, such as the oxides pitchblende and carnotite. It finds use in luminous paint and neutron sources, and (progressively less) in radiotherapy.

Symbol : Ra: m.p. 700°C; b.p. 1140°C; r.d. 5 (?): p.n. 88; r.a.m. 226.02.

Radon. It is a colourless monatomic radioactive element of the rare-gas group. It is now thought to form compounds. It has 19 short-lived radioisotopes: the most stable, radon-222 is a decay product of radium-226 and itself disintegrates into an isotope of polonium with a half-life of 3.82 days. ^{222}Rn is used in radiotherapy.

Symbol: Rn: m.p.—71°C; b.p.—61.8°C; d. 9.73 kg m^{-3} p.n. 86.

R.A.M. Relative Atomic Mass.

Raney Nickel. A spongy form of nickel which is obtained by the action of sodium hydroxide on a nickel—aluminium alloy. The sodium hydroxide dissolves the aluminium leaving a highly active form of nickel with a large surface area. The material is a black pyrophoric powder saturated with hydrogen. It has been an extremely efficient catalyst, especially for hydrogenation reactions at room temperature. It was discovered in 1927 by the American chemist M. Raney.

Rank. Applied to coal, rank is an indication of development or age of formation, and is generally denoted by the carbon content of the coal. E.g. anthracite is coal of high rank while lignite is low rank material.

Ranksite. A mineral consisting of a mixed sodium carbonate, sodium sulphate, and potassium chloride, $2Na_2CO_3 . 9Na_2SO_4 . KCl$.

Rare-earth Elements. See lanthanoids.

Rare Gases. Noble gases.

Realgar. A red mineral form of arsenic(II) sulphide, As_2S_2.

Reciprocal Proportions. See chemical combination.

Recombination Process. Refers to the process in which a neutral atom or molecule is formed by the combination of a positive ion and a negative ion or electron; *i.e.* a process of the type:

$$A^+ + B^- \rightarrow AB.$$

or

$$A^+ + e^- \rightarrow A$$

In recombination, the neutral species formed has been usually in an excited state, from which it can decay with emission of light or other electromagnetic radiation.

Red Lead. See dilead(II) lead(IV) oxide.

Redonda Phosphate. Naturally occurring $AlPO_4$ from the West Indies. Used in phosphorus manufacture.

Redox. An abbreviation for oxidation-reduction, *e.g.* redox potential—oxidation potential.

Redox. Relating to the processes of oxidation and reduction, which are intimately connected in that during oxidation by chemical agents the oxidizing agent itself becomes reduced, and vice versa. Thus an oxidation process is always accompanied by a reduction process.

Reducing Agent (reductant). Refers to a substance that brings about reduction in other substances. It achieves this by being itself oxidized. Reducing agents are having atoms with low oxidation numbers; that is the atoms have gained electrons. In reducing other substances, these atoms lose electrons.

Reductant. See reducing agent.

Reduction. Removal of oxygen from a compound or addition of hydrogen. More generally, an atom a molecule, or an ion is said to be reduced or to undergo reduction when it gains electrons. Reduction can be effected chemically, *i.e.* by the use of reducing agents, or electrically, in which case the reduction process occurs at the cathode. For example,

$$2Fe^{3+} + Cu \rightarrow 2Fe^{2+} + Cu^{2+}$$

where Cu is the reducing agent and Fe^{3+} is reduced, and

$$2H_2O + SO_2 + 2Cu^{2+} \rightarrow 4H^+ + SO_4^{2} + 2Cu^+$$

where SO_2 is the reducing agent and Cu^{2+} is reduced.

Stannous chloride, sulphur dioxide, the bisulphite ion, and hydroxylamine hydrochloride have been common reducing agents.

Refining. Refers to the process of purifying substances or extracting substances from mixtures.

Refractory. Describing a compound (*e.g.* an inorganic oxide) that has a very high melting point.

Refractory Materials, Refractories. A general term for materials which are not damaged by heating to at least 1500°C in a clean (oxidizing) atmosphere.

Acid refractory materials include fireclays, flint clays, china clays (kaolins), silica, flint, chalcedony, ganister and titanium dioxide.

Neutral refractory materials include graphite, charcoal, coke, chromite and various carbides.

Basic refractory materials include lime, magnesia, various materials composed chiefly of alumina (bauxite, diaspore, laterite, gibbsite, etc.), dolomite and most of the rare refractory oxides, particularly zirconia.

To be satisfactory when in use, a refractory material must not only be undamaged by heating it to or maintaining it at a high temperature, it must also resist the abrasive action of flue-dust, the contents of the charge of the furnace, the corrosion of slag, hot gases, flame etc. and movement of the various parts of the structure.

Relative Atomic Mass, A. The ratio of the average mass of an element to 1/12 of the mass of ^{12}C. Formerly known as atomic weight.

Relative Molecular Mass. Symbol : M_r. Refers to the ratio of the average mass per molecule of the naturally occurring form of an element or compound to 1/12 of the mass of an atom of nuclide ^{12}C. This was formerly called molecular weight. It is not to be used only for compounds that have discrete molecules; for ionic compounds (*e.g.* NaCl) and giant-molecular structures (*e.g.* BN) the formula unit is used.

Rem (radiation equivalent man). Refers to a unit for measuring the effects of radiation dose on the human body. One rem has been equivalent to an average adult male absorbing one rad of radiation. The biological effects depend on the type of radiation as well as the energy deposited per kilogram. The statutory limit for occupational exposure to radiation has been 5 rem per year.

Resonance. The term resonance refers to the fact that many molecules cannot be represented by any single valency bond structure, but must be considered as intermediate between two or more such structures. Such molecules are called resonance hybrids. This term does not imply the molecule resonances (in the physical sense) between the structures; the molecule is not well represented by one structure during part of its life time, by another during another part; rather the structure is unique, but is most conveniently described by the mathematical expedient called resonance. Thus benzene, for example, cannot be represented by any one of the structures I to V alone but is intermediate between all of them, and many others resembling.

I II III IV V

The canonical forms have no physical significance because these are merely a convenient way of picturing a molecule in terms of familiar Lewis structures.

Mathematically this is expressed by saving that the wave function describing the molecule, ψ_g, is a linear combination of the wave functions, ψ_1, ψ_2, ψ_3...describing the various structures.

$$\psi_g = C_1\psi_1 + C_2\psi_2 + \ldots$$

The classic difficulty over the concept of resonance is the disposition to regard the molecule as oscillating from one to another of the assigned forms. The fault is not with the molecule but rather with the human inability to describe properly this hybrid by the single structural formula that characterises it.

Reverberatory Furnace. In this furnace fuel is burnt in a separate part of the furnace (fireplace) and only the flames and hot gases come in contact with the charge being heated. The fireplace is closed at the bottom by iron bars on which the fuel

Reverberatory furnace.

is stacked and burnt. The rest of the horizontal part of the furnace is known as the bed or hearth. The material to be heated (charge) is placed on the hearth. The roof of the furnace is concave and takes a bend towards the end away from the fireplace and communicátes with the chimney (flue).

The charge is placed on the hearth and fuel is burnt in the fireplace. Hot gases and the flames are deflected or reverberated by the concave roof on the bed and heat the charge lying there.

Since the fuel does not come in direct contact with the charge, a reverberatory furnace can be used for reduction as well as oxidation processes. For reduction the material is mixed with a reducing agent like coke and heated while for oxidation it is heated in a current of air. Thus calcination and roasting are generally carried out in a reverberatory furnace.

Reverse Osmosis. Refers to a process of obtaining pure water from water containing a salt, as in desalination. Pure water and the salt water are separated by a semipermeable membrance and the pressure of the salt water is raised above the osmotic pressure, causing water from the brine to pass through the membrance into the pure water. This process needs a pressure of some 25 atmospheres, which makes it difficult to apply on a large scale.

Roasting The process of heating the ore strongly in excess of air. During roasting the volatile impurities are removed and the ore changed to oxide. Sulphide ores are generally roasted when they change to oxides. For example, zinc sulphide is roasted to get zinc oxide.

$$2ZnS + 3O_2 \longrightarrow 2ZnO + 2SO_2$$

Rhenium. Re. At. no. 75, at. wt. 186 2, m p. 3180°C, b.p. 5627°C (est.), d 20.53. Re is a rare metal extracted from flue dusts after roasting Mo or Cu ores. The metal is obtained by reduction of Re compounds with H_2, it has a hcp structure; is silvery grey in colour and dissolves in H_2O_2 and reacts with hot Cl_2, S or O_2. Used as an additive in W-and Mo-based alloys, in Re—W thermocules and in catalysts. World production 1968 4100 lbs. ^{187}Re with a half-life of 7×10^{10} years can be used in determining the age of the universe.

Rhodium. Rh. At. no. 45, at. wt. 102.9055, m.p. 1966°C, b.p. 3727°C, d 12.4. One of the platinum group metals. Found with other platinum metals and often in Ni—Cu deposits. After reduction to the mixed metals Rh is extracted from the aqua regia insoluble material with fused $KHSO_4$, extracted with water and precipitated as hydrated Rh_2O_3 which dissolves in HCl to give $(NH_4)_3RhCl_6$. The metal can be obtained by reduction with H_2, it is ccp and is very inert. It is mainly used as an alloying agent for Pt or Pd. U.S. production 1970 400 kg.

Rinmann's Green, $ZnCo_2O_4$. A spinel formed when cobalt nitrate solution is placed on zinc oxide and the mixture heated to redness. The green colour forms a delicate test for Zn.

Rock. Refers to an aggregate of mineral particles that makes up part of the earth's crust. It may be consolidated or unconsolidated (*e g.* sand, gravel, mud, shells, coral, and clay).

Rock Asphalt. Naturally occurring mixtures of bitumen with mineral matter. Rock asphalts generally contain 5—15% by weight of bitumen.

Rock Crystal. See quartz.

Rock Salt. A transparent naturally occurring mineral form of sodium chloride.

Roentgen. The former unit of dose equivalent (see radiation units). It is named after the discoverer of X-rays, W. K. Roentgen (1845—1923).

Roman Cement. A hydraulic cement made from clay and calcium carbonate. It is a crude Portland cement.

Rose's Metal. An alloy of low melting point (about 100°C) consisting of 50% bismuth, 25—28% lead, and 22—25% tin.

Roussin Salts. Iron nitrosyls. There are two series black (iron(II) plus NO plus sulphides) $M(Fe_4S_3(NO)_7]$ and red (black plus OH^-), $M_2[Fe_2S_2(NO)_2]$.

Rubidium Symbol Rb. It is a soft silvery-white metallic element belonging to group 1 of the periodic table; a.n. 37; r.a.m. 85.47; r.d. 153; m p. 38 4°C; b.p 688°C. It occurs in a number of minerals (*e.g.* lepidolite) and in certain brines. The metal is produced by electrolysis of molten rubidium chloride. The naturally occurring isotope rubidium—87 is radioactive (see rubidium—strontium dating). The metal is highly reactive, with properties similar to those of other group I elements, igniting spontaneously in air. It was discovered spectroscopically by R.W. Bunsen and G.R. Kirchhoff in 1861.

Rubidium—Strontium Dating. Refers to a method of dating geological specimens which is based on the decay of the radioisotope rubidium—87 into the stable isotope strontium—87. Natural rubidium contains 27.85% of rubidium—87, which has a half-life of 4.7×10^{11} years. The ratio $^{87}Rb/^{87}Sr$ in a specimen gives an estimate of its age (up to several thousand million years).

Ruby. The transparent red variety of the mineral corundum, the colour being due to the presence of traces of chromium. It is a valuable gemstone, more precious than diamonds. Rubies have been produced synthetically by the Verneuil flame-fusion process. Industrial rubies are used in lasers, watches, and other precision instruments.

Rusting. The corrosion of iron in air to form an oxide. Both oxygen and moisture must be present for rusting to occur. The process is electrolytic, involving electrochemical reactions on the wet ion :

$$Fe(s) \rightarrow Fe^{2+}(aq) + 2e^-$$
$$H_2O + O_2(aq) + 2e^- \rightarrow 2OH^-(aq)$$

Iron(II) hydroxide precipitates and is oxidized to the red hydrated iron(II) oxide $Fe_2O_3.H_2O$.

Ruthenium. A transition metal that occurs naturally with platinum. It forms alloys with platinum that are used in electrical contacts. Ruthenium is also used in jewellery alloyed with palladium.

Symbol : Ru; mp. 2430°C; b.p. 3700°C; r.d. 12.4; p.n. 44; r.a.m. 101.07.

Rutherfordium. Element 104.

Rutile. Mineral TiO_2, a reddish brown material used as a source of Ti and in ceramics. The tetragonal structure has each Ti octahedrally co-ordinated by O and each O co-ordinated by three Ti in an approximate plane. This structure is adopted by many oxides and fluorides, *e.g.* PbO_2 and MnO_2.

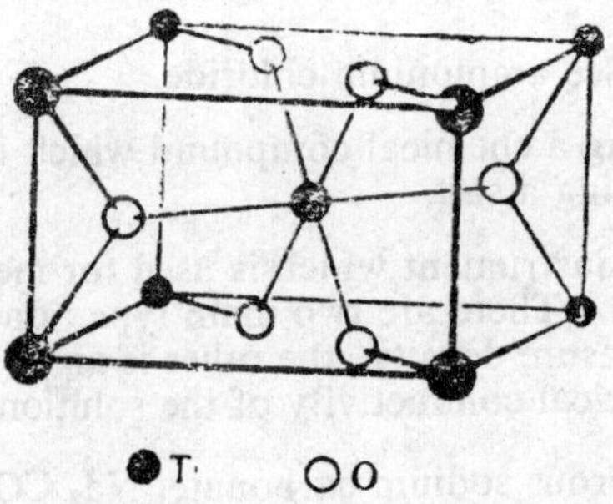

Crystal structure of rutile

Rutile. A mineral form of titanium(IV) oxide, TiO_2.

Rydberg's Constant. The wave number of any line in a spectrum can be expressed as the difference of two terms of the form R/x^2, that is

$$v = \frac{R}{x_1^2} - \frac{R}{x_2^2}$$

where x_1 and x_2 are small integers and R is a constant, the Rydberg's constant. The value of R which has only a small variation in all terms of all such formulae, is 109678.

Rydberg Spectrum. An absorption spectrum of a gas in the ultra-violet region, consisting of a series of lines that become closer together towards shorter wavelengths, merging into a continuous absorption region. The absorption lines correspond to electronic transitions to successively higher energy levels. The onset of the continuum corresponds to photoionization of the atom or molecule, and can thus be used to determine the ionization potential.

S

Sacrificial Protection. Refers to a method of protection against electrolytic corrosion (especially rusting). In protecting steel pipelines, for instance, zinc or magnesium rods are buried in the ground at points along the line and connected to the pipeline.

Sal Ammoniac. See ammonium chloride.

Saline. Describing a chemical compound which is a salt, or a solution containing a salt.

Salinometer. An instrument which is used for measuring the salinity of a solution. There are two main type : one is a type of hydrometer to measure density; the other is an apparatus for measuring the electrical conductivity of the solution.

Sal Soda. Anhydrous sodium carbonate, $Na_2 CO_3$.

Salt. A compound having an acidic and a basic radical, or a compound which is formed by total or partial replacement of the hydrogen in an acid by a metal. In general terms a salt has been a material that has identifiable cationic and anionic components.

Salt Cake. Industrial sodium sulphate.

Salt Hydrates. Clathrates formed by frame-works of hydrogen-bonded water and anions with cations occupying holes in the lattice. Examples have very high water content, such as $(Pr^{n}_{3}S)\,F,20H_2O$.

Saltpetre. See potassium nitrate.

Samarium Symbol Sm. A soft silvery metallic element which belongs to the lanthanoids; a n. 6.2; r.a.m. 150 35; r.d. 7.52 (20°C); m.p. 1073°C; b.p. 1791°C. It is found in monazite and bastnatite. There are seven naturally occurring isotopes, all of which are stable except samarium −147, which is weakly radioactive (half-life 2 5 × 10^{11} years). The metal finds use in special alloys for making nuclear-reactor parts as it is a neutron absorber. Samarium oxide (Sm_2O_3) is used in small quantities in special optical glasses. The largest use of the element has been in the ferromagnetic alloy $SmCo_5$, which produces permanent magnets five times stronger than any other material. The element was discovered by Francois Lecoq de Boisbaudran in 1879.

Sand. Particles of rock having diameters in the range 0.06 −2.00 mm Most sands are composed chiefly of particles of quartz, which are derived from the weathering of quartz bearing rocks.

Sandwich Compound. A type of complex which is formed between transition-metal ions and aromatic compounds, in which the metal ion is 'sandwiched' between the rings. The bonding is between the d-orbitals of the metal and the π electrons on the ring. Ferrocene ($Fe(C_5H_5)_2$) is the best known example.

Sapphire. As a gemstone a blue form of aluminium oxide (corundum) containing some metal impurities. Now prepared artificially by fusing Al_2O_3. Used as a very inert material, *e.g* for handling HF.

S-block Elements. The elements of the first two main groups of the periodic table; *i.e.* groups IA (Li, Na, K, Rb, Cs, Fr) and IIA (Be, Mg, Ca, Sr, Ba, Ra). The outer electronic configurations of these elements all are having inert-gas structures plus outer ns^1 (IA) or ns^2 (IIA) electrons.

Scandium Symbol Sc. A rare soft silvery metallic element which belongs to group IIIB of the periodic table; a.n. 21; r.a.m. 44.956; r.d. 2.985 (alpha form), 3.19 (beta form); m.p. 1540°C;

b p. 2850°C. Scandium often occurs in lanthanoid ores, from which it can be separated on account of the greater solubility of its thiocyanate in either. The only natural isotope, which is not radioactive, is scandium-45, and there are nine radioactive isotopes, all with relatively short half-lives. Predicted in 1869 by Mendeleev, and then called ekaboron, the oxide (called scandia) was isolated by Nilson in 1879.

Scheelite. A mineral form of calcium tung-state, $CaWO_4$, which is used as an ore of tungsten. It is found in contact metamorphosed deposits and vein deposits as colourless or white tetragonal crystals.

Schomaker-Stevenson Equation. The equation $r_{A-B}=r_A+r_B-0.09(X_A-X_B)$ relating the bond length r_{A-B} to the individual radii r_A and r_B of the two atoms concerned and the electronegativities X_A and X_B of the two atoms concerned in the bond. This relation is only empirical and is not very accurate.

Schrodinger Equation. An equation which is used in wave mechanics (see quantum mechanics) for the wave function of a particle. The time-independent Schrodinger equation has been as follows :

$$\Delta^2\psi+8\pi^2m(E-U)\psi/h^2=0$$

where ψ is the wave function; Δ^2 the Laplace operator, h the Planck constant, m the particle's mass, E its total energy, and U its potential energy.

Screening Rule. The electrons in complete inner shells screen the outer electrons against attraction by the nucleus. This is known as screening effect. It has been observed that this screening effect is less operative for the more penetrating orbitals than for the less penetrating ones. As a result of the screening effect, every ns orbital is filled earlier than the corresponding (n—1) d orbital. Thus, 4s is filled earlier than 3d. For similar reasons 5s, 5p and 6s orbitals are filled before the 4f orbital.

Seed. A crystal which is used to induce other crystals to form from a gas, liquid, or solution.

Selenides. Refer to the binary compounds of selenium with other more electropositive elements. Selenides of nonmetals are covalent (*e.g* H_2Se). Most metal selenides can be prepared by direct combination of the elements. Some are well-defined ionic compounds (containing Se^{2-}), while others are nonstoichiometric interstitial compounds (*e.g.* Pd_4Se, $PdSe_2$).

Selenium. A metalloid element which occurs in several allotropic forms and belonging to group VI of the periodic table. It is found in minute quantities in sulphide ores and industrial sludges. The common grey allotrope has been very light-sensitive and is used in photocells, solar cells, and in xerography.

Symbol : Se; m.p. 217°C (grey); b.p. 684.9°C (grey); r.d. 4.79 (grey); p.n. 34; r.a.m. 78.96.

Self-ionization. The process of partial dissociation into cations and anions envisaged for self-ionizing solvents. E.g.

$$2H_2O \rightleftharpoons [H_3O]^+ + OH^-$$

$$3NH_3 \rightleftharpoons [NH_4]^+ + NH_2^-$$

Ionic species containing the characteristic cation—*e.g.* NH_4Cl—are acids; species containing the characteristic anion are base, *e.g.* KNH_2.

Semi-b : Bituminous Coals. Non coking coals containing about 90–93% C, 4–4.5% H and 3–4.5% O. They have a volatile matter of about 15–20% and a calorific value of about 36 MJ/kg. This type of coal is important in steam-raising. See also bituminous coals, anthracites.

Semipolar Bond. A coordinate bond.

Semi-steel. Cast iron to which has been added steel scrap. The product is still cast iron since the carbon content is between 2.5–3.2%.

Septivalent (Heptavalent). Having a valency of seven.

Sequestering Agent, Complexones. Compounds which are very effective in forming complexes with metal cations (and thus with salts) and thus preventing them acting as simple hydrated cations. Ethylenediaminetetra-acetic acid (EDTA) and related derivatives from complexes with most M^{2+} and M^{3+} species. Gluconic acid and other hydroxy acids act similarly. Sequestering agents are used extensively, *e.g.* for preventing deleterious catalytic effects of traces of Fe or Cu and in analysis.

Sequestration. Refers to the process of forming coordination complexes of an ion in solution. Sequestration often involves the formation of chelate complexes, and is used to prevent the chemical effect of an ion without removing it from the solution (*e.g.* the sequestration of Ca^{2+} ions in water softening). It is also used as a way of supplying ions in a protected form, *e.g.* the use of sequestered iron solutions for plants in regions having alkaline soil.

Sesqui. Prefix which indicates a ratio of 2 : 3 in a chemical compound. For example, a sesqui-oxide has the formula M_2O_3.

Sexivalent (Hexavalent). Having a valency of six.

Shell. A group of electrons that share the same principal quantum number. Early work on X-ray emission studies used the terms K, L, M, and these have been still sometimes used for the first three shells : n=1 K-shell; n=2 L-shell; n=3 M-shell.

Sherardizing. Refers to the process of coating iron or steel with a zinc corrosion-resistant layer by heating the iron or steel in contact with zinc dust to a temperature slightly below the melting point of zinc. At a temperature of about 371°C the two metals amalgamate to yield internal layers of zinc-iron alloys and an external layer of pure zinc. The process was invented by Sherard Cowper-Coles (d. 1935).

Shift Conversion, Shift Reaction. An important reaction in which carbon monoxide reacts with steam to produce hydrogen and carbon dioxide.

$$CO + H_2O \rightarrow CO_2 + H_2$$

Carbon monoxide and excess steam are normally passed over a cobalt catalyst at about 250-300°C resulting in greater than 99% conversion of CO to CO_2. The carbon dioxide is often removed by absorption, *e.g.* by a hot carbonate process. This conversion reaction is widely used in oil or solid fuel gasification processes for the production of town gas or substitute natural gas.

Sial. Refer to the rocks that form the earth's continental crust. These are granite rock types rich in silica (SiO_2) and aluminium (Al), hence the name. Compare sima.

Siderite. Refers to a brown or grey-green mineral form of iron(II) carbonate, $FeCO_3$, often with magnesium and manganese substituting for the iron. It occurs in sedimentary deposits or in hydrothermal veins and is an important iron ore. It is found in England, Greenland, Spain, N Africa, and the USA.

Sigma Bond, σ-bonds. Covalent bond containing two electrons with the major overlap between the orbitals forming the bond being in one region in space directly between the nuclei. E.g. H_2.

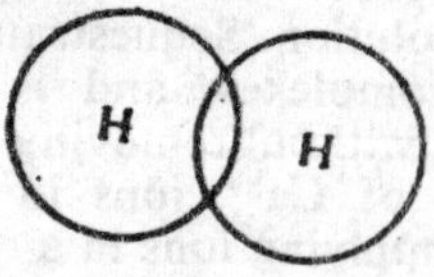

Sigma Electron. An electron in a sigma orbital. See orbital.

Silane. Any of a small number of hydrides of silicon : SiH_4, Si_2H_6, Si_3H_8, etc. Silanes can be made by the action of acids on magnesium silicide (Mg_2Si), They are unstable compounds and ignite spontaneously in air. Only the first few members of the series are known and there is not the range of hydrosilicon' compounds to compare with the hydrocarbons.

Silatranes. Internal complexes of silicon. Some of these complexes have physiological activity. E.g.

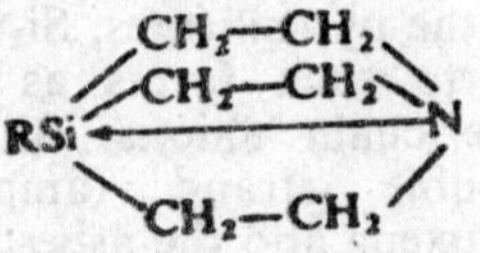

Silicate. Any of a group of substances containing negative ions composed of silicon and oxygen. The silicates are a very exten-

SiO_4^{4-} as in Be_2SiO_4 (phenacite)

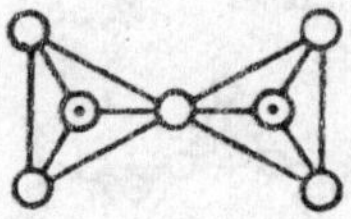

$Si_2O_5^{2-}$ as in $Sc_2Si_2O_7$ (thortveitite)

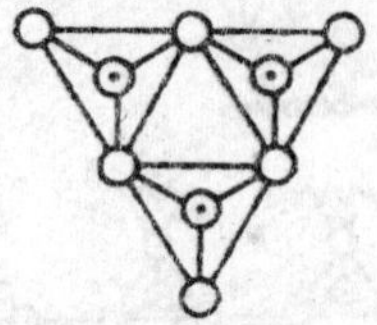

$Si_3O_9^{6-}$ as in $BaTiSi_3O_9$ (bentonite)

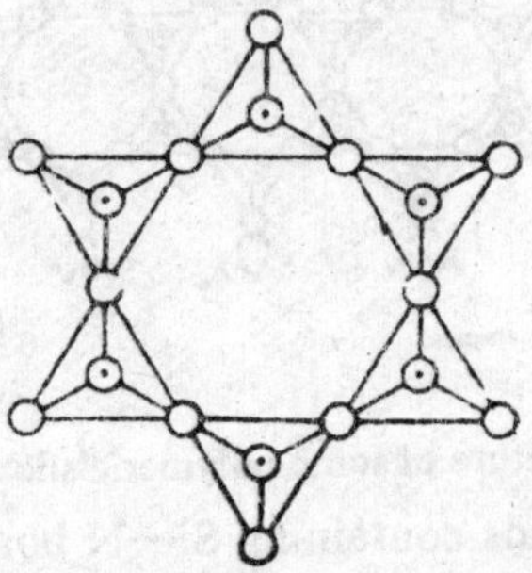

$Si_6O_{18}^{12-}$ as in $Be_3Al_2Si_6O_{18}$ (beryl)

Structure of some discrete silicon ions

sive group and natural silicates from the major component of most rocks (see silicate minerals). The basic structural unit is the tetrahedral SiO_4 group. This may occur as a simple discrete SiO_4^{4-} anion as in the orthosilicates, *e.g.* phenacite (Be_3SiO_4) and willemite (Zn_2SiO_4). Many larger silicate species are also found. These are composed of SiO_4 tetrahedra linked by sharing oxygen atoms as in the pyrosilicates, $Si_2O_7^{6-}$, *e.g.* $Sc_2Si_2O_7$. The linking can extent to such forms as benitoite, $BaTiSi_3O_9$, or alternatively infinite chain anions, which are single strand (pyroxenes) or double strand (amphiboles). Spodumene, $LiAl(SiO_3)_2$, is a pyroxene and the asbestos minerals are amphiboles. Infinite two-dimensional sheets are also possible, as in the various micas and the linking can extend to full three-dimensional framework structures, often with substituted trivalent atoms in the lattice. The zeolites are examples of this.

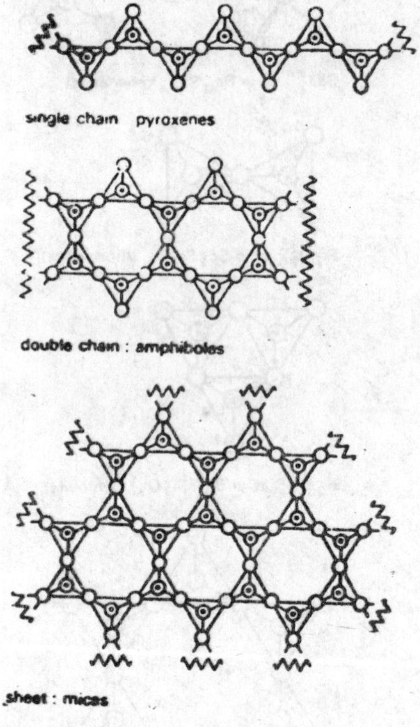

Structure of some polymeric silicate ions

Silazanes. Compounds containing Si—N bonds.

Silica. See silicon(IV) oxide.

Silicane. See silane.

Silicide. A compound of silicon with a more electropositive element. The silicides are structurally similar to the interstitial carbides but the range encountered is more diverse. They react with mineral acids to form a range of silanes.

Silicon. A hard brittle grey metalloid element; the second element in group IV of the periodic table. It has the electric configuration of neon with four additional outer electrons; *i.e.*,

Symbol : Si; m.p. 1414°C; b.p. 2355°C; r.d. 2.32-2.34; p.n. 14; r.a.m. 28.

Silicon Carbide (carborundum). A black solid compound, SiC, insoluble in water and soluble in molten alkali; r.d. 3.217; m.p. c. 2700°C. Silicon carbide is made by heating silicon(IV) oxide with carbon in an electric furnace (depending on the grade required sand and coke may be used). It is extremely hard and is widely used as an abrasive. The solid exists in both zinc blende and wurtzite structures.

Silicon Dioxide. See silicon (IV) oxide.

Silicone Rubbers, $(Me_2SiO)_n$ (also other alkyl derivatives). Silicone derivatives. Retains elastic properties over very wide temperature range, inert to acids, alkalis, oils and oxidizing agents.

Silicones. Polymeric organosilicon derivatives containing Si—O—Si links. Organosilicon halides are prepared RX plus Cu—Si or Ag—Si alloy; RX plus $SiCl_4$ over Zn or Al; $SiCl_4$ or Si—H compounds plus olefine over catalyst (hydrosilylation). (These derivatives also formed $SiCl_4$ plus, *e.g.*, LiR, RMgX on laboratory scale). The halides are hydrolysed to give polymers with properties depending upon the degree of polymerization (extent of hydrolysis). Silicones are used in greases, hydraulic fluids, sealing compounds, enamels, varnishes, resins, synthetic rubbers. Resins are used in electrical insulation and laminates. They are odourless, colourless and inert, immiscible with water and have high flash points.

Silicon Hydride. Silane.

Silicon(IV) Oxide (silicon dioxide, silica, SiO_2). A hard crystalline compound occurring naturally in three crystalline forms (quartz, tridymite, and crystobalite). Sand is mostly silicon(IV) oxide. Fused silica is a glassy substance used in laboratory apparatus. Silica is used in the manufacture of glass.

Siloxanes. Molecular compounds containing SiOSi groups, *e.g.* hexachlorosiloxane, $Cl_3SiOSiCl_3$ (controlled hydrolysis of $SiCl_4$:

$SiCl_4$ plus air at high temperature). Poly (organosiloxanes) are silicones.

Siloxene. Polymeric reducing substance of formula $(Si_2H_2O)_n$ formed from Ca_2Si and EtOH/HCl.

Silver. A transition metal that occurs native and as the sulphide (Ag_2S) and chloride (AgCl). It is extracted as a by-product in refining copper and lead ores. Silver darkens in air due to the formation of silver sulphide. It is used in coinage alloys, tableware, and jewellery. Silver compounds are used in photography.

Symbol : Ag; m.p. 960°C, b.p. 2180°C; r.d. 10.5; p n. 47; r.a.m. 107.87.

Silver(I) Bromide. A yellowish solid compound, AgBr; r.d. 65; m.p. 432°C. It can be precipitated from silver(I) nitrate solution by adding a solution containing bromide ions. It dissolves in concentrated ammonia solutions (but, unlike the chloride, does not dissolve in dilute ammonia). The compound is used in photographic emulsions.

Silver(I) Chloride (AgCl) A compound prepared as a curdy white precipitate by the addition of a soluble chloride solution to a solution of silver(I) nitrate. It occurs in nature as the mineral 'horn silver'. Silver(I) chloride is insoluble in water but dissolves in concentrated hydrochloric acid and concentrated ammonia solution. It is rather unreactive but is affected by sunlight, being used extensively in photography. Silver(I) chloride crystallizes in the sodium chloride lattice pattern. It absorbs ammonia gas forming ammines, *e.g.* $AgCl.2NH_3$ and $AgCl.3NH_3$.

Silver(I) Iodide (AgI) A pale yellow precipitate prepared by the addition of a soluble iodide solution to a solution of silver(I) nitrate. Silver(I) iodide occurs in nature as iodoargyrite in the form of hexagonal crystals. It is virtually insoluble in ammonia solution and is used in the photographic industry. Silver(I) iodide will absorb ammonia gas. It is trimorphic and is found to contract when heated and expand on cooling.

Silver (I) Nirate. A colourless solid, $AgNO_3$; r.d. 4.3; m.p. 212°C. It is an important silver salt because it is water-soluble. It is used in photography. In the laboratory, it is used as a test for chloride, bromide, and iodide ions and in volumetric analysis of chlorides using an absorption indicator.

Silver (I) Oxide (argentous oxide, Ag_2O). A brown amorphous solid formed by the addition of sodium or potassium hydroxide solution to a solution of silver nitrate. Silver (I) oxide turns moist red litmus blue and decomposes (at 160°C) to give silver and oxygen. Silver (I) oxide dissolves in concentrated ammonia solution, forming the complex ion $[Ag(NH_3)_2]^+$. In organic chemistry it is used when moist to convert alkyl halides to the corresponding alcohol and when dry to convert alkyl halides into ethers.

Silver (II) Oxide (argentic oxide, AgO). A black diamagnetic substance prepared by the oxidation of either silver or silver (I) oxide using ozone. Alternatively it can be prepared as a precipitate by the addition of a solution of potassium persulphate to one of silver (I) nitrate. Silver (II) oxide has a structure similar to that of zinc sulphide.

Sodium Chloride (common salt, salt, NaCl) A white solid prepared by the neutralization of hydrochloric acid with aqueous sodium hydroxide. It occurs in sea water and natural brines. Natural solid deposits of rock salt are also found in certain places. The compound dissolves in water with the absorption of heat, its solubility changing very little with temperature. It is used to season and preserve food. Industrially, it is used as a raw material in the manufacture of sodium carbonate (Solvay process), sodium hydroxide (electrolysis), and soap. Sodium chloride is almost insoluble in alcohol.

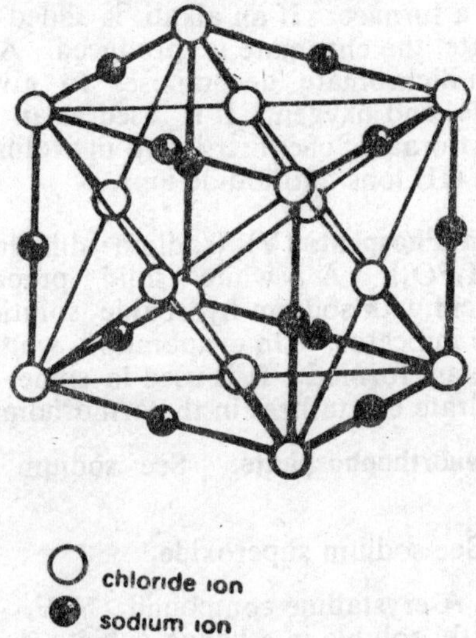

Sodium-chloride structure.

Sodium chloride Structure. A form of crystal structure hat consists of a face-centred cubic arrangement of sodium ions with chloride ions situated at the middle of each edge and in the centre of the cube. Electrostatic attraction holds the oppositively charged ions together.

The lattice can also be thought of as two interpenetrating face-centred cubes, one composed of sodium ions and the other of chloride ions. Other compounds (*e.g.*, sodium bromide and potassium chloride) having their ions arranged in the same positions, are also described as having the sodium-chloride structure.

Sodium Cyanide. A white or colourless crystalline solid, NaCN, deliquescent, soluble in water and in liquid ammonia, and slightly soluble in ethanol; cubic; m.p. 564°C; b.p. 1496°C. Sodium cyanide is now made hy absorbing hydrogen cyanide in sodium hydroxide or sodium carbonate solution. The compound is extremely poisonous because it reacts with the iron in haemoglobin in the blood, so preventing oxygen reaching the tissues of the body. It is used in the extraction of precious metals and in electroplating industries. Aqueous solutions are alkaline due to salt hydrolysis.

Sodium Dichromate ($Na_2Cr_2O)^-$. An orange deliquescent solid that is very soluble in water. It is made from finely powdered chromite, which is heated with calcium oxide and sodium carbonate in a furnace. If an alkali is added to a solution of the dichromate, the chromate is produced. At high temperatures, sodium dichromate decomposes to give the chromate, chromic oxide, and oxygen. It is used as an oxidizing agent, particularly in organic chemistry and in volumetric analysis to estimate iron (II) ions and iodide ions.

Sodium Dihydrogen Phosphate (V) (sodium dihydrogen orthophosphate, NaH_2PO_4). A white solid prepared by titrating phosphoric acid with sodium hydroxide solution using methyl orange as the indicator. On evaporation white crystals of the monohydrate are formed. It is used in some baking powders. The monohydrate crystallizes in the orthorhombic system.

Sodium Dihydrogenorthophosphate. See sodium dihydrogenphosphate (V).

Sodium Dioxide. See sodium superoxide.

Sodium Fluoride. A crystalline compound, NaF, soluble in water and very slightly soluble in ethanol; cubic; r.d. 2.56; m.p. 993°C; b.p. 1695°C. It occurs naturally as villiaumite and may be prepared by the reaction of sodium hydroxide or of sodium carbonate with hydrogen fluoride. The reaction of sodium fluoride with concentrated sulphuric acid may be used as a source of hydrogen fluoride. The compound is used in ceramic enamels and as a preservative agent for fermentation. It is highly toxic but in very dilute solution (less than 1 part per million) it is used in the fluoridation of water for the prevention of tooth decay on account of its ability to replace OH groups with F groups in the material of dental enamel.

Sodium Hexafluoraluminate. A colourless monoclinic solid, Na_3AlF_6, very slightly soluble in water; r.d. 2.9; m p. 1000°C. It changes to a cubic form at 580°C. The compound occurs naturally as the mineral cryolite but a considerable amount is manufactured by the reaction of aluminium fluoride with alumina and sodium hydroxide or directly with sodium aluminate. Its most important use is in the manufacture of aluminium in the Hall Heroult cell. It is also used in the manufacture of enamels, opaque glasses, and ceramic glazes.

Sodium Hydride (NaH). A white crystalline solid prepared by passing a pure stream of dry hydrogen over sodium at 350°C; the sodium is usually suspended in an inert medium. Electrolysis of fused sodium hydride yield hydrogen at the anode. Sodium hydride reacts with water to form sodium hydroxide solution and hydrogen. It is used as a powerful reducing agent to convert water to hydrogen, concentrated sulphuric acid to hydrogen sulphide, and iron (III) oxide to iron. Sodium hydride bursts into flames spontaneously on contact with the halogens at room temperature. It dissolves in liquid ammonia to give sodamide, $NaNH_2$.

Sodium Hydrogencarbonate (bicarbonate of soda; sodium bicarbonate). A white crytalline solid, $NaHCO_3$, soluble in water and slightly soluble in ethanol; monoclinic; r.d. 2.159; loses carbon dioxide above 270°C. It is manufactured in the Solvay process and may be prepared in the laboratory by passing carbon dioxide through sodium carbonate or sodium hydroxide solution. Sodium hydrogencarbonate reacts with acids to give carbon dioxide and, as it does not have strongly corrosive or strongly basic properties itself, it is employed in bulk for the treatment of acid spillage and in medicinal applications as an antacid.

Sodium Hydrogen Sulphate (sodium bisulphat, $NaHSO_4$). A white solid formed either by the partial neutralization of sodium hydroxide solution with sulphuric acid, by the action of concentrated sulphuric acid on sodium nitrate, or by heating equimolar quantities of sodium chloride and concentrated sulphuric acid. In aqueous solution sodium hydrogensulphate is strongly acidic. It crystallizes as the monohydrate ($NaHSO_4.H_2O$), which dehydrates on warming. If heated strongly the pyrosulphate ($Na_2S_2O_7$) is formed, which decomposses to give the sulphate and sulphur trioxide. Sodium hydrogen sulphate is used as a cheap source of sulphuric acid and in the dyeing industry.

Sodium Hydrogensulphite (sodium bisulphite). A white solid, $NaHSO_3$, which is very soluble in water (yellow in solution) and slightly soluble in ethanol; monoclinic; r.d. 1.48. It decomposes on heating to give sodium sulphate, sulphur dioxide and sulphur. It is formed by saturating a solution of sodium carbonate with sulphur dioxide. The compound is used in the brewing industry and in the sterilization of wine casks. It is a general antiseptic and bleaching agent.

Sodium Hydroxide (caustic soda, NaOH). A white delinquescent slightly translucent solid with a fibrous texture. In solution it is a strong alkali and electrolyte. In the laboratory it can be prepared by reacting sodium, sodium monoxide, or sodium peroxide with water. Industrially it can be prepared by the electrolysis of sodium chloride using a mercury cathode (Castner-Kellner) or 'diaphragm' cell. It can also be manufactured by the Gossage process. Sodium hydroxide dissolves readily in water with the evolution of heat. Its solutions have a soapy feel and are very corrosive. It is used to absorb acidic gases, such as carbon dioxide and sulphur dioxide. Industrially it is used in soap and paper manufacture and in the purification of bauxite.

Sodium Iodide. A white crystalline solid, NaI, very soluble in water and solube in both ethanol and ethanoic acid It is known in both the anhydrous form (cubic; r.d. 3.67; m.p. 661°C; b.p. 1304°C) and as the dihydrate (monoclinic; r.d. 2·45). It is prepared by the reaction of hydrogen iodide with sodium carbonate or sodium hydroxide in solution. Like potassium iodide, sodium iodide in aqueous solution dissolves iodine to form a brown solution containing the I_3 ion. It finds applications in photography and is also used in medicine as an expectorant and in the administration of radioactive iodine for studies of thyroid function and for treatment of diseases of the thyroid.

Sodium Monoxide (Na_2O). A white solid formed by burning sodium in a deficiency of oxygen or alternatively by reducing sodium peroxide or sodium hydroxide with the requisite amount of sodium. It reacts violently with water to form sodium hydroxide and with acids to form solutions of their salts. Sodium monoxide forms cubic crystals. It dissolves in liquid ammonia to give sodamide and sodium hydroxide.

Sodium Nitrate (Chile saltpetre). A white solid, $NaNO_3$, soluble in water and in ethanol; trigonal; r.d. 2.261; m.p. 306°C; decomposes at 380°C. A rhombohedral form is also known. It is obtained from deposits of caliche or may be prepared by the reaction of nitric acid with sodium hydroxide or sodium carbonate. It was previously used for the manufacture of nitric acid by heating with concentrated sulphuric acid. Its main use is in nitrate fertilizers.

Sodium Nitrate ($NaNO_2$) A yellowish-white solid formed by the thermal decomposition of sodium nitrate. It is readily soluble in water. The anhydrous salt forms orthorhombic crystals. When treated with cold dilute hydrochloric acid, sodium nitrite forms nitrous acid. It is used in organic chemistry in the process of diazotization and industrially as a corrosion inhibitor.

Sodium Orthophosphate. See trisodium phosphate (V)

Sodium Peroxide (Na_2O_2). A yellowish-white ionic solid formed by the direct combination of burning sodium with excess oxygen. It reacts with water to form sodium hydroxide and hydrogen peroxide the latter decomposes rapidly in the alkaline solution to give oxygen. Sodium peroxide is used as a bleaching agent and an oxidizing agent for such materials as wool and wood pulp. Sodium peroxide can convert nitrogen (II) oxide to sodium nitrate and iodine to sodium iodate.

Sodium Sesquicarbonate. A white crystalline hydrated double salt, Na_2CO_3 $NaHCO_3$ $2H_2O$, soluble in water but less alkaline than sodium carbonate; r.d. 2 12; decomposes on heating. It may be prepared by crytallizing equimolar quantities of the constituent materials; it also occurs natùrally as trona and is Searless Lake brines. It is widely used as a detergent and soap builder and, because of its mild alkaline properties, as a water-softening agent and bath-salt base.

Sodium Sulphate (Na_2SO_4). A white soluble solid formed by heating a mixture of sodium chloride and concentrated sulphuric acid. Industrially it is prepared by the first stage of the Leblanc process. It forms two hydrates, the decahydrate (Na_2SO_4 $10H_2O$) and the heptahydrate (Na_2SO_4 $7H_2O$). Sodium decahydrate, known as Glauber's salt, is used in the manufacture of glass and as a purgative in medicine. It effloresces on exposure to air to form the anhydrous salt, which is used as a drying agent in organic chemistry. At room temperature sod um sulphate crystallizes in the orthorhombic system; around 250°C there is a transition to the hexagonal system. The decahydrate crystallizes in the monoclinic form.

Sodium Sulphide. A yellow-red solid, Na_2S, formed by the reduction of sodium sulphate with carbon (coke) at elevated temperatures. It is a corrosive and readily oxidized material of variable composition and usually contains polysulphides of the type Na_2S_2 Na_2S_3, and Na_2S_4, which cause the variety of colours. It is known in an anhydrous form (r.d. 1.85; m.p 1180°C) and as a non-hydrate, $Na_2S.9H_2O$ (r.d. 1.43; decomposes at 920°C). Other hydrates of sodium sulphide have been reported. The compound is deliquescent, soluble in water with extensive hydrolysis, and slightly soluble in alcohol. It is used in wood pulping, dyestuffs manufacture, and metallurgy on account of its reducing properties. It has also been used for the production of sodium thiosulphate (for the photographic industry) and as a depilatory agent in leather preparation. It is a strong skin irritant.

Sodium Sulphite (Na_2SO_3). A white solid formed by reacting the exact amount of sulphur dioxide with either sodium carbonate or sodium hydroxide. It is readily soluble in water and crystallizes as colourless crystals of the heptahydrate ($Na_2SO_3.7H_2O$). When treated with dilute mineral acids, sulphur dioxide is evolved. At high temperatures, sodium sulphite undergoes thermal decomposition to give sodium sulphate and sodium sulphide.

Sodium Superoxide (sodium dioxide). A whitish-yellow solid, NaO_2, formed by the reaction of sodium peroxide with excess oxygen at elevated temperatures and pressures. It reacts with water to form hydrogen peroxide and oxygen.

Sodium Thiosulphate ($Na_2S_2O_3$). A white solid prepared either by boiling sodium sulphite with flowers of sulphur or by passing sulphur dioxide into a suspension of sulphur in boiling sodium hydroxide. Sodium thiosulphate is readily soluble in water and crystallizes as large colourless crystals of the pentahydrate ($Na_2S_2O_3.5H_2O$). It reacts with dilute acids to give sulphur and sulphur dioxide. It is used in photography as 'hypo' and industrially as an antichlor. In volumetric analysis solutions sodium thiosulphate are usually prepared from the pentahydrate. On heating it disproporticnates to give sodium sulphate and sodium sulphide.

Sodium Tungstate. $Na_2WO_4,2H_2O$.

Soft Acids and Bases. See hard and soft acids and bases.

Soft Water. See hardness of water.

Solder. An alloy used in joining metals. The molten solder alloys with (wets) the surfaces to be joined, without melting them, and solidifies on cooling to form a hard joint. The surfaces must be clean and free of oxide and a flux is often used to achieve this. Soft solders consist of lead with up to 60% tin and melt in the range 183—250°C. Soft-soldering is used, for example, in plumbing and making electrical contacts. Brazing solders are copper-zinc alloys that have higher melting points and produce stronger joints than soft solders; silver can be added to produce silver solders.

Solid State Reactions. Reactions occurring between two or more solid compounds or decompositions occurring from the solid state. Gases may be evolved in such reactions. Most ternary oxides are formed by solid state reactions.

Solvation. The attraction of an ion in solution for molecules of solvent. In water, for example, a positive ion will be surrounded by water molecules, which tend to associate around the ion because of attraction between the positive charge of the ion, and the negative part of the polar water molecule. The energy of this solvation (hydration in the case of water) is the 'force' overcoming the attraction between positive and negative ions when an ionic solid dissolves. The attraction of the dissolved ion for solvent molecules may extend for several layers. In the case of transition metal elements, ions may also form complexes by coordination to the nearest layer of molecules.

Solvay Process (ammonia—soda process). An industrial method of making sodium carbonate from calcium carbonate and sodium chloride. The calcium carbonate is first heated to give calcium oxide and carbon dioxide, which is bubbled into a solution of sodium chloride in ammonia. Sodium hydrogencarbonate is precipitated:

$$H_2O + CO_2(g) + NaC(aq) + NH_3(aq) \rightarrow$$
$$NaHCO_3(s) + NH_4Cl(aq)$$

The sodium hydrogencarbonate is heated to give sodium carbonate and carbon dioxide. The ammonium chloride is heated with calcium oxide (from the first stage) to generate the ammonia. The process was patented in 1861 by the Belgian chemist Ernest Solvay (1838-1922).

Solvolysis. A reaction between a compound and the solvent in which it is dissolved.

Sommerfeld Model. Inspite of many successes, Bohr's theory was found to be inadequate to explain certain details in the spectrum of hydrogen. For instance, the H_α line of the Balmer series was found to contain several components. This fine structure of spectral lines could not be explained by Bohr's theory which assumed that there was only one orbit for each quantum number n, whereas the observed fine structure suggested that for any given quantum number n there might be several orbits of slightly different energies.

Sommerfeld in 1915, guided by the above suggestion, modified Bohr's theory by introducing the following modifications :

(a) Concept of ellipitical orbits, and

(b) Relativistic variation of the mass of electron.

We will discuss these modifications one by one.

(a) Ellipitical Orbits: In order to explain the multiplicity of spectral lines, Sommerfeld introduced the concept of elliptical orbits. He postulated that :

(i) Since the electron is moving around and under the influence of a massive nucleus, like a planet around the central massive sun, it might describe elliptical orbits as well.

(ii) While retaining the first circular orbit suggested by Bohr, Sommerfeld assumed one additional ellipitical orbit in the case of Bohr's second orbit and added two additional elliptical orbits to Bohr's third orbit and so on (Fig.).

(iii) The nucleus is one of the foci for all these orbits.

(iv) In an elliptical orbit, the major and minor axes will differ in lengths, but as the orbit broadens, they will approach each other and become equal when the orbit becomes circular. Circular orbit is only a special case of the ellipitical orbit. The orbit is defined by two quantum numbers, *i.e.*, n and k which correspond to the major and minor axes of the ellipse. They are related as follows :

$$\frac{\text{Principal quantum number}}{\text{Azimuthal quantum number}} = \frac{n}{k} = \frac{\text{length of major axis}}{\text{length of minor axis}} = \frac{a}{b}$$

It is clear from the above that for any given value of n, k cannot be zero as in that case the ellipse would degenerate into a straight line passing through the nucleus. Further, k cannot be more than n since b is always less than a. When $n=k$, the path becomes circular. There is a limit to the number of different orbits that an electron may have in any energy level. Such possible paths are referred to as sub-levels. For a given value of n, k can have only n different values which means that there may be only n elliptical orbits or sub-levels with different eccentricity. For elliptical orbits the value of k would be $(n-1)$, $(n-2)$ etc. down to $k=1$. When $k=0$, the ellipse would be a straight line. When $n=3$, $k=3$ (circular) 2, 1 (two elliptical), Thus for $n=3$, there are three different orbits. Thus, the greatest contribution of Sommerfeld model lies in its sub-division of the original Bohr stationary levels into various sub-levels of slightly differing energies as given by differences in orbit shapes.

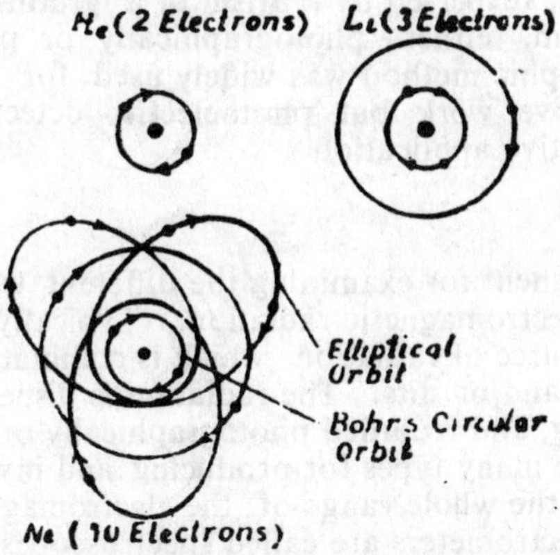

Sommerfeld ellipitical orbits.

Sorel's Cement. A cement made from 2MgO (ignited) to 1$MgCl_2$ (concentrated solution). Sets to a solid on stirring. The solid contains hydroxy chlorides, *e.g.* $5Mg(OH)_2,MgCl_2,7H_2O$, $Mg_2(OH)_3Cl,4H_2O$.

Solvolysis. The reaction when a compound reacts with a solvent. In the case of water the special term hydrolysis is used, *e.g.* the reaction of a salt with water giving the free acid and base.

$$BA+H_2O \longrightarrow HA+BOH$$

Species. A chemical entity, such as a particular atom, ion, or molecule.

Spectral Line. A particular wavelength of light emitted or absorbed by an atom, ion, or molecule.

Spectral Series. A group of related lines in the absorption or emission spectrum of a substance. The lines in a spectral series occur when the transitions all occur between one particular energy level and a set of different levels.

Spectrochemical Series. Ligands arranged in order of their tendency to cause splitting of the d orbitals in complexes (see crystal field theory). The order is obtained from changes in magnetism and the frequencies of d–d transitions in the u.v.-vis-ible-near i.r. spectra. The series is $I^- < Br^- < Cl^- < F^- < OH^- < H_2O <$ pyridine $< NH_3 <$ en $< NO_2^- < CN^-$.

Spectrograph. An instrument for producing a photographic record of a spectrum.

Spectrographic Analysis A method of analysis in which the sample is excited electrically (by an arc or spark) and emits radiation characteristic of its component atoms. This radiation is passed through a slit, dispersed by a prism or a grating, and recorded as a spectrum, either photographically or photoelectrically. The photographic method was widely used for qualitative and semiquantitative work but photoelectric detection also allows wide quantitative application.

Spectrometer

1. An instrument for examining the different wavelengths present in electromagnetic radiation. Typically, spectrometers have a source of radiation, which is collimated by a system of lenses and/or slits. The radiation is dispersed by a prism or grating, and recorded photographically or by a photocell. There are many types for producing and investigating spectra over the whole range of the electromagnetic spectrum. Often spectrometers are called spectroscopes.

2. Any of various other instruments for analysing the energies, masses, etc., of particles.

Spectrophotometer. A form of spectrometer able to measure the intensity of radiation at different wavelengths in a spectrum, usually in the visible, infrared, or ultraviolet regions.

Spectroscope. An optical instrument that produces a spectrum for visual observation. The first such instrument was made by R. W. Bunsen; in its simplest form it consists of a hollow tube with a slit at one end by which the light enters and a collimating lens at the other end to produce a parallel beam, a prism to disperse the light, and a telescope for viewing the spectrum (see illustration). In the spectrograph, the spectroscope is provided with a camera to record the spectrum. For a broad range of spectroscopic work, from the ultraviolet to the infrared, a diffraction grating is used instead of a prism.

Spectroscopy. The study of methods of producing and analysing spectra using spectroscopes, spectrometers, spectrographs, and spectrophotometers. The interpretations of the spectra so produced can be used for chemical analysis, examining atomic and molecular energy levels and molecular structures, and for determining the composition and motions of celestial bodies.

Spectrum (plural spectra)

1. A range of electromagnetic radiation emitted or absorbed by a substance under particular circumstances. In an emission spectrum, light or other radiation emitted by the body is analysed to determine the particular wavelengths produced. The emission of radiation may be induced by a

variety of methods; for example, by high temperature, bombardment by electrons, absorption of higher-frequency radiation, etc. In an absorption spectrum a continuous flow of radiation is passed through the sample. The radiation is then analysed to determine which wavelengths are absorbed. See also band spectrum, continuous spectrum, line spectrum.

2. In general, any distribution of a property. For instance, a beam of particles may have a spectrum of energies. A beam of ions may have a mass spectrum (the distribution of masses of ions). See mass spectrometer.

Speculum Metal. A 66% Cu and 34% Sn alloy, silver-white in colour. Used in mirrors and optical grantings.

Speiss. Crude cobalt arsenide obtained by smelting As-containing ores of Ni, Cu, Pb.

Speltar. Commercial zinc containing about 3% impurities, mainly lead.

Spence Metal. An iron sulphide, m.p. 160°C (FeS plus S). Used for medallions, busts, etc.

Sphalerite, (Zn,Fe) S. Zinc blende containing some iron. An important zinc ore.

Sphene, $CaSiTiO_5$. A constituent of granites. Contains discrete SiO_4 groups.

Spiegel (spiegeleisen) A form of pig iron containing 15-30% of manganese and 4-5% of carbon. It is added to steel in a Bessember converter as a deoxidizing agent and to raise the manganese content of steel.

Spin. A property of certain elementary particles whereby the particle acts as if it were spinning on an axis; *i.e.* it has an angular momentum. Such particles also have a magnetic moment. In a magnetic field the spins line up at an angle to the field direction and process around this direction. Certain definite orientations to the field direction occur such that $m_s h/2\pi$ is the component of angular momentum along this direction. m_s is the spin quantum number. For an electron it has values $+1/2$ and $-1/2$.

Spin Moment. An electron spinning on its own axis produces a magnetic moment, termed the spin moment. For a multielectron system the magnitude of the spin moment, μ_s, is given by

$$\mu_s = g\sqrt{S(S+1)}$$

where g is the gyromagnetic ratio and S is the sum of the spin quantum numbers (m_s) of the individual electrons.

Spin Multiplicity. For an electronic state containing n unpaired electrons, the total spin quantum number, S, is equal to (n/2). The spin multiplicity of the state is given by (2S+1). When this is equal to unity, it is called a singlet state; when it is equal to three it is a triplet state, etc.

Spin-orbit Coupling. The interaction of spin and angular momentum.

Spin Quantum Number. This quantum number which is designated as s arises from the spectral evidence that an electron in its motion around the nucleus in an orbit also rotates or spins about its own axis. The electron can spin clockwise or counter clockwise. The spin of electron contributes to its overall angular momentum and since it can either be forward (clockwise) or backward (anticlockwise) relative to the direction of the path of electron, it may cause angular momentum to increase or decrease (Fig. [b]). This leads to two sub-levels of slightly different energy for each value of m, thus corresponding to two possibilities of s. Assuming the total energy difference between these two sub-levels to be one quantum, s level will have values $+\frac{1}{2}$ and $-\frac{1}{2}$. The two values are often conveniently put as ↑ and ↓.

Mathematically this quantum number is defined by the equation

$$mur = \frac{\pi\sqrt{[s(s+1)]}}{2\pi}$$

where mur denotes the angular momentum an s may have values of either $+\frac{1}{2}$ or $-\frac{1}{2}$.

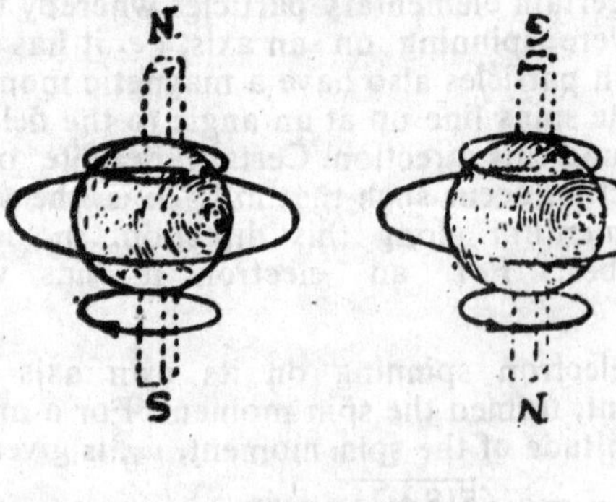

Spinning electron.

The concept of electron spin is important in applying the Pauli's principle which limits only two electrons in an orbital provided they have opposite spins. Such electrons are known as paired electrons. When an orbital has only one electron, it is known as odd or unpaired electron.

The spin quantum number gives rise to magnetic properties of materials. If there are two electrons in any orbital with opposite spins (paired electrons), their magnetic moments will be cancelled. Presence of magnetic properties indicates the presence of one or more unpaired electrons.

Spin-spin Coupling. The process of interaction between nuclear spins giving rise to fine structure in n.m.r. spectra.

Spinel. A group of oxide minerals with the general formula $F^{2+}R_2{}^{3+}O_4$, where F^{2+}=Mg, Fe, Zn, Mn, or Ni and R^{3+}=Al, Fe, or Cr, crystallizing in the cubic system. The spinels are divided into three series : spinel ($MgAl_2O_4$), magnetite, and chromite. They occur in high-temperature igneous or metamorphic rocks.

Spirits of Salt. Hydrogen chloride, so-called because it can be made by adding sulphuric acid to common salt (sodium chloride).

Square Antiprismatic Co ordination. Co-ordination by eight neighbours in a square antiprismatic arrangement (*i.e.* the two square faces at 45°). Found in $[Mo(CN)_8]^{4-}$. A square antiprism has 2 square and 8 triangular faces.

Square planar. Describing a coordination compound in which four ligands positioned at the corners of a square coordinate to a metal ion at the centre of the square.

Stabilization Energy. The amount by which the energy of a delocalized chemical structure is less than the theoretical energy of a structure with localized bonds. It is obtained by subtracting the experimental heat of formation of the compound (in kJ mol^{-1}) from that calculated on the basis of a classical structure with localized bonds.

Stability Constant. When a complex is formed in solution between a metal ion and a ligand the equilibrium may be expressed by a constant which is related to the free energy change for the process :

$$M+A \rightleftharpoons MA;\ K_1=\frac{[MA]}{[M][A]}$$

K_1 is the stability constant, the terms in the brackets representing activities. Charges have been dropped for the sake of clarity and

$$-\Delta G=RT\ln K_1$$

Step-wise equilibria may also be expressed in this way, *e g.*

$$M+A\rightleftharpoons MA;\ K_1=\frac{[MA]}{[M][A]}$$

$$MA+A\rightleftharpoons MA_2;\ K_2=\frac{[MA_2]}{[MA][A]}$$

$$MA_{(n-1)}+A\rightleftharpoons MA_n;\ K_n=\frac{[MA_n]}{[MA_{(n-1)}][A]}$$

$K_1, K_2 \ldots K_n$ being called the step-wise stability constants. Overall stability constants, designated β, may be evaluated by multiplication of the individual stability constants, *e.g.*

$$M+2A\rightleftharpoons MA_2;\quad \beta_2=\frac{[MA_2]}{[M][A]^2}=K_1\,K_2$$

Stainless Steel. A form of steel containing at least 11-12% of chromium, a low percentage of carbon, and often some other elements, notably nickel and molybdenum. Stainless steel does not rust or stain and therefore has a wide variety of uses in industrial, chemical, and domestic environments. A particularly successful alloy is the steel known as 18·8, which contains 18% Cr, 8% Ni, and 0.08% C.

Stalactites And Stalagmites. Pillars of calcium carbonate found hanging from the ceiling (stalactites) and standing on the floor (stalagmites) in limestone caverns. They form when water containing dissolved carbon dioxide forms weak carbonic acid, which is able to react with limestone rock (calcium carbonate) to give a solution of calcium hydrogen-carbonate. The solution accumulates in drops on the roof of caves and as it evaporates, the reaction is reversed causing calcium carbonate to be deposited. Over a very long period, the deposits build up to give a stalactite hanging from the cavern roof. Water that drips onto the floor from the tip of the stalactite also evaporates to leave a deposit of calcium carbonate. This gradually grows to form a stalagmite. Stalagmites and stalactites can eventually meet to form a single rock pillar.

Stannane (tin(IV) hydride, SnH_4) A colourless poisonous gas prepared by the action of lithium aluminium hydride on tin(II) chloride.

It is unstable, decomposing immediately at 150°C. Stannane acts as a reducing agent.

Stannate. A compound formed by reaction of tin oxides (or hydroxides) with alkali. Tin oxides are amphoteric (weakly acidic) and react to give stannate ions. Tin(IV) oxide with molten alkali gives the stannate(IV) ion

$$SnO_2+2OH^- \rightarrow SnO_3^{2-}+H_2O$$

In fact, there are various ions present in which the tin is bound to hydroxide groups, the main one being the hexahydroxostannate(IV) ion, $Sn(OH_6^{2-}$. This is the negative ion present in crystalline 'trihydrates' of the type $K_2Sn_2O_3.3H_2O$.

Tin(II) oxide gives the trihydroxostannate(II) ion in alkaline solutions

$$SnO(s)+OH^-(aq)+H_2O(l) \rightarrow Sn(OH)_3^-(aq$$

Stannate(IV) compounds were formerly referred to as orthostannates (SnO_4^{4-}) or metastannates (SnO_3^{2-}). Stannate(II) compounds were called stannites.

Stannic Compounds. Compounds of tin in its higher (+4) oxidation state; *e.g.* stannic chloride is tin(IV) chloride.

Stannite. See stannate.

Sima. The rocks that form the earth's oceanic crust and underlie the upper crust. These are basaltic rock types rich in silica (SiO_2) and magnesium (Mg), hence the name. The sima is denser and more plastic than the sial that forms the continental crust.

Single Bond. A covalent bond between two elements that involves one pair of electrons only. Often referred to as a sigma bond, (σ bond), it is presented by a single line, for example H-Br.

Sintered Glass. Porous glass made by sintering powdered glass, used for filtration of precipitates in gravimetric analysis.

Sintering. The process of heating and compacting a powdered material at a temperature below its melting point in order to weld the particles together into a single rigid shape. Materials commonly sintered include metals and alloys, glass, and ceramic oxides. Sintered magnetic materials, cooled in a magnetic field, make especially retentive permanent magnets.

Slag. Material produced during the smelting or refining of metals by reaction of the flux with impurities (*e.g.* calcium silicate formed by reaction of calcium oxide flux with silicon dioxide impurities). The liquid slag can be separated from the liquid metal because it floats on the surface. See also basic slag.

Slaked Lime. See calcium hydroxide.

Slurry. A thin paste of suspended solid particles in a liquid.

Smalt. A blue pigment obtained by grinding a potassium cobalt silicate glass under water and used for tinting glazes and enamels. The term smalt was once used to describe any vitrified pigment used for painting on glass.

Smelting. The melting of an ore with chemical reaction with other materials. A part of the process of winning an element from its ores.

Soda. Any of a number of sodium compounds, such as caustic soda (NaOH) or, especially, washing soda ($Na_2CO_3.10H_2O$).

Soda Ash. Anhydrous sodium carbonate, Na_2CO_3.

Soda Lime. A grey solid produced by adding sodium hydroxide solution to calcium oxide, to give a mixture of $Ca(OH)_2$ and NaOH. It is used in the laboratory as a drying agent and to absorb carbon dioxide.

Sodamide ($NaNH_2$) A white ionic solid formed by passing dry ammonia over sodium at 300-400°C. The compound reacts with water to give sodium hydroxide and ammonia. It reacts with red-hot carbon to form sodium cyanide and with nitrogen(I) oxide to form sodium azide. Sodamide is used in the Castner process and in the explosives industry.

Sodium Symbol Na. A soft silvery reactive element belonging to group I of the periodic table (see alkali metals); a.n. II; r.a.m. 22.9898; r.d. 0.97; m.p 97.5°C; b.p. 892°C. Sodium occurs as the chloride in sea water and in the mineral halite. It is extracted by electrolysis in a Downs cell. The metal is used as a reducing agent in certain reactions and liquid sodium is also a coolant in nuclear reactors. Chemically, it is highly reactive, oxidizing in air and reacting violently with water (it is kept under oil). It dissolves in liquid ammonia to form blue solutions containing solvated electrons. Sodium is a major element required by living organisms. The element was first isolated by Humphry Davy in 1807.

Sodium Acetate. See sodium ethanoate.

Sodium Aluminate ($NaAlO_2$). A white solid produced by adding excess aluminium to a hot concentrated solution of sodium hydroxide. Once the reaction has been initiated, sufficient heat is liberated to keep the reaction going; hydrogen is also produced. In solution the aluminate ions have the structure $Al(OH)_4$ and $NaAl(OH)_4$ is known as sodium(I) tetrahydroxoaluminate(III). Addition of sodium hydroxide to an aluminium salt solution produces a white gelatinous precipitate of aluminium hydroxide, which dissolves in excess alkali to give a solution of sodium(I) tetrahydroxoaluminate(III).

Sodium Amide (sodamide). A white crystaline powder, $NaNH_2$, which decomposes in water and in warm ethanol, and has an odour of ammonia; m.p. 210°C; b.p. 400°C. It is produced by passing dry ammonia over metallic sodium at 350°C. It reacts with red-hot carbon to give sodium cyanide and with nitrogen (I) oxide to give sodium azide.

Sodium Azide (NaN_3). A white solid prepared by passing nitrogen(I) oxide over heated sodamide On heating, sodium azide decomposes to give sodium and nitrogen. It is used as a reagent in organic chemistry and in the preparation of lead azide (for use in detonators).

Sodium Bicarbonate. See sodium hydrogen-carbonate.

Sodium Bisulphate. See sodium hydrogen-sulphate.

Sodium Bisulphite. See sodium hydrogensulphite.

Sodium Bromide (NaBr). A white solid formed by the action of bromine on hot sodium hydroxide solution. Alternatively, it can be made by the action of dilute hydrobromic acid on sodium carbonate or hydroxide. Sodium bromide is soluble in water and forms crystals similar in shape to sodium chloride. It is used in medicine and photography. Sodium bromide forms a hydrate, $NaBr \cdot 2H_2O$.

Sodium Carbonate (soda ash, Na_2CO_3). A white amorphous powder, which aggregates on exposure to air owning to the formation of hydrates. Industrially it is prepared by the ammonia-soda (Solvay) process. On crystallizing from aqueous solution large translucent crystals of the decahydrate ($Na_2CO_3. 10H_2O$) are formed (washing soda). These effloresce forming the monohydrate, $Na_2CO_3H_2O$. Large quantities of sodium carbonate are used in the manufacture of sodium hydroxide. Washing soda is used as a domestic cleanser. The salt produces an alkaline solution in water by hydrolysis.

$Na_2CO_3 + 2H_2O \rightarrow 2NaOH + H_2CO_3$. It is used in volumetric analysis to standardize strong acids.

Sodium Chlorate(V). A white crystalline solid, $NaClO_3$; cubic; r d. 2 49; m p. 250°C. It decomposes above its melting point to give oxygen and sodium chloride The compound is soluble in water and in ethanol and is prepared by the reaction of chlorine on hot concentrated sodium hydroxide. Sodium chlorate is a powerful oxidizing agent and is used in the manufacture of matches and soft explosives, in calico printing, and as a garden weedkiller.

Stannous Compounds. Compounds of tin in its lower (+2) oxidation state; *e.g*, stannous chloride is tin(II) chloride.

Stassfurt Deposits. Deposits of various salts mined at Stassfurt chiefly for potassium (carnallite). Kainite, kienerite and gypsum are also present, Br_2 is obtained as a by-product.

State Symbols. Symbols which designate the spin multiplicity and electronic state of a species. *E.g.* the ground state of a d^2 species is ^3F, the spin multiplicity symbol (3) designates 2 unpaired electrons, the symbol F designates an angular momentum quantum number of 3.

Stationary States of Electrons. The stable electronic arrangements in atoms or molecules. Transitions between stationary states give rise to absorptions in the u.v., visible or near i.r. regions of the electromagnetic spectrum.

Steam Reforming. The conversion of a methane-steam mixture at 900°C with a nickel catalyst into a mixture of carbon monoxide and hydrogen. The mixture of gases (synthesis gas) provides a starting material in a number of processes, *e.g.* the the manufacture of methanol.

Steel. Any of a number of alloys consisting predominantly of iron with varying proportions of carbon (up to 1.7%) and, in some cases, small quantities of other elements (alloy steels), such as

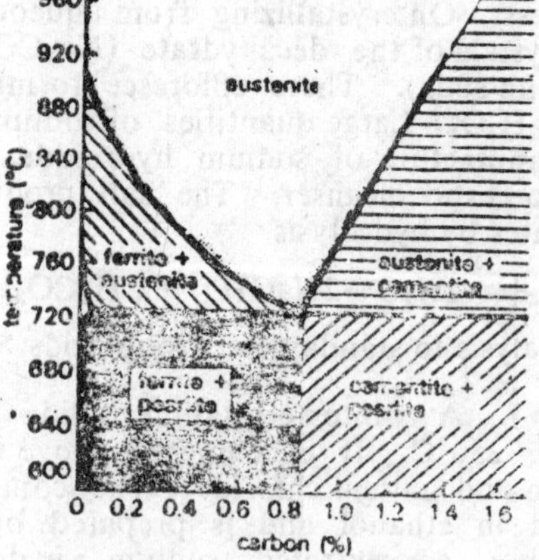

Phase diagram for steel.

manganese, silicon, chromium, molybdenum, and nickel. Steel containing over 11—12% of chromium are known as stainless steels.

Carbon steels exist in three stable crystalline phases: ferrite has a body-centred cubic crystals austenite has a face-centred cubic crystal, and cementite has an orthorhombic crystal. Pearlite is a mixture of ferrite and cementite arranged in parallel plates. The phase diagram shows how the phases form at different temperatures and composition.

Steels are manufactured by the basic-oxygen process (L—D process), which has largely replaced the Bessemer process and the open-hearth process, or in electrical furnaces.

Alloy steels are steels containing deliberate additions of alloying elements over and above the amounts which are residual from the steel-making process. Small quantities, 1 or 2% of Ni, Cr, Mn, Si are commonly added for improved quench hardening. Such steel may be quenched in oil in place of water, with less danger of cracking : 1—3% of Ni gives improved toughness especially at low temperatures. Carbide formers such as Cr (up to 18%), W (up to 18) and Mo (up to 10%) are added to high carbon steels to give wear resistance due to the alloy carbides. Cr in low C steels forms a solid solution with ion and in this condition it forms a protective invisible passive oxide layer on the surface giving corrosion and scaling resistance : 3—25% Cr is used.

Austenitic steels are steels which retain the ccp structure right down to room temperature. (Normally the structure of steel changes to body-centred cubic at around 700°C on cooling) For this reason these steels cannot be hardened by quenching. There are two main groups :

(1) For corrosion resistance; see stainless steel (below).

(2) For wear resistance (Hadfield steel). This steel contains 12% Mn and 1.2% C. The austenite is metastable and abrasion of the surface causes it to break down and form hard wear-resistant martensite. It is used for crushing plant, mechanical digger bucket edges and railway crossings.

Stainless steel. The name given to a group of Cr or Cr—Ni steels showing an unusually high resistance to corrosion by the atmosphere and many chemical reagents. In order to achieve this resistance to corrosion > 12% Cr is necessary.

Stellite. An alloy, 45—50% Co, 25—30% Cr 15—20% W, 2.5% C alloy. Used as a cutting tool tipping material, but differs from many other non-ferrous cutting alloys which are sintered and cemented as it can be cast and welded, and may be applied as a coating to a softer material to resist abrasion.

Step. A single stage in a chemical reaction. For example, the addition of hydrogen chloride to ethene involves three steps :

$$HCl \rightarrow H^+ + Cl^-$$

$$H^+ + C_2H_4 \rightarrow CH_3CH_2^+$$

$$CH_3CH_2^+ + Cl^- \rightarrow CH_3CH_2Cl$$

Stereochemistry. The branch of chemistry concerned with the shapes of molecules and the way these affect the chemical properties.

Steric Effect. An effect in which the shape of a molecule influences its reactions. A particular example occurs in molecules containing large groups, which hinder the approach of a reactant (steric hindrance).

Stibine. See antimony hydride.

Stibnite, Sb_2S_3. The only important antimony ore. A steel-grey material with a metallic lustre, the crystals are transparent to i.r. The crystal contains rather complex chains of linked Sb (co-ordination 3+2 weak links to other chains) and S atoms. Stibnite was used as an eyebrow pigment.

Stilbite. A zeolite, $Na_2Ca[Al_2Si_6)O_{16}],6H_2O$, used for softening hard water.

Stoichiometric. A stoicheiometric compound is one in which the ratio of the number of atoms to each other, as determined from the atomic weights, is a ratio of small whole numbers. In nonstoicheiometric compounds there are defects in the crystal lattice, or part replacement of atoms of one element by another.

Stoichiometric Compound. A compound in which atoms are combined in exact whole-number ratios. Compare nonstoichiometric compound.

Stoichiometry. The relative proportions in which elements from compounds or in which substances reacts.

Strass. A dense glass used for making imitation gems and also for jewels to be cemented to vases and other ornamental ware. It is chiefly characterized by its high but variable refractive index.

Strengths of Acids and Bases. The strength of an aqueous acid is its ability to give hydroxonium ions (H_3O^+) and that of an aqueous base its ability to give hydroxyl ions or to accept protons. For Lewis acids and bases, acidic and basic character are measured by the tendency to react with standard bases and acids respectively. Since the degree of dissociation varies with the dilution, it is necessary to compare the relative strengths of acids or bases in solutions of equivalent concentrations. This can be done by measuring the concentrations of hydroxonium ion or of hydroxyl ion. Those acids like hydrochloric and nitric which are largely dissociated into their ions are strong acid; those like ethanoic and tartaric which are only slightly dissociated are weak acids. Potassium, sodium and barium hydroxides which are largely dissociated are strong bases whilst the slightly dissociated ammonium hydroxide is a weak base.

Strong Acid. An acid that is completely dissociated in aqueous solution.

Strontia. See strontium oxide.

Strontianite. A mineral form of strontium carbonate, $SrCO_3$.

Strontium. A soft low-melting reactive metal; the fourth member of group II of the periodic table and classed as a typical alkaline-earth element. The electronic configuration is that of krypton with two additional outer 5s electrons. Strontium and barium are both of low abundance in the Earth's crust, strontium occurring as strontanite ($SrCO_3$) and celestine ($SrSO_4$).

The element is produced industrially by roasting the carbonate to give the oxide (800°C) and then reducing with aluminium.

$$3SrO + 2Al \rightarrow Al_2O_3 + 2Sr$$

Strontium Bicarbonate. See strontium hydrogencarbonate.

Strontium Carbonate. A white solid, $SrCO_3$; orthorhombic; r.d. 3.7; decomposes at 1770°C. It occurs naturally as the mineral strontianite and is prepared industrially by boiling celestine (strontium sulphate) with ammonium carbonate. It can also be prepared by passing carbon dioxide over strontium oxide or hydroxide or by passing the gas through a solution of strontium salt. It is a phosphor, used to coat the glass of cathode-ray screens, and is also used in the refining of sugar, as a slagging agent in certain metal furnaces and to provide a red flame in fireworks.

Strontium Chemistry. Strontium is an alkaline earth, electronic configuration $5s_2$. It is an electropositive metal ($E^\circ Sr^{2+}$ to Sr in acid -2.89 volts) which readily dissolves in acids. It forms a single range of Sr(II) compounds.

Strontium Chloride ($SrCl_2$). A white solid obtained directly from strontium and chlorine or by passing chlorine over heated strontium oxide. It is used in fireworks to give a red flame. The hexahydrate ($SrCl_2.6H_2O$) is prepared by the neutralization of hydrochioric acid by strontium hydroxide or carbonate.

Strontium Fluoride SrF_2. Almost insoluble in water.

Strontium Hydrogencarbonate (strontium bicarbonate). A compound, $Sr(HCO_3)_2$, which is stable only in solution. It is formed by the action of carbon dioxide on a suspension of strontium carbonate in water. On heating, this process is reversed.

Strontium Hydroxide ($Sr(OH)_2$. A solid that normally occurs as the octahydrate ($Sr(OH)_2.8H_2O$), which is prepared by crystallizing an aqueous solution of strontium oxide. Strontium hydroxide is readily soluble in water and aqueous solutions are strongly basic. It is used in the purification of sugar.

Strontium Nitrate, $Sr(NO_3)_2$. Formed $SrCO_3$ plus HNO_3. Forms 4 and 0 hydrates.

Strontium Oxide, SrO. M.p. 2430°C. Formed by ignition $SrCO_3$, $Sr(NO_3)_2$ or $Sr(OH)_2$. Forms $Sr(OH)_2$ in water.

Strontium peroxide, SrO_2, is formed from SrO and O_2 at dull red heat under pressure. $SrO_2,8H_2O$ is formed from $Sr(OH)_2$ plus H_2O_2 in cold aqueous solution.

Strontium Sulphate, $SrSO_4$. Occurs naturally as celestine. Prepared SrO, $Sr(OH)_2$ or $SrCO_3$ plus H_2SO_4. Sparingly soluble in water.

Strontium Sulphate ($SrSO_4$). A white sparingly soluble salt that occurs naturally as the mineral celestine. It can be prepared by dissolving strontium oxide, hydroxide, or carbonate in sulphuric acid. It is used as a substitute for barium sulphate in paints.

Sublimate. A solid formed by sublimation.

Sub-shell. A subdivision of an electron shell. It is a division of the orbitals that make up a shell into sets of orbitals, which are degenerate (*i.e.* have the same energy) in the free atom. For example, in the third shell or M-shell there are the 3s, 3p, and 3d sub-shells of which the latter two are 3—and 5—degenerate respectively.

Sugar of Lead. See lead(II) ethanoate.

Sulphamic Acid, $H_3N^+.SO_3^-$. A colourless crystalline solid normally existing as a zwitterion: prepared by the reaction of sulphur dioxide with acetoxime, or commercially, by the treatment of urea with fuming H_2SO_4. Sulphamic acid is very soluble in water and is a strong acid, forming a series of salts, the sulphamates. It is an excellent acidimetric primary standard. It is used in weed-killers, flame-proofing agents, electrolytes in metal deposition processes and sweeteners.

Sulphamide, $SO_2(NH_2)_2$ and sulphimide, $(SO_2NH)_n$. Colourless crystalline solids formed by the action of ammonia on a solution of sulphuryl chloride in benzene; free sulphimide exists only in the polymerized form Both compounds are soluble in water and are readily hydrolysed to sulphamic acid, $H_3N^+.SO_3^-$ and ammonia; the hydrogen atoms are in each case replaceable by metals to form salts. Many derivatives of sulphamide and cyclic derivatives of sulphimide are known.

Sulphanes. The hydrogen sulphides H_2S_2 to H_2S_6 and beyond (for H_2S see hydrogen sulphide). Formed S in aqueous Na_2S and addition to HCl or S_2Cl_2 and H_2S. Decompose into H_2S and S.

Sulphanes. Compounds of hydrogen and sulphur containing chains of sulphur atoms. They have the general formula H_2S_n The simplest is hydrogen sulphide, H_2S; other members of the series are H_2S_2, H_2S_3, H_2S_4, etc.

Sulphate A salt or ester of sulphuric (VI) acid. Organic sulphates have the formula R_2SO_4, where R is an organic group. Sulphate salts contain the ion SO_4^{2-}.

Sulphides. Compounds of other elements with sulphur. Generally prepared by direct interaction of the elements. Metal sulphides are often very in soluble occur, as minerals and can be used for separation of the elements. Sulphides of the most electropositive elements give H_2S with water. Covalent derivative formed with the lees electropositive elements, *e.g.* liquid CS_2 S_2Cl_2.

Sulphate (dithionite; hyposulphite). A salt containing the negative ion $S_2O_4^{2-}$, usually formed dy the reduction of sulphites with excess SO_2 Solutions are not very stable and decompose to give thiosulphate and hydrogensulphite ions. The structure is $^-O_2S{\cdot}SO_2$.

Sulphininic Acid (dithionous acid; hyposulphurous acid). An unstable acid, $H_2S_2O_4$. known in the form of its salts (sulphinates).

Sulphinyl. Containing the > SO group.

Sulphites. Salts of sulphurous acid, *i.e.,* sulphates (IV) containing $[SO_3]^{2-}$ species.

Sulphur. A low melting non-metallic solid, yellow coloured in its common forms; the second member of group VI of the periodic table. It has the electronic configuration [Ne]$3s^2p^4$.

Symbol : S; m p. 118.9°C ;b.p. 444.6°C; r.d. 2.07; p.n. 16; r.a.m. 32.06.

Sulphur Dichloride. See disulphur dichloride.

Sulphur Dichloride Dioxide (sulphuryl chloride). A colourless liquid, SO_2Cl_2; r.d. 1.67; m.p. —54.1°C; b.p. 69°C. It decomposes in water but is soluble in benzene. The compound is formed by the action of chlorine on sulphur dioxide in the presence of an iron (III) chloride catalyst or sunlight. It is used as a chlorinating agent and a source of the related fluoride, SO_2F_2.

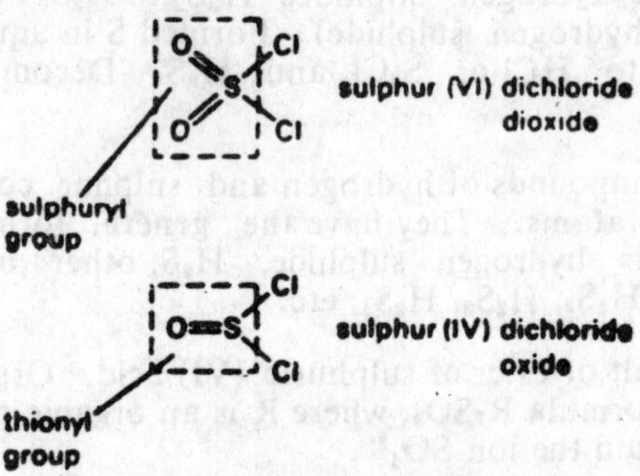

Oxychlorides of sulphur

Sulphur Dioxide (sulphur (IV) oxide). A colourless liquid or pungent gas, SO_2, formed by sulphur burning in air. r.d. 1.43 (liquid), m.p. —72.2°C. b.p. —10°C. It can be made by heating iron sulphide (pyrites) in air. The compound is a reducing agent and is used in bleaching and as a fumigant and food preservative. Large quantities are also used in the contact process for manufacturing sulphuric acid. It dissolves in water to give a mixture of sulphuric and sulphurous acids.

Sulphuretted Hydrogen, See hydrogen sulphide.

Sulphuric Acid (oil of vitriol). A colourless oily liquid, H_2SO_4; r d. 1.84; m.p. 10 36°C, b.p- 338°C. The pure acid is rarely used; it is commonly available as a 96·98% solution (m.p. 3 0°C). The compound also forms a range of hydrates: $H_2SO_4H_2O$ m.p. 8.62°C), $H_2SO_42H_2O$ (m.p. —38/39°C); $H_2SO_46H_2O$ (m p. —54°C), $H_2SO_4.8H_2O$.

Until the 1930s, sulphuric acid was manufactured by the lead-chamber process, but this has now been replaced by the contact process (catalytic oxidation of sulphur dioxide). More sulphuric acid is made in the UK than any other chemical product; production levels (UK) are commonly 12000 to 13000 tonnes per day. It is extensively used in industry, the main applications being fertilizers (32%), chemicals (6%), paints and pigments (15%), detergents (11%), and fibres (9%).

In concentrated sulphuric acid there is extensive hydrogen bonding and several competing equilibria, to give species such as H_3O^+, HSO_4^-, $H_3SO_4^+$, and $H_2S_2O_7$. Apart from being a powerful protonating agent (it protonates chlorides and nitrates producing hydrogen chlorine and nitric acid (m.p.—(2°C). Its full systematic name is tetrooxosulphuric (VI) acid, the com-

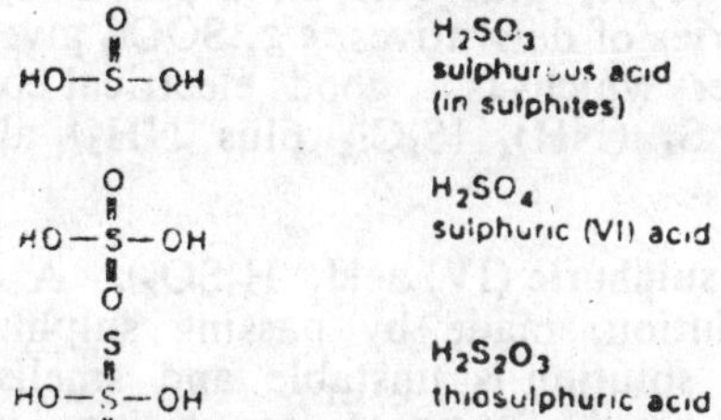

St uctures of some oxo acids of sulphur.

pound is a moderately strong oxidizing agent. Thus, is will dissolve copper :

$$Cu(s) + H_2SO_4(l) \rightarrow CuO(s) + H_2O(l) + SO_2(g)$$

$$CuO(s) + H_2SO_4(l) \rightarrow CuSO_4\,(aq) + H_2O\,(l)$$

It is also a powerful dehydrating agent, capable of removing H_2O from many organic compounds (as in the production of acid anhydrides). In dilute solution it is a strong dibatic acid forming two series of salts, the sulphates acid the hydrogen-sulphates.

Sulphuric (IV) Acid. See sulphurous acid.

Sulphuric Acid, fuming, Nordhausen Sulphuric Acid; Oleum. A solution of SO_3 in H_2SO_4 containing some $H_2S_2O_7$, etc. SO_3 distils from the acid on heating. Used in organic synthesis.

Sulphur Monochloride. See disulphur dichloride.

Sulphur Nitrogen Derivatives. There is an extensive chemistry of sulphur nitrogen derivative many of them being cyclic derivatives. S_4N_4 (S_2Cl_2 plus NH_4Cl, S plus NH_3 in CCl_4) gives an extensive series of derivatives, *e g.* $SOCl_2$ gives $(S_4N_3)Cl$. SN)x is a polymer which is a good electrical conductor. S_7NH, S_5 $(NH)_3$ and S_4 $(NH)_4$ (S_2Cl_2 plus NH_3) all have structures related to S_8.

Sulphurous Acid (sulphuric (IV) acid, H_2SO_3). A weak acid found only in solution, made by passing sulphur (IV) oxide into water. The solution is unstable and smells of sulphur (IV) oxide. It is a reducing agent, converting ion (III) ion to iron (II) ions, chlorine to chloride ions, and orange dichromate (VI) ions to green chromium (III) ions.

Sulphur (IV) Oxide. See sulphur dioxide.

Sulphur (VI) Oxide (sulphur trioxide, SO_3). A fuming volatile white solid prepared by passing sulphur (IV) oxide and oxygen over hot vanadium (V) oxide (acting as a catalyst) and cooling the product in ice. Sulphur (VI) oxide reacts vigorously with water to form sulphuric acid.

Sulphur Oxyacids. Sulphur forms an extensive range of oxyacids which are dealt with under separate headings. The acids are :

$S(OH)_2$, HS(O)OH	sulphoxylic acid
$OS(OH)_2$	sulphurous acid
$O_2S(OH)_2$	sulphuric acid
$S(O)S(OH)_2$	thiosulphuric acid
O(HO)SS(OH)O	dithionous acid
O(HO)SSnS(OH)O	Polythionic acids
O_2(HO)SS(OH)O	pyrosulphurous acid. disulphurous acid
O_2(HO)SS(OH)O_2	dithionic acid
O_2(HO)SOS(OH)O_2	pyrosulphuric acid disulphuric acid

$H_2S_3O_{10}$	trisulphuric acid
$H_2S_4O_{13}$	tetrasulphuric acid
$O_2S(OH)(OOH)$	peroxymonosulphuric acid
$O_2(HO)SOOS(OH)O_2$	peroxydisulphuric acid

Sulphur Trioxide (sulphur (VI) Oxide. A colourless fuming solid, SO_3, which has three crystalline modifications. In decreasing order of stability 'these are: α, r.d. 1.97; m p. 16.83°C; b.p. 44°C; β, m.p. 16.24°C; sublimes at 50°C, r.d. 2.29,γ m.p. 16.8°C; b.p. 44.8°C. All are polymetric, with linked SO_4 tetrahedra: the γ-form has an icelinked structure and is obtained by rapid quenching of the vapour, the β-form has infinite helical chains; and the α-form has infinite chains with some cross-linking; of the SO_4 tetrahedra. Even in the vapour, there are polymeric species, and not discrete sulphur trioxide molecules (hence the compound is more correctly by its systematic name sulphur (VI) oxide).

Sulphur trioxide is prepared by the oxidation of sulphur dioxide with oxygen in the presence of a vanadium (V) oxide catalyst. It may be prepared in the laboratory by distilling a mixture of concentrated sulphuric acid and phosphorus (V) oxide. It reacts violently with water to give sulphuric (VI) acid and is an important intermediate in the preparation of sulphuric acid and oleum.

Sulphuryl Group. The group$=SO_2$, as in sulphur dichloride oxide.

Sulphuryl Halides, SO_2X_2. Sulphuryl chloride, SO_2Cl_2, b.p. 69°C (SO_2 plus Cl_2 plus light) is used as a chlorinating agent. SO_2F_2 and SO_2Br_2 are also known.

Sulphydryl. The SH group, also called thiol.

Sulphydryl Group. See thiols.

Super Acids. Mixtures, *e.g.*, $HF-SbF_5$, HSO_3F-SbF_5, which act as very strong acids and, *e.g*, induce carbenium ion formation, protonation, hydride abstraction.

Superadditive Developer Mixtures. Photographic developer mixtures containing two or more developing agents the combined power of which is much greater than either agent used separately. *E.g.* M.Q. developers (aminophenols plus hydroquinone). Also known as synergistic developers.

Superoxides. A group of inorganic compounds that contain the O_2^- ion. They are formed in significant quantities only for sodium potassium rubidium, and caesium. They are very powerful oxidizing agents and a react vigorously with water to give oxygen gas and OH^- ions. The superoxide ion has an unpaired electron and is paramagnetic and coloured (orange).

Superphosphates. Insoluble calcium phosphate, $Ca_3(PO_4)_2$, treated with concentrated sulphuric acid gives calcium hydrogen phosphate, $Ca(H_3PO_4)_2$, which is soluble in water and is a very effective fertilizer. Commercially, one of the available forms of calcium phosphate is treated with sulphuric acid of carefully controlled concentration, preferably in a closed autoclave under pressure ('Oberphos' process) The dry product consists of the soluble calcium salt together with calcium sulphate and sulphates of other metals present in the raw material, and is known as 'superphosphate'. Care is taken in the manufacturing process to avoid a product containing excess phosphoric acid, or unattacked insoluble calcium phosphate which results if too much or too little acid is used.

The grades of superphosphate most widely used have about 13.7% of phosphoric acid (30% tricalcium phosphate) and 16-17% of phosphoric acid. Higher grades are also manufactured.

Sylvine, Sylvite, KCl. Occurs in salt deposits an important source of potassium salts.

Sylvinite, KCl—NaCl. A source of potassium salts.

Sylvite (sylvine). A mineral form of potassium chloride, KCl.

Symmetry Elements. The operators which give repetition and pattern in molecules and lattices. Symmetry elements comprise.

1. Axes of symmetry. An axis about which rotation of the body through an angle of $2\pi/n$ (where n is an integer) gives an identical pattern; 2-fold, 3-fold, 4-fold and 6-fold axes are known in crystals, 5-fold axes are known in molecules. In lattice the rotation may be accompanied by a lateral movement parrallel to the axis (screw axis).
2. Planes of symmetry, Planes through which there is reflection to an identical point in the pattern. In a lattice there may be a lateral movement parallel to one or more axes (glide plane).
3. Identity operator. Present in all lattices and molecules.
4. Centre of symmetry. A point through which there is reflection to an identical point in the pattern.

Symproportionation, Conproportionation. The opposite of disproportionation.

Synchrotron. A device used for accelerating electrons and protons. The particles, which are kept in a circular orbit by means of a magnetic field, are accelerated by a radio-frequency oscillator. By this method protons with energies of the order of 100 Mev can be produced.

Syngenite Mineral K_2SO_4, $CaSO_4$,HO. It is decomposed by water to give insoluble $CaSO_4$. Commonly occurs in association with rock salt anhydrite and gypsum.

Synthesis. In a chemical sense the building up of complex molecules from smaller ones.

Synthesis Gas. The term used to describe a mixture of hydrogen and carbon monoxide, normally produced from the reaction of methane and steam under pressure or, less usually, by the partial oxidation of methane. Synthesis gas is used in the manufacture of methanol. Oxo alcohols and synthetic gasoline.

Synthesis Gas. A mixture of carbon monoxide and hydrogen produced by steam reforming of natural gas.

$$CH_4 + H_2O \rightarrow CO + 3H_2$$

Synthesis gas is a useful starting material for the manufacture of a number of organic compounds.

T

Ta. Tantalum.

Tachydrite. $CaCl_2 \cdot 2MgCl_2, 12H_2O$.

Talc, $Mg_3(OH)_2 Si_4O_{10}$. A silicate with layers of linked SiO_4 tetrahedra, the layers are electrically neutral so that there is only weak interaction between layers. The crystals are soft and used as a lubricant (French chalk). Also used as an extender and filler in paint, rubber, insecticides, and tar and asphalt roofing materials, as a constituent of toilet powders and in some types of porcelain. Steatite is a pure compact variety, soap stone is a massive impure form.

Tantalite (Fe, Mn) (Nb, Tb, Ta)$_2$O$_6$ with excess Ta. The principle ore of tantalum.

Tantalum Ta. At.no. 73, at.wt. 180.95, m.p. 2996°C, b.p. 5425°C. d 16.6. An element of group V, electronic configuration $6s^2 5d^3$, Occurs with Nb in tantalite (Fe, Mn) $(Nb, Ta)_2O_6$. The ores are treated by NaOH fusion followed by acid washing to give mixed pentoxides. Ta is separated by liquid extraction from HF with isobutylmethyl-ketone. The metal is prepared from Ta_2O_5 by reduction with alkali metals or by electrolysis of fused fluorides, it is boc, remaining lustrous in air but reacts with oxygen or steam at high temperatures and dissolves in HNO_3—HF. Used in resistant alloys particularly construction for the chemical industry and in surgical appliances. World production Nb—Ta minerals 1970 15400 tonnes.

Tantalum Chemistry. Largely that of the +5 oxidation state and very similar to niobium chemistry. The major difference is the absence of stoicheiometric lower oxides for Ta and minor differences in complex species, *e g.* $TaCl_6^-$ and $NbOCl_5^{2-}$ in HCl solutions.

Tantalum Oxide Ta_2O_5 is the most important oxide and is prepared by dehydration of the dydrated oxides formed by dehydrolysis of halides. It is a dense white inert material, reduced to $Ta_{2.5}O_4$. The pentoxide forms tantalates, *e.g.*, K_3TaO_4, as mixed metal oxides by fusion with oxides or carbonates, reduction gives bronzes. Isopoly-tantalates, *e.g.*, $[H_xTa_6O_{19}]^{(8-x)-}$ (x=0, 1 or 2) are formed by fusing Ta_2O_5 with alkali and dissolving the melt in water.

Tantiron. An Fe—Si (15%) alloy containing small quantities of C, Mn, P and S. Very resistant to acids but brittle.

Tapiolite. $FeTa_2O_6$. A tantalum mineral. Has the trirutile (tapiolite) structure with three rutile unit calls with regular occupation of the cation positions by Fe and Ta.

Technetium Symbol Tc. A radioactive metallic transition element; a.n. 43; m.p. 2171°C; b p. 4876°C. The element can be detected in certain stars and is present in the fission products of uranium. It was first made by Perrier and Segre by bombarding molybdenum with deuterons to give technetium-97. The most stable isotope is techmetium-99 (half-life 2.12×10^5 years); this is used to some extent in labelling formedical diagnosis. There are sixteen known isotopes. Chemically, the metal has properties intermediate between manganese and rhenium.

Tellurates. Formally salts containing oxyanions of tellurium. Tellurates(VI) contain octahedral $[TeO_6]^{6-}$ species and derivatives of these. Tellurates(IV), tellurites, contain $[TeO_3]^{2-}$ species.

Telluric Acid, $Te(OH)_6$ See tellurium oxyacids.

Tellurides Binary derivatives of Te with other elements. Similar structurally to sulphides.

Tellurites. Tellurates (IV).

Tellurium. A brittle silvery metalloid element which belongs to group VI of the periodic table. It is found native and in combination with metals. Tellurium is used mainly as an additive to improve the qualities of stainless steel and various metals.

Symbol : Te; m.p. 449.5°C; b.p. 989.8°C; r.d. 6.24; p.n. 52; r.a.m. 127.6

Tellurium Bromides. Tellurium dibromide, $TeBr_2$, m.p. 210°C, o p. 339°C. Unstable compound formed Te plus $TeBr_4$ in Et_2O.

Tellurium tetrabromide, $TeBr_4$ m.p. 380°C, b.p. 420°C. Red solid, Te plus excess Br_2. The vapour is dissociated to $TeBr_2$, and Br_2. Forms many complexes, including $[TeBr_6]^{2-}$ ions.

Tellurium Chlorides. Tellurium dichloride, $TeCl_2$. Unstacle solid $TeCl_4$ plus Te.

Tellurium tetrachloride, $TeCl_4$. Colourless solid, m.p. 225°C, b.p. 390°, obtained Te plus excess Cl_2. Hydrolysed by water, forms complexes including $[TeCl_6]^{2-}$ ions.

Temper Brittleness. A serious fall in stock resistance occurring in some alloy steels when tempered in the range 300-600°C. Mo (<0.5%) prevents temper brittleness particularly in-Cr steels.

Tempering. The process of increasing the toughness of an alloy, such as steel, by heating it to a predetermined temperature, maintaining it at this temperature for a predetermined time, and cooling it to room temperature at a predetermined rate.

Temporary Hardness. Refers to a type of water hardness which is caused by the presence of dissolved calcium, iron, and magnesium hydrogencarbonates. This form of hardness can be removed by boiling the water. Temporary hardness arises because rainwater combines with carbon dioxide from the atmosphere to form weak carbonic acid. This reacts with carbonate rocks, such as limestone, to produce calcium hydrogencarbonate, which then goes into solution. When this solution is heated, the complete reaction sequence is reversed so that calcium carbonate is precipitated. If this is not removed, it will accumulate in boilers and hot water pipes and reduce their efficiency.

Terbium. Tb. At.no. 65, at.wt. 158.93, m.p. 1360°C, b.p. 3041°C, d 8.28. A typical lanthanide. The metal is hcp. The oxide has potential as a green phosphor for colour television. Sodium terbium borate has been used as a laser material.

Terbium Compounds. Terbium shows +3 and +4 oxidation states. Tb (III) compounds are typical lanthanide compounds

$$Tb^{3+}\ (f^8 \text{ pale pink}) \longrightarrow Tb\ (-2\ 39 \text{ volts in acid})$$

TbF_4 (TbF_3+F_2 at 300°C) and $Cs_3\ TbF_7$ are known and Tb oxides intermediate between Tb_2O_3 and TbO_2 are formed by heating Tb compounds in air. These higher oxides liberate oxygen from water. Tb_2Cl_3 and TbCl ($TbCl_3$ plus Tb) have metal-metal bonding.

Ternary Compound. A chemical compound formed from three elements *e.g.*, Na_2SO_4 or $LiAlH_4$.

Tervalent (trivalent). Having a valency of three.

Tetraethyl Lead. See lead tetraethyl.

Tetraflouoroberyllates. Containing the $[BeF_4]^{2-}$ ion.

Tetrafiuoroborates. Containing the $[BF_4]^-$ ion.

Tetrafluorohydrazine, N_2F_4. See nitrogen fluorides.

Tetragonal System. The crystal system with a 4-fold axis as principal axis of symmetry. The unit cell has its z-axis parallel to the 4-fold axis and the other two equal axes mutually perpendicular at right angles to this axis, E.g., rutile, TiO_2.

Tetrahedral Angle. The angle between the bonds in a tetrahedral compound (approximately 109° for a regular tetrahedron).

Tetrahedral Co-ordination. Regular co-ordination by four ligands situated at the corners of a tetrahedron. Methane, CH_4, has

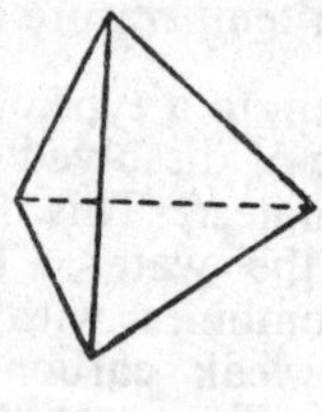

Tetrahedron

tetrahedral co-ordination about the carbon. A tetrahedron has four triangular faces.

Tetrahedral Compound. A compound in which an atom has four bonds direcred towards the corners of a regular tetrahedron. The angles between the bonds in such a compound are about 109°.

Tetrahydrate. A crystalline hydrated compound containing four molecules of water of crystallization per molecule of compound.

Tetrahedrite, Fahl Ore, Cu_3SbS_3. A common copper ore with a zinc blende type structure with Cu and Sb in the Zn position and S occupying 3/4 of the S positions. Sb can be randomly replaced by As, Cu_3AsS_3 with the same structure is tennantite.

Tetraoxophosphoric (V) Acid. See phosphoric (V) acid.

Tetraoxosulphuric (V) Acid. See sulphuric acid.

Tetravalent (quadrivalent). Having a valency of four.

Th. Thallium.

Thallates. Thallium (III) compounds.

Thallium. It is a soft malleable greyish metallic element belonging to group III of the periodic table. It is found in lead and cadmium ores, including pyrites (FeS_2). Thallium is highly toxic and was used previously as a rodent and insect poison. Various compounds have been now used in photocells, infrared detectors, and low-melting glasses.

Symbol : Tl; m.p. 303.5°C; b.p. 1457°C; r.d. 11.85; p.n. 81; r.a.m. 204.37.

Thermite (thermit). It is a mixture of aluminium powder and iron (III) oxide. When ignited the oxide is reduced; the reaction is strongly exothermic and molten iron is formed :

$$2Al + Fe_2O_3 \longrightarrow Al_2O_3 + 2Fe$$

Thermite is used in incendiary bombs and for welding steel.

Thermite Reaction. A reduction process which involves Al powder and a metal oxide (generally) Fe_2O_3) used to produce heat for welding or incendiary devices. Generally ignited with Mg ribbon; the reaction gives, *e.g.*, Al_2O_3 and Fe.

Thiocyanates. Salts of thiocyanic acid, HSCN. Complexes and esters can be N or S bonded. Formed cyanides plus S. Used in estimation of Ag^+ (precipitate of AgSCN) and Fe^{3+} (red colouration). Photographic sulphur sensitizers. With oxidizing agents (MnO_2. Br_2) give thiocyanogen, $(SCN)_2$, which polymerizers above 0°C.

Thiocyanic Acid, HSCN. Prepared $KHSO_4$ plus KSCN Gas at room temperature but rapidly polymerized. Forms thiocyanic esters RSCN and isothiocyanic esters or mustard oils, RNCS.

Thiocyanogen $(SCN)_2$. See thiocyanates.

Thiomolybdates. Species such as $[MoS_4]^{2-}$, $[MoO_2S_2]^{2-}$. Can act as bidentate ligands. Thiotungstates are similar.

Thionic Acid, $H_2S_nO_6$. See polythionic acids.

Thionitrosyls. Complexes containing NS groups, analogous to nitrosyls.

Thionyl. Containing >SO groups.

Thionyl Chloride, $SOCl_2$ M.p.−105°C, b.p. 79°C. Formed SO_2 plus PCl_5. Hydrolysed by water to HCl, SO_2 and H_2O. Used to convert COH groups to CCl and for the dehydration of metal chloride hydrates.

Thionyl Halides. SOF_2($SOCl_2$ plus SbF_3 or anhydrous HF); SOClF (as for SOF_2); $SOBr_2$ ($SOCl_2$ plus HBr); $SOCl_2$.

Tniosulphate. A salt having the ion $S_2O_3^{2-}$ formally derived from thiosulphuric acid. Thiosulphates readily decompose in acid solution to give elemental sulphur and hydrogensulphite (HSO_3^-) ions.

Thiosulphuric Acid, $H_2S_2O_3$. The free acid is only known (SO_3 plus H_2S) at−78°C as an etherate. Thiosulphates, containing the $[S_2O_3]^{2-}$ ion are formed from sulphites (sulphates(IV)) plus S. From complexes—particularly with Ag^+. Thiosulphates are used in photography. React quantitatively with I_2; reaction used in volumetric analysis. Acids on $[S_2O_3]^{2-}$ give SO_2 and S.

Thiotungstates. See thiomolybdates.

Thoria. See thorium.

Thorium. Refers to a toxic radioactive element of the actionoid series that is a soft ductile silvery metal. It has several long-lived radioisotopes occurring in a variety of minerals including monazite. Thorium is used in magnesium alloys, in the electronics industry, and is a potential nuclear fuel.

Symbol : Th; m.p. 1750°C; b.p. 4750°C (approx.); r.d. 11.72; p.n. 90; r.a.m. 232.04.

Thorium Halides. All ThX_4 are known

$$(ThO_2+CCl_2F_2 \text{ or } HF \xrightarrow{400^\circ C} ThF_4,$$

$$ThCl_4aq. \xrightarrow{SOCl_2} ThF_4,$$

$$Th \xrightarrow{Br\,700^\circ C} ThBr_4,$$

$$Th \xrightarrow{I_2\,400^\circ C} ThI_4)$$

ThF_4 is insoluble in water (but ThF_4, $4H_2O$ can be formed); the other halides are soluble in water. Halide oxides. $ThOX_2$, are formed from ThO_2 and $ThXb_4$. Complex halides, *e.g.* K_5ThF_9, Na_4ThF_8,Rb_2ThCl_6 are formed from acid solution and have high co-ordination numbers about the metal. For lower halides.

Thorium Oxide. ThO_2, Thorium salts are hydrolysed in aqueous solution and the hydrated oxide is precipitated by alkalis. Ignition of this hydrated oxide, not other oxyacid salts, gives white ThO_2.

Thorium Oxyacid Salts. Commercially available thorium salts include the nitrate ($4H_2O$ and H_2O), the oxalate ($6H_2O$) and sulphate (0, 4, 5, 8 and $9H_2O$). The anion is generally co-ordinated to the metal; the salts are eventually hydrolysed in aqueous solution.

Thorium Series. See radioactive series.

Thortveitite $Sc_2Si_2O_7$. An important sourcc of Sc. Contains $Si_2O_7^{6-}$ ions.

Thrse-centre Bond. A bonding orbital formed by two electrons between three atoms.

Threnardite. A mineral form of sodium sulphate, Na_2SO_4.

Thulium. Symbol Tm. A soft grey metallic element belonging to the lanthanoids; a.n. 69; r.a.m. 168.934; r d. 9.321 (20°C); m p. 1545°C: b.p. 1950°C. It occurs in apatite and xenotime. There is one natural isotope thulium—169, and seventeen artificial isotopes have been produced. There are no uses for the element, which was discovered by P.T. Cleve in 1879.

Thulium Compounds. Thulium forms a series of +3 compounds which are typical lanthanide compounds.
Tm^{3+} (f^{12} pale green)$\rightarrow$Tm(—2.28 volts in acid).

$TmBr_2$ ($TmBr_3$+Tm) is known but is probably Tm^{3+}, $2Br^-$,e^-.

Tin. Sn. At. no. 50, at. wt. 118.69, m.p. 232°C, b.p. 2270°C, d, 7.28—7.31. Group IV metal easily obtained from its ores and hence known since prehistoric times. Occurs as cassiterite or tinstone, SnO_2, which is reduced with C (also recovered electrolytically). The element has three allotropic forms, metallic tin is silvery in colour, soft and malleable, tetragonal (distorted cp lattice). Below 13.5°C the stable form is grey tin (α) diamond lattice, brittle powder (the phase change is slow in the absence of grey tin). λ-Sn, another brittle modification is stable above 161°C. The metal is attacked by halogens (Cl_2

used in Sn recovery) but because of high over voltage is not attacked by cold dilute acids. Concentrated acids will dissolve Sn. Used in coating steel and in alloys (solder bronze) (Sn—Nb alloys are superconductors). Compounds are used in glass contains and as fungicides. World production 1976 178000 tonnes.

Tin(II) Chloride A white solid, $SnCl_2$, soluble in water and ethanol. It exists in the anhydrous form (rhombic; r.d. 3 95: m.p. 246°C; b p. 65.2°C) and as a dihydrate, $SnCl_2 \cdot 2H_2O$ (monoclinic; r.d. 2.71; m p. 37.7°C). The compound is made by dissolving metallic tin in hydrochloric acid and is partially hydrolysed in solution.

$$Sn^{2+} + H_2O = SnOH^+ + H^+$$

Excess acid must be present to prevent the precipitation of basic salts. In the presence of additional chloride ions the pyramidal ion $[SnCl_3]^-$ is formed; in the gas phase the $SnCl_2$ molecule is bent. It is a reducing agent in acid solutions and oxidizes slowly in air :

$$Sn^{2+} \rightarrow Sn^{4+} + 2e$$

Tin(IV) Chloride ($SnCl_4$). A colourless fuming liquid. Tin(IV) chloride is soluble in organic solvents but is hydrolysed by water. It dissolves sulphur, phosphorus, bromine, and iodine, and it dissolves in concentrated hydrochloric acid to give the anion $SnCl_6^{2-}$.

Tin(IV) Hydride (stannane) A highly reactive and volatile gas (b p.—53°), SnH_4, which decomposes on moderate heating (150°C). It is prepared by the reduction of tin chlorides using lithium tetrahydridoaluminate(III) and is used in the synthesis of some organo-tin compounds. The compound has reducting properties.

Tin(II) Oxide (SnO). A dark green or black solid. It can also be obtained in an unstable red form, which turns black on exposure to air. Tin(II) oxide can be made by precipitating the hydrated oxide from a solution containing tin(I) ions and dehydrating the product at 100°C.

Tin(IV) Oxide (tin dioxide). A white solid, SnO_2, insoluble in water; tetrahedral; r.d. 6.95; m.p. 1630°C; sublimes at about 1850°C. Tin(IV) oxide is trimorphic : the common form, which occurs naturally as the ore cassiterite, has a rutile lattice but hexagonal and rhombic forms are also known. There are two so-called dihydrates, $SnO_2 \cdot 2H_2O$, known as α-and β-stannic acid. These are essentially tin hydroxides. Tin(IV) oxide is amphoteric, dissolving in molten alkalis to form stannate; in the presence of sulphur, thiostannates are produced.

Tin Plague. See tin.

Tin(II) Sulphide (SnS). A grey solid that can be prepared from tin and sulphur. Above 265°C tin(II) sulphide slowly turns into tin(IV) sulphide and tin :

$$2SnS(s) \rightarrow SnS_2(s) + Sn(s)$$

Tin(IV) Sulphide (mosaic gold). A bronze or golder yellow crystalline compound, SnS_2, insoluble in water and in ethanol; hexagonal; r.d. 4.5; decomposes at 600°C. It obtained by the reaction of hydrogen sulphide with a soluble tin(IV) salt or by the action of heat on thiostannic acid, H_2SnS_3. The golden-yellow form used for producing a gilded effect on wood is prepared by heating tin, sulphur, and ammonium chloride.

Titania. See titanium(IV) oxide.

Titanium. Ti At. no. 20, at. wt. 47, 90, m.p. 1660°C, b.p. 3287°C, d 4.5. Common element, the most important minerals are ilmenite, $FeTiO_3$, and rutile, TiO_2. The metal is obtained by reducing $TiCl_4$ with Mg. The metal is used as a light weight construction material; TiO_2 is used extensively as a white pigment and in catalysts. The massive metal has a protective layer of oxide in air.

Titanium Carbide, TiC. Steel grey solid of high m.p., formed TiO_2 plus charcoal in an electric furnace. Very resistant to chemical attack. Used in tool tips and as a deoxidizer in steel manufacture.

Titanium(IV) Chloride (titanium tetrachloride, $TiCl_4$). A volatile colourless liquid formed by heating titanium(IV) oxide with carbon in a stream of dry chlorine at 700°C. The product is purified by fractional distillation. Titanium(IV) chloride fumes in moist air forming the oxychlorides of titanium. Crystallization of such a solution gives crystals of the di-and pentahydrates. Titanium(IV) chloride finds use in the preparation of pure titanium and as an intermediate in the production of titanium compounds from minerals.

Titanium Dioxide. See titanium(IV) oxide.

Titanium(IV) Oxide (titania; titanium dioxide). A white oxide, TiO_2, occurring naturally in various forms, particularly the mineral rutile. It is used as a white pigment and as a filler for plastics, rubber, etc.

Topaz. A variably coloured aluminium silicate mineral, $Al_2(SiO_4)(OHF)_2$, that forms orthorhombic crystals. It is found mainly is acid igneous rocks, such as granites and pegmatities. Topaz is valued as a gemstone.

Tourmaline, $(Na,Ca)(Li,Al)_3Al_6(OH)_4(BO_3)_3Si_6O_{18}$. Contains cyclic Si_6O_{18} species. Occurs as black or blue-black crystals in granites, pegmatites and various metamorphic rocks.

Tracer. An isotope of an element used to investigate chemical reactions or physical processes (*e.g.* diffusion).

Tracing (radioactive tracing). See labelling.

Trans. Term used in designating isomers to indicate isomer with non-adjacent (opposite) likeligands. Compare cis.

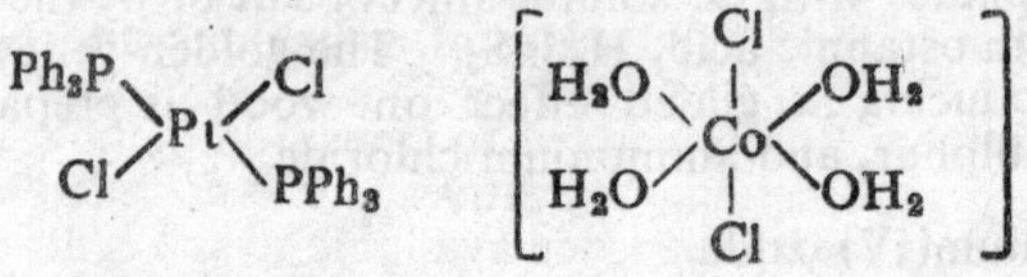

trans-planar trans-octahedral

Trans-actinides. See post actinides.

Trans Effect. In substitution in complexes the proportions of cis and trans isomers formed depends markedly upon the other ligands present. Thus in square-planar platinum complexes the proportion of the two products A and B depends upon the relative trans-directing effects of L and L′. Ligands can be placed in a

A B

series of ability to direct trans substitution; some common ligands in order are $H_2O < NH_3 < Cl^- < CO$.

The order of ligands in decreasing trans-directing power is :

$$CN^- > NO_2 > I^- > Br^- > Cl^- > NH_3 > H_2O.$$

Transition Elements. Refers to a class of elements occurring in the periodic table in three series : from scandium to zinc; from yttrium to cadmium; and from lanthanum to mercury. The transition elements are all metals. They owe their properties to the presence of d electrons in the atoms. In the first transition series, calcium has the configuration $3s^2 3p^6 4s^2$. The next element scandium has $3s^2 3p^6 3d^1 4s^2$, and the series is formed by filling the d levels up to zinc at $3d^{10} 4s^2$. The other two transition series occur by filling of the 4d and 5d levels. Often the elements scandium, yttrium, and lanthanum are considered with the lanthanoids rather than the transition elements, and zinc, cadmium, and mercury are also treated as a separate group.

Transition Point (transition temperature). The temperature at which one crystalline form of a substance changes to another form.

Transmutation. A change of one element into another by radioactive decay or by bombardment of the nuclei with particles.

A proton fired with a high velocity at an atom of lithium, may yield two atoms of helium.

$$7/3Li + 1/1p \rightarrow 2\ 4/2He$$

Transuranic Elements. Refer to those elements of the actinoid series that have higher atomic numbers than uranium. The transuranic elements are obtained by adding neutrons to the lower actinoids by high energy bombardment and they generally have such short half-lives that macroscopic isolation is impossible. There has been a considerable amount of speculation about other 'islands of stability' at much higher mass numbers, as high as 184, and the possible existence of a 'super-actinoid' series.

Tremolite, $(OH)_2Ca_2Mg_5(Si_4O_{11})_2$. An amphibole, an abestos mineral. Used in acid-resistant filters.

Triatomic. Denoting a molecule, radical, or iron consisting of three atoms For example, O_3 and H_2O are triatomic molecules.

Triclinic System. The crystal system without symmetry planes or axes. A unit cell has three unequal axes at mutually oblique angles. E.g. $CuSO_4,5H_2O$.

Tricresyl Phosphate. See phosphate esters.

Tridymite, SiO_2. The form of silica stable between 573°C and 1470°C. It is a desirable constituent of silica bricks because of its low coefficient of expansion.

Trigonal Bipyramidal Co-ordination. Co-ordination of five ligands at the concerns of a trigonal bipyramid. Occurs in PF_5. A trigonal pyramid has six triangular faces. The three equatorial ligands are not equivalent to the two apical but many trigonal bipyramidal species are fluxional and the apical and equatrial ligands cannot be identified separately by n.m.r. spectroscopy.

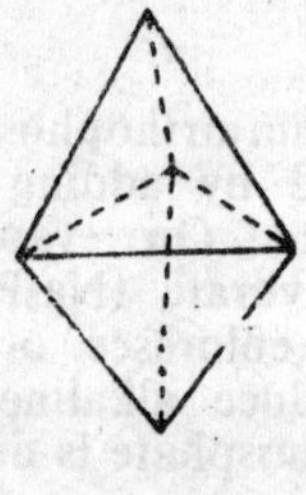

Trigonal Prismatic Co-ordination. Co-ordination of six ligands at the corners of a trigonal prism. Occurs for Mo in MoS_2. A trigonal prism has two triangular and three rectangular faces.

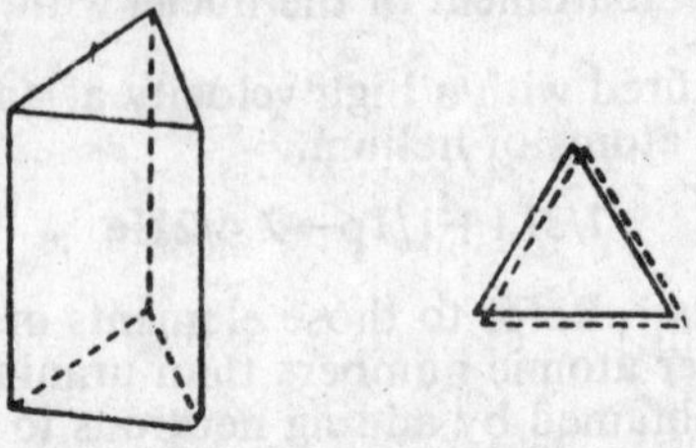

trigonal prism

Trigonal System. The crystal system with a 3-fold axis as principle axis. The unit cell is generally taken with the z-axis parallel to the 3-fold axis and the other two axes mutually inclined at 60° and perpendicular to the 3-fold axis. Rhombohedral unit cells (three equal axes equally at an angle [not 90°]) can be expressed in terms of trigonal cells. E g. trigonal $CaCO_3$.

Trihydrate. A crystalline hydrated compound that contains three molecules of water of crystallization per molecule of compound.

Tri-iron Tetroxide (ferrosoferric oxide). A black magnetic oxide, Fe_3O_4; r.d. 5.2. It is formed when iron is heated in steam and also occurs naturally as the mineral magnetite. The oxide dissolves in acids to give a mixture of iron(II) and iron(III) salts.

Trioxoboric(III) Acid. See boric acid.

Trioxosulphuric(IV) Acid. See sulphurous acid.

Trioxygen. See ozone.

Triple Bond. A covalent bond which is formed between two atoms in which three pairs of electrons contribute to the bond. One pair forms a sigma bond (equivalent to a single bond) and two pairs give rise to two pi bonds. It is conventionally represented as three lines.

Trisilane. See silane.

Trisodium Phosphate(V) (sodium orthophosphate, Na_3PO_4) A white solid which is obtained by adding sodium hydroxide to disodium hydrogenphosphate. On evaporation white hexagonal crystals of the dodecahydrate ($Na_3PO_4 \cdot 12H_2O$) are obtained. These crystals do not effloresce or deliquesce. They dissolve readily in water to produce alkaline solutions owing to salt hydrolysis. Trisodium phosphate is used as a water softener.

Tritium, 3/1H. The heaviest isotope of hydrogen, occurring in a proportion of less than 1 in 10^{17} in natural hydrogen. It is β-active and is used as a radioactive tracer in thermonuclear reactions. It may be prepared by the deuterium bombardment of various light atoms or by the thermal neutron reaction $^6Li(n\alpha)^3H$.

Tritiated Compound. A compound in which one or more ^1H atoms have been replaced by tritium (^3H) atoms.

Trivalent (tervalent). Having a valency of three.

Trona. A mineral form of sodium sesqui-carbonate, consisting of a mixed hydrated sodium carbonate and sodium hydrogen-carbonate, $Na_2Co_3 \cdot NaHCO_3 \cdot 2H_2O$.

Troostite. The very fine structures in heat-treated steels. The term is also used for a variety of zinc silicate, Zn_2Sio_4

Tungstates. Formed by dissolving WO_3 in NaOH solution and subsequently by cation exchange. The normal tungstates, *e.g.*, Na_2WO_4, $2H_2O$, contain tetrahedral WO_4^{2-} ions Most metal salts other than alkali metal salts are insoluble. Acidification of a solution of WO_4^{2-} ions gives polymeric anions—isopoly-tungstates—containing WO_6 octahedra joined corner to corner or edge to edge; $[HW_6O_{21}]^{5-}$, $[H_2W_{12}O_{42}]^{10-}$ and $]W_{12}O_{39}]^{6-}$ are well established. If the acidification is carried out in the presence of another oxyanion, heteropoly-tungstates in which MO_6 or MO_4 groups (M=heteroatom) are incorporated into the polyanion are formed. E g. $[Fe_2W_{12}O_{42})^{6-}$ and $[PW_{12}O_{40}]^{3-}$. It seems unlikely that free tungstic acids of any of these forms are stable. Further acidification of isopolytungstates gives WO_3, $2H_2O$. Heteropolytungstates are used to form W-containing catalysts by heating.

Tungsten Symbol W. A white or grey metallic transition element (formerly called wolfram); a.n. 74; r.a.m. 183.85; r.d. 19.3; m p. 3410°C; b.p. 5660°C. It is found in a number of ores, including the oxides wolf-ramite, (Fe, Mn) WO_4, and scheelite, $CaWO_4$. The ore is heated with concentrated sodium hydroxide solution to form a soluble tungstate. The oxide WO_3 is precipitated from this by adding acid, and in reduced to the metal using hydrogen. It is used in various alloys, especially high-speed steels (for cutting tools) and in lamp filaments Tungsten forms a protective oxide in air and can be oxidized at high temperature. It does not dissolve in dilute acids. It forms compounds in which the oxidation state ranges from +2 to +6. The metal was first isolated by J.J. and F. d'E'huyer in 1783.

Tungsten Blue. Partially reduced WO_3 or tungstates.

Tungsten Bronzes. Coloured compounds M_nWO_3 (M=unipositive metal 0<n<1) prepared by reducing tungstates with, *e.g.*, Na. The compounds contain W^{VI} and W^{V}.

Tungsten Carbide (W_2C and WC) Either of two carbides produced by heating powered tungsten with carbon. The carbides are extremely strong and are used in industry to make cutting tools or as an abrasive. WC has a very high melting point (2770°C) and will conduct electricity. W_2C also has a very high melting point (2780°C) but is a less efficient conductor of electricity. It is very resistant to chemical attack and behaves in a manner very similar to that of tungsten. It is strongly attacked by chlorine to give tungsten hexachloride, WC_6.

Tungten Carbonyl, $W(CO)_6$. Very similar to molybdenum carbonyl but less reactly substituted.

Turquoise. A blue basic hydrated aluminium phosphate, $Al_2(OH)_3PO_4$, H_2O containing some Cu. Usually microcrystalline but amorphous in appearance with a waxy lustre; used as a gemstone.

Tutton Salts. The salts $M^{I}_2SO_4$, $M^{II}SO_4$, $6H_2O$ where M^{I} is an alkali metal and M^{II} is a dipositive transition metal.

Tysonite. The naturally occurring mixed fluorides (La, Ce,...) F_3. The tysonite structure, which is adopted by many lanthanide and actinide trifluorides, has the metal in trigonal bipyramidal plus other near neighbour co-ordination.

U

Ulexite $NaCaB_5O_9$, $8H_2O$. An important source of borates.

Ultraviolet (UV) A form of electromagnetic radiation, which is shorter in wavelength than visible light. Ultraviolet wavelengths range between about 1 nm and 400 nm. Ordinary glasses have been not transparent to these waves; quartz has been a much more effective material for making lenses and prisms for use with ultraviolet. Like light, ultraviolet radiation is obtained by electronic transitions between the outer energy levels of atoms. However, having a higher frequency, ultraviolet photons carry more energy than those of light and can induce photolysis of compounds.

Uncertainty Principle (heiserbeṛg uncertainty principle; principle of indeterminism). The principle which states that it is not possible to know with unlimited accuracy both the position and momentum of a particle. This principle, discovered in 1927 by Werner Heisenberg (1901—76), may be usually stated in the form : $\Delta x \Delta p_x > h/2\pi$, where Δx is the uncertainty in the x-coordinate of the particle, Δp_x is the uncertainty in the x-component of the particle's momentum and h is the Planck constant. An explanation of the uncertainty has been that in order to locate a particle exactly, an observer must be able to bounce off it a photon of radiation; this act of location itself changes the position of the particle in an unpredictable way.

Uniaxial Crystal. A double-refracting crystal which is having only one optic axis

Unit Cell. In a crystal there is a regular repetition of a three-dimensional arrangement of the constituent atoms. In order to describe the structure of the crystal it is usually sufficient to describe the characteristic portion of the crystal which, if repeated indefinitely in the directions of the crystal axes, reproduces the crystal. This portion of the crystal is contained within a parallelopiped called the unit cell, whose sides are parallel to the crystal axes.

United Atom Method. In the LCAO method, atoms are brought from infinite distance gradually near one another till they take up a position of equilibrium in one molecule. Thus, the individual atomic orbitals yield molecular orbitals which may be bonding if the spins are paired or antibonding is the spins are unpaired.

In the united atom method, we start with an united atom in which the nuclei aresuprimposed over one another. Then, the individual nuclei are gradually separated till they acquire a position of equilibrium. It is important to remember that the united atom is purely hypothetical and has no actual existence.

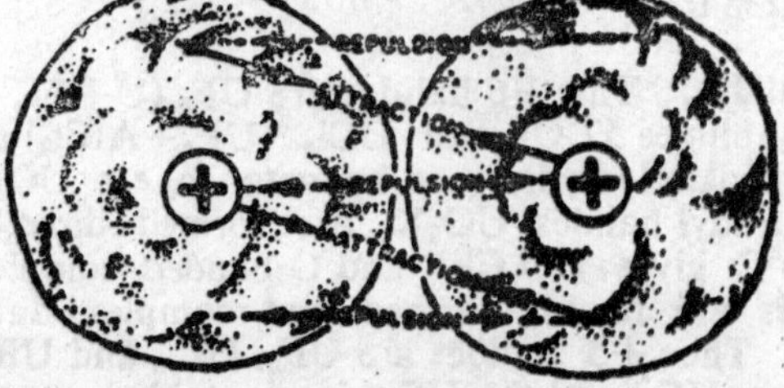

Let us illustrate the formation of hydrogen molecule by the united atom method. In this treatment, we consider the structure of H_2 where the hydrogen nuclei are superimposed over each other as shown in Fig. In this structure there are no interatomic forces of attraction and repulsion. But there are naturally a great deal of repulsion between the two nuclei and electrons. Thus, the system will have higher energy and it will not be stable. If we now separate the nuclei very slowly, the repulsive forces will decrease and the energy of the system will be lowered. At equilibrium position, the two nuclei will be at a distance at which the forces of repulsion are minimum. This is also the position at which the energy of the system is lowest. This is the position at which the molecular orbital is formed. If we remove the atoms to infinity, the energy of individual atomic orbitals increases. Thus, we can represent the relationship between the orbitals in the united atom, molecular orbitals when the atoms are at equilibrium distance, and the individual atomic orbitals when are at infinity as follows.

Univalent (monovalent). Having a valency of one.

Uraninite. Refers to a mineral form of uranium(IV) oxide, which is having minute amounts of radium, thorium, polonium, lead, and helium. When uraninite occurs in a massive form with a pitchy lustre it is termed as pitch-blende, the chief ore of uranium Uraninite is found in Saxony (East Germany), Romania, Norway, the UK (Cornwall). E. Africa (Zaire), USA and Canada (Great Bear Lake).

Uranium. A toxic radioactive silvery element of the actinoid series of metals. Its three naturally occurring radioisotopes, uranium 238 (99.283% in abundance), uranium-235 (0.711%), and uranium-234 (0.005%), occur in numerous minerals including the uranium oxides pitchblende, uraninite, and carnotite. The readily fissionable ^{235}U has been a major nuclear fuel and nuclear explosive, while ^{238}U has been a source of fissionable plutonium-239.

Symbol : U : m.p. 1132°C; b.p. 3818°C; r.d. 18.95; p.n 92; r.a.m. 238.03.

Uranium Halides. The +6 halides are UF_6 ($U+F_2$, UF_4+O_2, m.p. 64°C, sublimes 57°C) and UCl_6 (UF_6+AlCl_3) both of which are immediately hydrolysed by water to, *e.g* UOF_4 and UO_2F_2. Other uranyl halides, UO_2X_2, are formed direcly (*e.g* UCl_4+O_2/350°C gives UO_2Cl_2) and O-bonded complexes, including hydrates are readily formed and complex fluorides are also known. The +5 halides are UF_5, UCl_5 and UBr_5 (*e g.*, UF_6+HBr gives UF_5) and the UF_6^- ion is stable in aqueous HF. All of the tetrahalides, UX_4, are known (UO_2+HF/500°C gives UF_4;U_3O_8+C_3Cl_6) and all form O-and N-bonded complexes and a wide range of complex fluorides is known. The +3 halides UF_3, UCl_3 and UBr_3 are formed by H_2 reduction of UX_4; hydrates, ammines and other N-bonded complexes are known.

Uranium Hexafluoride. See uranium (VI) fluoride.

Uranium (VI) Fluoride (uranium hexafluoride). A volatile white solid, UF_6; r.d. 5.1; m.p. 6.45°C. It finds use in the separation of uranium isotopes by gas diffusion.

Uranium-lead Dating. Refers to a group of methods of dating certain rocks that depends on the decay of the radioisotopes uranium—238 to lead—206 (half-life 4.5×10^9 years) or the decay of uranium—235 to lead—207 (halife $7.1 \times 10_8$ years). One form of uranium—lead dating is depending on measuring the ratio of the amount of helium trapped in the rock to the amount of uranium present (since the decay $^{238}U \rightarrow {}^{206}Pb$ releases eight alpha-particles). Another method of calculating the age of the rocks to has been measure the ratio of radiogenic lead (^{206}Pb, ^{207}Pb, and ^{208}Pb) present to nonradiogenic lead (^{204}Pb). These methods provide reliable results for ages of the order $10^7 - 10^9$ years.

Uranium (IV) Oxide. A black solid, UO_2; r.d. 109; m.p. 3000°C. It occurs naturally as uraninite and is used in nuclear reactors.

UraniumOxides. Uranium forms an extensive series of oxides including UO (semi-metallic), UO_2, U_3O_8 (the stable oxide in air), U_4O_9 $U_{2.25}$⁻ $U_{2.40}$ (two forms) and UO_3 (five forms). The hydrated peroxide UO_4,$2H_2O$ is precipitated from an aqueous solution of a U(VI) salt by H_2O and on heating gives UO_3 which may be reduced with hydrogen. Complex insoluble uranates (VI), generally polymeric mixed oxides, are precipitated from solutions of uranium (VI) salts or are formed by solid state reactions, many uranatest (VI) contain linear UO_2 groupings. U(V) and U(IV) mixed metal oxides are formed by solid state reactions (*e.g.*, $BaUO_3$+UO_2/550°C gives $Ba(UO_3)_2$). So called 'sodium di or pyro-nranate, uranium yellow, Na_2UO_7,$6H_2O$, is an insoluble sodium salt used in glass.

Uranium Series. See radioactine series.

Uranium Yellow. See uranium oxides.

Uranocene. See uranium compounds.

UV. See ultraviolet radiation

V

Vacuum Arc Furnace Refining. Crude metal made into as electrode is progressively melted in an arc furnace under vacuum. As the electrode melts, volatile impurities boil off. The molten metal is then chilled by pouring into a water cooled copper crucible. Refractory metals (with high m. pt.) like Ti, Zr and Mo are refined by this method.

In a recent development of the method, the metal to be melted is bombarded by electrons from a heated tungsnten element under conditions of high vacuum.

Valence (valency). The combining power of an element or radical which is equal to the number of hydrogen atoms that will combine with or displace one atom of the element.

The valence of an element has been generally equal to either the number of valence electrons or eight minus the number of valence electrons. Transition metal ions display variable valence.

Valence Bond Method. It is due to the work of Heitler, London, Slater and Pauling. In this treatment, a molecule is formed by the overlapping of atomic orbitals and a molecule is composed of atoms which retain their individual character when linked to the other atoms.

Valence Electron. An electron in one of the outer shells of an atom that takes parts in forming chemical bonds.

Valence Shell Electron-pair Repulsion Theory (VSEPR Theory). This theory was developed by Gillespie and Nyholm. According to this theory, the geometry of a molecule depends upon the number of bonding and non-bonding electron-pairs (lone pairs) in the central atom which arrange themselves in such a way that there is a minimum repulsion between them so that the molecule has minimum energy and maximum stability. Since there can only be one orientation of orbitals corresponding to minimum energy, a molecule of a given substance has invariably a definite shape, *i.e.* a definite geometry.

Gillespie has proposed the following rules for explaining the shapes of molecules and ions.

Rule 1. When the central atom in a molecule is surrounded only by bonded electron-pairs but not by lone-pairs, the molecule will have a regular geometry or shape. The geometry depends on the number of bonded electron pairs as given in the following Table.

Number of electron pairs	*Geometry*	*Bond angle*	*Examples*
2	Linear	180°	BeF_2, $HgCl_2$, CdI_2
3	Trigonal plane	120°	BF_3, BI_3, GaI_3
4	Tetrahedral	109° 28,	CH_4. $SnCl_4$,$SnBr_4$,BF_4^{-1}
5	Trigonal bipyramid	90° and 110°	PCl_5, $SbCl_5$. VO_3^{-1}
6	Octahedral	90°	SF_6, $[SiF_6]^{2-}$$[PbCl_6)]^{2-}$
7	Pentagonal bipyramid	72° and 90°	IF_7

Rule 2. Some molecules may have distorted or irregular geometry if the electron-pairs surrounding the central atom are not equivalent, *i.e.*, some are bonded while remaining are lone pairs. Lone pair electrons repel adjacent electron pairs more strongly than do bonding electron pairs. Also, the repulsion is expected to increase in the following order.

bond pair-bond pair < bond pair-lone pair < lone pair-lone pair.

This rule is justified because a bonding electron is under the influence of two nuclei whereas a lone pair is under the influence of only one nucleus. Therefore, a lone pair will occupy a greater space on the surface of an atom than a bonding pair. Consequently, the presence of lone pairs will alter the bond angle to a small extent which does not pertain to any regular geometry. The molecule in such a case is said to have an irregular geometry. One can justify this rule by comparing the bond angles which decrease in the series.

$$CH_4,\ 109.5° > NH_3,\ 107.3° > H_2O,\ 104.5°$$

Rule 3. The greater the number of lone pairs of electrons on the central atom, the greater, in general, will be the tendency of bond angles to contract and, hence, greater will be the distortion in the geometry of the molecule. This rule is understood by considering the following series;

NO_2^+, 180°; $\dot{N}O_2$, 104°; $\ddot{N}O_2^-$, 115°.

Rule 4. The smaller the electronegativity of the central atom, the smaller is the bond angle and hence, smaller is the distortion in the regular shape of molecules. For example,

(i) PI_3. 102° > PBr_3 101.5° > PCl_3, 100°.

(ii) AsI_3, 101.5° > $AsBr_3$, 100.5° > $AsCl_3$, 98.4°.

Rule 5. Bond angles involving multiple bonds are generally larger than those involving only single bonds. However, multiple bonds do not affect the geometry of a molecule.

Rule 6. Re,ulsion between electron pairs in field shells is greater than that in incomplete shells. For example, the bond angle decreases.

OH_2 ⩾	SH_2 >	SeH_2 >	TeH_2
104 5	92.2	91.0	89.5

Valency. See valence.

Vanadates. Anionic derivatives of vanadium. Practically vanadates(V) although vanadates(IV), mixed metal oxides, are also known. Free vanadic acids are not known.

Vanadic Acids. See vanadates(V).

Vanadium. V. At. no. 23, at.wt. 50.9415, m.p. 1890°C, b.p. 3380°C, d 5.96. Group V transition element occurring naturally as patronite (V_2S_5 plus S), carnotite (KUO_2VO_4, $1.5H_2O$—also uranium ore), and vanadinite ($Pb_5(VO_4)_3Cl$—apatite structure). Also occurs in some crude oils. Obtained from the above ores or from ores worked for Fe or Ti production. The ore is converted to vanadate(V) by roasting with NaCl, leached and acidified to V_2O_5 which is purified via NH_4VO_3. V_2O_5 is reduced with Ca in a $CaCl_2$ flux and refined by electrolysis of fused NaCl–LiCl—VCl_2 or by a van-Arkel process. Often used as ferrovanadium (Al reduction in the presence of Fe). The metal is bcc; soft grey or silver. It burns on heating in O_2 or Cl_2 and dissolves in HNO_3 and slowly in hot conc. H_2SO_4 and fused alkalis. Used mainly as a steel additive (80%) and in V—Al alloys (10%). Used in catalysts. World production 1976 23620 tonnes.

Vanadium Bromides. Vanadium(II) bromide, VBr_2. Reddish brown crystals formed from VBr_3 and H_2.

Vanadium(III) bromide, VBr_3, Dark green or black (V plus Br_2) gives green solution in water and green crystalline VBr_3, $6H_2O$. Forms many complexes.

Vanadium oxide tribromide, $VOBr_3$, Dark red deliquescent liquid formed by heating V_2O_3 plus Br_2.

Vanadium oxide dibromide, $VOBr_2$. Yellow powder obtained by heating $VOBr_3$ or Br_2 plus S_2Br_2 over heated V_2O_5.

Vanadium Carbides. VC (NaCl structure) and V_4C_3 are formed from the elements and are important in the constitution of vanadium steels.

Vanadium Carbonyl, $V(CO)_6$. Green-black paramagnetic carbonyl formed VCl_3 plus Na plus CO in diglyme to give [Na diglyme$_2$] [$V(CO)_6$] which gives $V(CO)_6$ with HCl—Et_2O. Many substituted carbonyl derivatives and organometallic derivatives, *e g.* $C_5H_5V(CO)_4$ are known.

Vanadium(V) Oxide (vanadium pentoxide, V_2O_5). An oxide of vanadium which is extensively used as a catalyst in oxidation processes, as in the contact process.

Vanadium Pentoxide. See vanadium(V) oxide.

Vanadium Sulphides. V_2S_5(V_2S_3 plus S), V_2S_3(V_2O_5 plus CS_2) and VS (V_2S_3 plus H_2) are known. Patronite (V_2S_5 plus S) is a vanadium mineral.

Vanadyl Species. Compounds containing the blue VO^{2+} group, *e g.* $VOSO_4$, $VOCl_2$.

Van Arkel—de Boer Process. Process for purification of metals, *e g.* Ti, V, where TiI_4 is vaporized and decomposed on a hot wire.

Van Arkel—de Boer Method. Ultra-pure metals are being prepared by the van Arkel-de Boer method. Metal is converted to a volatile stable compound taking care that the impurities are not affected during compound formation. The compound so obtained is then decomposed to get the pure metal. This method is being used for the purification of metals like titanium and zirconium.

Van Der Waals Force. An intermolecular force of attraction which is considerably weaker than chemical bonds and arising from weak electrostatic interactions between molecules (the energies are often less than 1 J mol^{-1}).

The van der Waals interaction includes contributions from three effects : permanent dipole-dipole interactions found for any polar molecule; dipole-induced dipole interactions, where one dipole causes a slight charge separation in bonds that have a high polarizability; and dispersion force, which result from temporary polarity arising from asymmetrical distribution of electrons around the nucleus. Even atoms of the noble gases exhibit dispersion forces.

Venetian Red. Native high grade Fe_2O_3, more usually Fe_2O_3 with some $CaSO_4$ prepared by heating $FeSO_4$ in the presence of calcium oxide.

Van Der Waal's Radii or Collision Radii. It is essentially the distance between two non-bonded atoms of two adjacent molecules. For example, the van der Waal's radius of hydrogen is 1.2A°; nuclei of two hydrogen atoms cannot approach nearer than 2×1 2 or 2.4A° from each other without overlapping of their electron charge clouds to form a bond. For example, in solid chlorine, each atom is found to have one close neighbour at a distance of 1.98A° and another at a distance of 3.60A°.

Thus, the covalent radius of chlorine atom is 0.99A° whereas the van der Waal's radius is 1.80A°.

Variation Method. According to the variation method, the wave function which gives the lowest energy of the system is close to the accurate wave function of the system. This method has been mainly used for the approximate determination of the lowest or ground state energy level of any system. The variation method is by far the most useful in the treatment of chemical bonding.

Verdigris. The green compound to which copper or bronze is converted on exposure to the atmosphere. It is usually a basic copper carbonate, but near the sea will be a basic chloride. The term is also used for basic copper ethanoate which is used as a pigment.

Veridian, Cr_2O_3. See chromium(III) oxide.

Vermiculite. A hydrated silicate clay. Used as a thermal insulator and in agriculture as a rooting medium and soil additive.

Vermilion Red, HgS. Occurs naturally as cinnabar. The red pigment in Chinese lacquer painting, red ink and temple paint are due to vermilion. Eventually Pb_3O_4 replaced HgS as a red pigment.

Villiaumite. A mineral form of sodium fluoride, NaF.

Volatile Matter. The % loss in weight when a standard weight of powdered coal or coke is heated under controlled conditions for seven minutes at 925°C in a crucible from which air is excluded. The loss consists of gas, tar and water from decomposition of the material. The surface or inherent moisture, separately determined, must be deducted from the VM figure obtained.

Volhard Method. Titration of Ag^+ with NCS^- in the presence of Fe_3^+. A deep red colour is formed at the end-point.

Vitreous. Resembling glass, or having the structure of a glass.

Vulcanite (ebonite). A hard black insulating material which is obtained by the vulcanization of rubber with a high proportion of sulphur (up to 30%).

W

Wackenroder's Solution. Solution containing colloidal sulphur and polythionic acids formed by passing H_2S into aqueous SO_2 solution.

Washing Soda. Sodium carbonate dedahydrate, $Na_2CO_3.10H_2O$.

Water (H_2O). It is a colourless liquid that freezes at 0°C and, at atmospheric pressure, boils at 100°C. In the gaseous state water contains single H_2O molecules. The atoms do not lies in a straight line, the angle between the central oxygen atom and the two hydrogen atoms being 105°; the distance between each hydrogen atom and the oxygen atom has been 0.099 nm. When ice forms, hydrogen bonds some 0.177 nm long develop between the hydrogen atom and oxygen atoms in adjacent molecules, giving ice its tetrahedral crystalline structure with a density of 916.8 kg m^{-3} at STP. Different ice structures develop under higher pressures. When ice melts to form liquid water, the tetrahedral structure breaks down, but some hydrogen bonds continue to exist; liquid water consists of group of associated water molecules, $(H_2O)_n$, mixed with some monomers and some dimers. This mixture of molecular species is having a higher density than the open-structured crystals. The maximum density of water, 999.97 kg m^{-3}, occurs at 3.98°C. This accounts for the ability of ice to float on water and for the fact that water pipes burst on freezing.

Although water has been predominantly a covalent compound, a very small amount of ionic dissociation takes place ($H_2O = H^+ + OH^-$). In every litre of water at STP there are approximately 10^{-7} moles of each ionic species. It is for this reason that, on the pH scale, a neutral solution has a value of 7.

As a polar liquid, water is the most powerful solvent known.

Water is decomposed by reactive metals (*e.g.* sodium) when cold and by less active metals (*e.g.* iron) when steam is passed over the hot metal. It is also decomposed by electrolysis.

Water Gas. A mixture of carbon monoxide and hydrogen which is obtained produced by passing steam over hot carbon (coke) :

$$H_2O(g)+C(s)\rightarrow Co(g)+H_2(g)$$

The reaction is strongly endothermic but the reaction can be used in conjunction with that for producer gas for making fuel gas. The main use of water gas before World War II was in producing hydrogen for the Haber process. Here the above reaction was combined with the water-gas shift reaction to increase the amount of hydrogen :

$$CO+H_2O\rightleftharpoons CO_2+H_2$$

Most hydrogen for the Haber process has been now made from natural gas by steam reforming.

Water Glass. A viscous colloidal solution of sodium silicates in water which is used to make silica gel and as a size and preservative.

Water of Crystallization. Water present in definite proportions in crystalline compounds. Compounds having water of crystallization are known as hydrates. Examples are copper(II) sulphate pentahydrate ($CuSO_4.5H_2O$) and sodium carbonate decahydrate ($Na_2CO_3.10H_2O$). The water can be removed by heating.

When hydrated crystals are heated the water molecules may get lost in stages For example, copper(II) sulphate pentahydrate changes to the monohydrate ($CuSO_4.H_2O$) at 100°C, and to the anhydrous salt ($CuSO_4$) at 250°C. The water molecules in crystalline hydrates may be held by hydrogen bonds (as in $CuSO_4$ H_2O) or, alternatively, may be coordinated to the metal ion as a complex aquo ion.

Water Softening. The removal from water of dissolved calcium, magnesium, and iron compounds, thereby reducing the hardness of the water. The compounds have been potentially damaging because they can accumulate in pipes and boilers and they react with, and therefore waste, soap. Temporary hardness could be removed by boiling the water. Permanent hardness can be removed in a number of ways : by distillation; by the addition of sodium carbonate, which causes dissolved calcium, for example, to precipitate out as calcium carbonate; and by the use of ion-exchange products such as Permutit and Calgon.

Wave Function. A function $\psi(x, y, z)$ appearing in Schrodinger's equation in wave mechanics. The wave function is a mathematical expression involving the coordinates of a particle in space. If the Schrodinger equation can be solved for a particle in a given system (*e.g.* an electron in an atom) then, depending on the boundary conditions, the solution is a set of allowed wave functions (eigenfunctions) of the particle, each corresponding to an allowed energy level (eigenvalue). The physical significance of the wave function has been that the square of its absolute value, $|\psi|^2$, at a point is proportional to the probability of finding the particle in a small element of volume, dxdydz, at that point. For an electron in an atom, this provides to the idea of atomic and molecular orbitals.

Wavelength Symbol : λ. May be defined as the distance between the ends of one complete cycle of a wave. Wavelength is related to the speed (c) and frequency (v) thus :

$$c = v\lambda$$

Wave Mechanics. The electronic theory of matter is confronted by certain difficulties which have only been overcome by assumptions of a very arbitrary nature. To overcome these defficulties by a general theorem, de Brogile and Schrodinger independently devised systems of mechanics which are essentially mathematical, and may only be very approximately described in terms of physical pictures. According to de Broglie, every kind of particle is associated with a wave motion, phase waves or Φ-waves, which is different from any of the radiations with which we are familiar. The interaction of these waves gives rise, however, to the familiar forms of radiation such as light Wave-mechanics has led to a considerable clarification of many difficulties inherent in the older quantum theory and in the electronic theory of matter, but it is still necessary to use these conceptions be cause of their more concrete nature.

Wave Number Symbol : σ. May be defined as the reciprocal of the wavelength of a wave. It is the number of wave cycles in unit distance, and is often used in spectroscopy. The unit is the metre^{-1} (m^{-1}) The circular wave number (Symbol : k) is given by :

$$k = 2\pi\sigma$$

Weberite Structure. A structure adopted by compounds of stoicheiometry $A_2B_2O_7$ (*e g.* $Ca_2Sb_2O_7$. Na_2MgAlF_7) which contain linked octahedra

Werner's Theory of Coordination. Werner, in 1893, put forward his famous theory of coordination to explain the formation of complex compounds. Taking the ammines like $PtCl_4.6NH_3$, $CoCl_3.6NH_3$, $CuSO_4.4NH_3$, $AgCl.2NH_3$, which were the best known at that time, he postulated the existence of two types of valencies possessed by many metals. These are : (i) the primary valencies and (ii) the secondary valencies.

(i) Primary valencies are those which a metal exercises in the formation of its simple salts. Thus, primary valencies of Pt, Co, Cu and Ag in the formation of their simple salts, *e g*, $PtCl_4$, $CoCl_3$, $CuSO_4$, and AgCl, are 4, 3, 2, and 1, respectively.

(ii) Secondary valencies are those which a metal cation exercises towards a neutral molecule or an anion in the formation of its complex ions. This valency, for example, as indicated by the composition of the ammines mentioned above, is 6 for Pt^{4+}, 6 for Co^{3+}, 4 for Cu^{2+} and 2 for Ag^{+} cations. The secondary valency is also termed as the coordination number of the metal cation under reference. Thus, coordination number gives the total number of neutral molecules or anions which may be linked to the metal cation in the formation of its complexes.

(iii) Every metal has a fixed number of secondary valencies, *i.e.*, it has a fixed coordination number.

White Arsenic. Crude arsenic (III) oxide, As_2O_3.

White Gold. A gold-palladium alloy.

Whiteheart Iron. See cast irons.

White Lead, $Pb_3(OH)_2(CO_3)_2$. See lead carbonate.

White-metal Alloys. Alloys of Sn, Pb and Sb. Solders are Pb—Sn alloys. Sn—Sb alloys include Britannia metal. Pb—Sb alloys include type-metal alloys (some Bi is sometimes added) which give precise castings. Fusible metal, used for boiler safety plugs is approx. 50%Bi, 25% Pb, 25% Sn. Anti friction alloys and bearing metals are also Pb—Sn—Sb alloys and show plasticity and low coefficients of friction. They include Babbitt metal (88% Sn, 8% Sb, 4% Cu), magnolia metal (80% Pb, 20% Sb) and Parson's white bronze (68% Sn, 30% Zn, 1% Cu, 1% Pb).

White Mica. See muscovite.

Whiting. A form of $CaCO_3$ made by grinding chalk and collecting the finer sediments from water. It is extensively used in the chemical and many other industries.

Whitherite. A mineral form of barium carbonate, $BaCO_3$.

Wolfram. A former alternative name for fungsten.

Wolframite, (Fe, Mn)WO_4. The principal ore of tungsten. It is crystallizing in the monoclinic system; the principal ore of tungsten. It commonly occurs as blackish or brownish tabular crystal groups. It is found chiefly in quartz veins associated with granitic rocks. China is the major producer of wolframite.

Wood's Metal. A fusible alloy having 50% bismuth, 25% lead, and 12.5% tin and cadmium. Its low melting point (70°C) leads to its use in fire-protection devices.

Work Hardening. Increase in strength and hardness, decrease in ductility and toughness of a metal alloy resulting from plastic deformation at a temperature below its recrystallization temperature (normally room temperature). Beneficial in production of high tensile wire by cold drawing and spring strip by cold rolling. Causes trouble in, *e.g.* pressing of metal sheet; cured by intermediate annealing. Results from mutual interference of dislocations moving on intersecting slip planes and also of grain boundary obstruction of dislocations.

Wrought Iron. A fairly pure iron with numerous slag inclusions. It was formerly much used by blacksmith because of its ease of welding and malleability.

Wurtzite. A form of ZnS which gives its name to an important structural type. Hexagonal with both Zn and S tetrahedrally co-ordinated. Has hcp S with Zn in tetrahedral holes. Other compounds with this structure include ZnO, BeO, AlN, GaN.

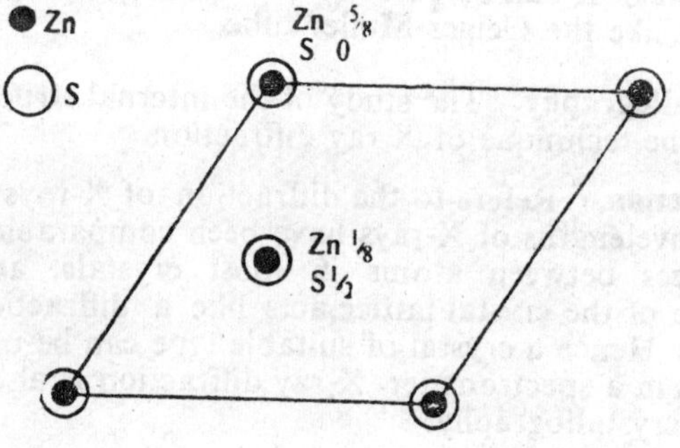

X

Xenon. Symbol Xe. A colourless odourless gas which belongs to group 0 of the periodic table (see noble gases); a.n. 54; r.a.m. 131.30; d. 5.887 g dm^{-3}; m.p. 111.9°C; b.p. -107.1°C. It occurs in the atmosphere (0.00087%) from which it is extracted by distillation of liquid air. There are nine natural isotopes with mass numbers 124, 126, 128—132, 134, and 136. Seven radioactive isotopes are also known. The element finds use in fluorescent lamps and bubble chambers. The element was discovered in 1898 by Ramsey and Travers.

Xenon Chemistry. Xenon has electronic configuration $5s^2 5p^6$ and shows oxidation states +2 [XeF_2 (Xe plus F_2 plus light), $XeCl_2$ (Xe plus Cl_2 plus electric discharge) both linear], +4 [XeF_4 (Xe plus F_2) square planar and its adducts with acid fluorides], +6 [XeF_6 (Xe plus F_2), explosive XeO_3 (XeF_6 plus H_2O)] and +8 [$(XeO_6)^{4-}$ ($(HXeO_4)^-$ plus base)] Other derivatives including oxyfluoro anions, *e.g.* $[XeO_3F]^-$, and oxide fluorides, *e.g.* $XeOF_2$ and $XeOF_4$, are also known.

X-radiation. Refers to an energetic form of electromagnetic radiation. The wavelength range has been 10^{-11} m to 10^{-8} m. X-rays are normally produced by absorbing high-energy electrons in matter. The radiation can pass through matter to some extent (hence its use in medicine and industry for investigating internal structures). It can be detected with photographic emulsions and devices like the Geiger-Muller tube.

X-ray Crystallography. The study of the internal structure of crystals using the technique of X-ray diffraction.

X-ray Diffraction. Refers to the diffraction of X-rays by a crystal. The wavelengths of X-rays have been comparable in size to the distances between atoms in most crystals, and the repeated pattern of the crystal lattice acts like a diffraction grating for X-rays. Hence a crystal of suitable type can be used to disperse X-rays in a spectrometer. X-ray diffraction is also the basis of X-ray crystallography.

X-rays. Electromagnetic radiation of shorter wavelength than ultraviolet radiation and longer wavelength than gamma radiation. The range of wavelengths is 10^{-11} m to 10^{-9} m. Atoms of all the elements emit a characteristic X-ray spectrum when they are bombarded by electrons.

Y

Ytterbium. A soft malleable silvery element having two allotropes and belonging to the lanthanoid series of metals. It is found in association with other lanthanoids. Ytterbium is used to improve the mechanical properties of steel.

Symbol : Yb; m,p. 819°C; b.p. 1194°C; r.d. 6.97 or 6.54; p.n. 70; r.a.m. 173.04.

Ytterbium Compounds. Ytterbium shows +3 and +2 oxidation states. Yb(III) species are typical lanthanide compounds.

$$Yb^{3+} (f^{13} \text{ colourless}) \rightarrow Yb \ (-2.27 \text{ volts in acid})$$

The dihalides YbX_2 are yellow (YbX_3+Yb) and are genuine Yb^{2+} salts.

$$Y^{3+} \rightarrow Yb^{2+} \ (-1.15 \text{ volts})$$

A lower oxide YbO probably exists.

Yttrium Symbol Y. A silvery-grey metallic element which belongs to group IIIB of the periodic table; a.n. 39; r.a.m. 88.905; r.d. 4.469 (20°C); m p. 1522°C; b.p. 3338°C It occurs in uranium ores and in lanthanoid ores, from which it can be extracted by an ion exchange process. The natural isotope is yttrium-89, and there are 14 known artificial isotopes. The metal finds use in superconducting alloys and in alloys for strong permanent magnets (in both cases, with cobalt). The oxide (Y_2O_3) is used in colour-television phosphors, neodymium-doped lasers, and microwave components.

Yttrium Compounds. Yttrium only forms compounds in the +3 oxidation state.

$$Y^{3+} (\text{colourless}) \rightarrow Y \ (-2.37 \text{ volts in acid})$$

Tne chemistry is similar to that of the lanthanides.

Z

Zeeman Effect. The term used for the spiltting of the lines in a spectrum when the source of the spectrum is exposed to a magnetic field. It was discovered in 1896 by P. Zeeman (1865-1943). In the normal Zeeman effect a single line is spilt into three if the field is prependicular to the light path or two lines if the field is parallel to the light path. This effect can be explained by classical electromagnetic principles in terms of the speeding up and slowing down of orbital electrons in the source as a result of the applied field. The anomolous Zeeman effect has been a complicated splitting of the lines into several closely spaced lines, so called because it does not agree with classical predictions. This effect is explained by quantum mechanics in terms of electron spin.

Zeise's Salt, $K[PtCl_3(C_2H_4)]$. A π-ethylene complex, the first such complex isolated.

Zeolites. Aluminosilicates containing $(Si, Al)_nO_{2n}$ frameworks with a negative charge which is balanced by cations present in the cavities. The cations are easily exchanged and water and gases can be selectively absorbed into the cavities.

Zinc. A transition metal occurring naturally as the sulphide (zinc blende) and carbonate (smithsonite). It is extracted by roasting the ore in air and then reduciny the oxide to the metal using carbon. Zinc is used to galvanize iron, in alloys (*e.g*, brass), and in dry batteries. It reacts with acids and alkalis but only corrodes on the surface in air. Zinc ions are identified in solution by forming white precipitates with sodium hydroxide or ammonia solution, both precipitates being soluble in excess reagent.

Symbol : Zn; m.p. 420°C; b.p. 908°C; r.d. 7.1; p.n. 30; r.a.m. 65.38.

Zinc Amalgam. A solid Zn—Hg solution, frequently used as a reducing agent.

Zincates. Ions such as $[Zn(OH)_4]^{2-}$, $[Zn(OH)_3H_2O]^-$ formed by the action of excess base on Zn^{2+} solutions.

Zinc Blende. This cubic form of zinc sulphide gives its name to an important crystal structure type. The *Zn* and S are tetrahe-

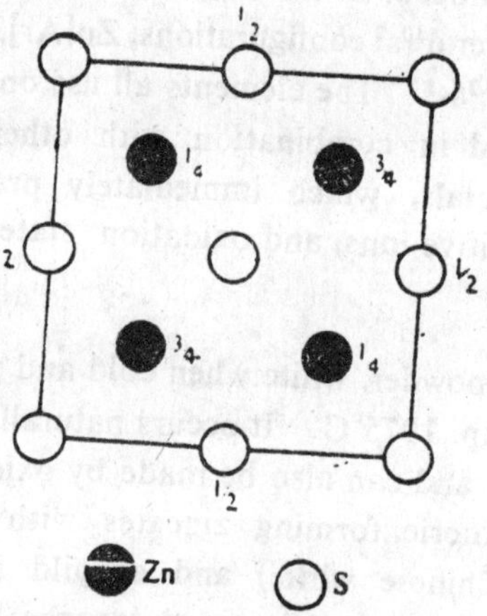

drally co-ordinated. Has ccp S with Zn in tetrahedral holes. Other compounds with this structure include BN, SiC, AlP, CuCl.

Zinc Borates. Compounds of rather indefinite composition prepared by treating a solution of borax with zinc sulphate. Widely used for fire-proofing textiles and in the ceramic and pharmaceutical industries.

Zinc Carbonate. A basic carbonate is precipitated by addition of an alkali carbonate to a solution containing Zn^{2+}. $ZnCO_3$ may result from the use of, *e.g.*, $NaHCO_3$. Hydrolysed to ZnO on boiling with Na_2CO_3 solution. The basic carbonate is used medicinally for external application.

Zinc Chloride. A white crystalline compound, $ZnCl_2$. The anhydrous salt, which is deliquescent, can be obtained by the action of hydrogen chloride gas on hot zinc; r.d. 2.9; m p. 290°C; b.p. 732°C. It has a relatively low melting point and sublimes easily, indicating that it is a molecular compound rather than ionic. Various hydrates also exist. Zinc chloride finds use as a catalyst, dehydrating agent, and flux for hard solder It was once known as butter of zinc.

Zinc Group. A group of metallic elements in the periodic table, consisting of zinc (Zn), cadmium (Cd), and mercury (Hg). The elements all occur at the end of the three transition series and all have outer $d^{10}s^2$ configurations; Zn[Ar]$3d^{10}4s^2$, Cd [Kr]$4d^{10}5s^2$, Hg [Xn]$5d^{10}6s^2$. The elements all use only outer s-electrons in reaction and in combination with other elements unlike the coinage metals, which immediately precede them. Thus all form dipositive ions, and oxidation states higher than +2 are not.

Zinc Oxide. A powder, white when cold and yellow when hot, ZnO; r.d. 5.5; m.p. 1975°C. It occurs naturally as a reddish orange ore zincite, and can also be made by oxidizing hot zinc in air. It is amphoteric, forming zincates with bases. It is used as a pigment (Chinese white) and a mild antiseptic in zinc ointments. An archaic name is philosopher's wool.

Zinc Sulphate. A white crystalline water-soluble compound made by heating zinc sulphide ore in air and dissolving out and recrystallizing the sulphate. The common form is theheptahydrate $ZnSO_4.7H_2O$; r.d. 1.9. This loses water above 30°C to give the hexahydrate and more water is lost above 100°C to form the monohydrate. The anhydrous salt forms at 450°C and this decomposes above 500°C. The compound, which was formerly called white vitriol, is used as a mordant and as a styptic (to check bleeding).

Zinc Sulphide. A yellow-white water-soluble solid, ZnS. It occurs naturally as sphalerite (see also zinc blende) and wurtzite. The compound sublimes at 1180°C. It is used as a pigment and phosphor.

Zone Refining. Metals of very high purity can be obtained by this method by removing an impurity which shows differences in solubility in the liquid and solid states of the metal.

A circular heater is fitted around a rod of impure metal and is slowly moved down the rod. At the heated zone the

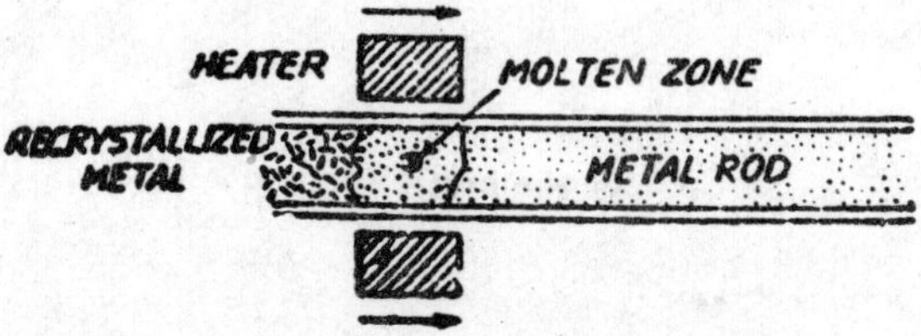

Zone refining.

rod melts and as the heater passes on, pure metal crystallizes while impurities pass into the adjacent molten part. In this way the impurities are swept into one end of the rod which is finally discarded. The heater may have to be moved from one end to the other more than once.

Ge, Si and Ga used as semi-conductors are refined in this manner; gallium arsenide and indium antimonide (also used as semi-conductors) are also zone refined.